Ganga Rejuvenation

Governance Challenges and Policy Options

Ganga Rejuvenation

Governance Challenges and Policy Options

Editors

Ora-orn Poocharoen
Chiang Mai University, Thailand

Robert James Wasson
National University of Singapore, Singapore

Xun Wu
Hong Kong University of Science and Technology, Hong Kong

World Scientific

NEW JERSEY • LONDON • SINGAPORE • BEIJING • SHANGHAI • HONG KONG • TAIPEI • CHENNAI

Published by

World Scientific Publishing Co. Pte. Ltd.
5 Toh Tuck Link, Singapore 596224
USA office: 27 Warren Street, Suite 401-402, Hackensack, NJ 07601
UK office: 57 Shelton Street, Covent Garden, London WC2H 9HE

British Library Cataloguing-in-Publication Data
A catalogue record for this book is available from the British Library.

GANGA REJUVENATION
Governance Challenges and Policy Options

ISBN 978-981-4704-57-1

Desk Editor: Sandhya Venkatesh

Typeset by Stallion Press
Email: enquiries@stallionpress.com

Printed in Singapore

PREFACE

This book is a collaborative project by graduate students of the Master of Public Administration (MPA) program (2015) at the Lee Kuan Yew School of Public Policy (LKY School), National University of Singapore. The project is a year-long module called the Governance Study Project (GSP), which is a flagship module of the MPA program. The module engages mid-career students to study a real-world complex problem as a cohort; and to provide analysis and policy recommendations from various angles.

We would like to sincerely thank all informants from the Indian public agencies, civil society groups, think tanks, and universities for sharing all valuable information to make this book possible. We would also like to thank Maitreyee Mukherjee, who worked tirelessly as a research assistant for this project and co-authored chapter 8; and Joost Buurman, senior research fellow of the Institute of Water Policy (IWP), who co-authored chapter 3.

We hope that this book is useful for policy-makers, activists, and the general public who are interested in the Ganga Rejuvenation project.

Ora-orn Poocharoen, Chiang Mai University
Robert James Wasson, National University of Singapore
Xun Wu, Hong Kong University of Science and Technology

CONTENTS

Chapter 1
INTRODUCTION

Ganga — India's "National River" — and its catchment have significant economic, cultural, social, environmental and religious importance for its more than 500 million inhabitants, or 43% of the population of the country. Across eleven states, the Ganga catchment accounts for more than a quarter of land mass and one third of its total surface water. The Ganga is also considered India's most sacred river, known as Ganga Ma (Mother Ganga), and is worshipped and revered as a living goddess by the hundreds of millions of Hindus in India and around the world.

Despite such iconic status, the Ganga River has been heavily polluted due to rapid urbanization, massive expansion of industrial development, and long-standing and increasingly intensive agriculture and pastoralism. The towns and cities located along the river generate some 3 billion litres of sewage every day, and less than half of this is treated before reaching the Ganga due to insufficient and often non-functional sewage treatment plants (STPs). Industrial waste is another pollutant source. Tanneries are one of the largest contributors to industrial pollution, followed by paper and pulp, chemical, sugar, and textile industries. The perennial flow of the river is limited by excessive withdrawals upstream into irrigation canals and by reservoirs constructed for irrigation and hydropower generation, which further aggravates the pollution problems by reducing dilution.

Indian governments have been trying to clean up the Ganga over the last three decades. The first Ganga conservation project, Ganga Action Plan-1 (GAP I) was launched in 1985 under the leadership of then Prime Minister, Mr. Rajiv Gandhi, to deal with water pollution. The project was initially expected to be completed in five years, but the Indian government extended it to 2000 after realizing the enormity of the problem. GAP Phase I was later joined by GAP Phase II to cover tributaries and cities exempted under GAP Phase I. In particular GAP II covered the Yamuna, Damodar and Gomati rivers and the big cities along these Ganga tributaries. The decades of effort through GAP I & II resulted in the addition of sewage treatment capacity but with limited visible change to the Ganga's health.

In May 2014, Prime Minister Narendra Modi brought the issue into the limelight by vowing to rejuvenate the Ganga by tackling the pollution problem. The National Ganga River Basin Authority (NGRBA) was merged into the Ministry of Water Resources (MWR) to become the Ministry of Water Resources, River Development and Ganga Rejuvenation, and a new platform named 'Namami Gange' (National Mission for Clean Ganga) was launched across multiple ministries with mandates for Ganga clean up. In 2015, the Union Cabinet approved an outlay of about US$ 3 billion for the first five years.

However, little progress has been made so far. It is estimated that only US$205 million had been spent between April 2015 and March 2017. The National Green Tribunal, a judicial body established in 2010 to deal with cases related to environmental protection and conservation of natural resources, observed that "Not a single drop of river Ganga has been cleaned so far" and criticised the government for "wasting public money" in the name of the Namami Gange project. The Supreme Court ridiculed the government for the slow progress by stating that "it would take 200 years to clean up the river at the current pace".

While much of the effort to clean the Ganga over the last 30 years, including the most recent attempt through Namami Gange, has been centred around building sewage treatment capacity in major urban centres along the river, other policy interventions can also be considered. For example, efforts to restore the flow of the river could be critical to enhance its pollutant transport and diluting capacities, and measures to reduce industrial pollution are essential along with building more wastewater treatment facilities.

In addition, the experience from GAP I & II highlights the importance of the quality of governance. A study conducted by the Central Pollution Control Board revealed that less than 60% of the installed capacity of existing sewage treatment plants along the Ganga was being used and 30% of the plants were not even functioning. A consortium of seven Indian Institutes of Technology (IITs) point out that the multiplicity of authorities, lack of assistance from state governments, and a dearth of monitoring are among the leading contributing factors for failure of past Ganga programs.

Policy interventions aiming at behavioral changes can play a key role. Behavior change is increasingly understood by public health experts around the world as critical to the success of water and sanitation projects. As Hans-Joachim Mosler has observed, efforts to provide people with water, sanitation, and hygiene "hardware" must be accompanied by programs that promote "software" changes. For example, no effort to reduce diarrheal disease and death has been successful without behavioral change to increase hand washing and reduce open defecation.

The book focuses on governance and management issues in the much-publicized 'Ganges Rejuvenation Project'. We argue that rejuvenation of the Ganga is more about governance than about technology, and that the major reasons for the lack of success in the past are — absence of long-term planning, poor co-ordination, and failure to sustain water and sewage treatment had been developed. The notion of governance implies that this is a problem that has solutions and that the solutions can only derive from proper policy design and the ability to connect all the dots related to what is a very difficult and complex problem. Although the Government of India has also appointed a task force for preparation of a long-term strategy, substantial knowledge gaps persist, especially with regard to governance.

Focusing on these broad aspects, chapters in the book explore spaces for better governance through active community participation, knowledge management, prospects of public-private-partnership, e-governance, youth education, lessons from past governance failures, comparative international analogies, and utilization of external aid and global expertise in successful implementation of a sustainable long-term plan for a river basin's integrated development of both the economy and environment. A host of activities, such as improving pollution monitoring systems, new development plans for tourism enhancement, river dredging and sewering riparian cities, are already being carried out with the aim of achieving quick results. But will they be successful? This book aims to address the governance challenge and is, therefore, a very timely contribution to cleaning one of the world's great rivers.

The Outline of the Volume

Contemporary public policies are often the product of older policies. It is, therefore, crucial to learn from the evolution of policies in order to detect successes and flaws in any given policy arena. Chapter 2 presents an assessment of earlier efforts in cleaning up the Ganga, namely GAP I and GAP II, with a view to drawing useful lessons for Namami Gange by identifying past gaps in policy design and implementation. Based on the stages model of policy making (agenda setting, policy formulation, decision-making, policy implementation and evaluation), a systematic review of GAP I and GAP II has been conducted to identify shortcomings corresponding to specific stages. Among 11 shortcomings identified in the chapter, the following are particularly important insufficient public buy-in, lack of clarity about policy outcomes, lack of ownership by state governments, and lack of sustained political interest. The authors argue that these shortcomings may undermine the performance of Namami Gange unless sufficient attention is paid to them and interventions are appropriate.. The chapter concludes with three main policy recommendations: 1) define a clear vision for Namami Gange; 2) make Namami Gangae a "People's Program", and 3) involve state and local governments more closely in policy design and implementation.

Similar to Chapter 2, the empirical base for Chapter 3 also derives from analysis of GAP I and GAP II. This chapter focuses on recommendations for coordination and collaboration strategies among key authorities and agencies. The problem of agencies working in silos is common in many governments, and the Indian case is not an exception. Based on key principles of Integrated Water Resources Management (IWRM), the chapter focuses on the effectiveness of interstate and intrastate coordination as a key contributing factor to the performance of the Ganga clean up projects. Factors impeding effective interstate coordination include institutional weaknesses such as the fragmentation, isolation, overlap and duplication of tasks and responsibilities, inadequate delegation of fiscal resources, imbalance of power between central and local governments, and conflicts of interest due to political affiliations and ideological differences. Factors affecting the effectiveness of intrastate coordination include lack of integration in the two-tiered system and at the project level, and perverse incentives and enforcement mechanisms. The chapter offers several specific measures to strengthen interstate and intrastate coordination, such as empowering state and local governments in the Ganga clean up, broader representation of different stakeholders in the NGRBA, making greater use of informal institutions, insulating public agencies from political changes, strengthening the Ganga Knowledge Centre, and enhancing the capacity and mechanisms for information collection and sharing.

Chapters 4 and 5 deal with an essential aspect of governance for sustainable transformation of the Ganga clean up. Chapter 4 focuses on the role of educating youth through environmental education (EE) for long-lasting impact on the health of the Ganga River, while Chapter 5 builds the case for social mobilization. Environmental education can help ensure that young people develop into individuals with the ambition, commitment and skills to become key drivers for change in the attitudes and behaviours of the community-at-large for sustainable transformation. The authors of this chapter carried out careful data collection among teachers and youths in Varanasi. Based on primary and secondary research, the findings of the study suggest that, though environmental education has a long history in India, it has failed to produce its intended objectives because of inefficient attention to EE, an overly broad EE curriculum, and ineffective teaching. The main recommendations include inducting the Minister responsible for school education into the NGRBA and establishing a youth division within the National Mission for Clean Ganga (NMCG). In addition, the chapter explicitly discusses the quality of curriculum design, pedagogy, teacher training and development, and the need to develop alliances between teachers.

The chapter on social mobilization (Chapter 5) highlights public apathy as one cause of the persistence of severe pollution in the Ganga, despite the existence of formal accountability mechanisms such as elections. This chapter investigates the possibility of large-scale social movement around the pollution problems in the Ganga River. With careful analysis of previous social movements targeted at the Ganga River, the authors explain reasons for public apathy and explore the possibility of igniting social mobilization to expedite the process of Ganga Rejuvenation. The chapter traces the evolution of social activism concerning Ganga pollution, and its success and failures in overcoming public apathy. The main contribution of the chapter is to offer a range of strategies and tactics for social movements based on theoretical frameworks and practices elsewhere. Practical and detailed advice is given on how to apply these frameworks, strategies and tactics, such as how to identify trigger issues, build coalitions, reach out to target groups, sustain support, and the staging of trigger events. The attention paid to the operational details of social movements makes the chapter an excellent manual for social and environmental activists who are motivated to harness the power of the people to deal with the social and environmental issues of both the clean up and the post clean up phases.

While the private sector has often been portrayed as the villain in environmental degradation, private sector players can play a key role in Ganga rejuvenation. The Indian Government has identified public-private partnerships (PPPs) as a key policy tool for Namami Gange, but India's experience with PPPs in water and wastewater has been mixed, and public opposition to the involvement of private sector players remains virulent. Chapter 6 presents the vicious cycle that has undermined the potential of PPP for restoring the Ganga: inadequate cost recovery for wastewater services led to low take-up of PPPs and low levels of investment in the sector overall, and the increase in untreated discharge further deteriorated water quality which, in turn, raised the costs of water treatment along the Ganga, making it more difficult to achieve cost-recovery. The chapter highlights several thought-provoking policy interventions to break this vicious cycle. First of all, STP construction should be unbundled from sewage network rollout to be more cost-effective. Second, water and wastewater tariffs as well as decision-making should be separated, instead of combined, to enable more extensive involvement of the private sector in wastewater treatment. Last, PPP contracts should be structured to recognize the high level of uncertainty in future developments.

Chapter 7 on river (or water) front development, focuses on Varanasi where the tensions among various competing interests, namely religious practices, tourism development, and environmental sustainability, culminated. Waterfront development is still a relatively new concept in many parts of the world. It combines urban planning, water accessibility, and environmental protection, and aims to achieve economic development while creating liveable cities. The chapter identifies waterfront development as a potential mechanism to better manage various competing interests through more systematic application of the principles of effective urban design and planning. Based on a comparative analysis of waterfront development in Varanasi and other cities around the world, specifically Minneapolis in the USA, Manado in Indonesia, Copenhagen in Denmark, and Singapore, the chapter offers several valuable suggestions for waterfront development in Varanasi, as well as concrete steps to address the current weaknesses in strategic choices, and institutional arrangements and processes. Given the importance of waterfront development, lessons drawn from the chapter based on Varanasi can have significant implications for other major cities along the Ganga where conflicts among competing interests have also intensified.

The availability and usage of data and knowledge can play a key role in Ganga rejuvenation due to the complex dynamics of the river pollution problems. The lack of authoritative data has led to differentiated views and contrasting arguments regarding past experiences of the Ganga River clean up projects. The chapter on data governance (Chapter 8) highlights the availability of accurate, updated data in fostering policy coherence and collaboration for Ganga rejuvenation. Applying the four-pillar framework of data governance, which consists of policy, people, process and technology, the chapter points to several key deficiencies in current and earlier efforts to clean up the Ganga, such as lack of policy coherence in data governance, fragmented approaches to data governance across different government agencies, lack of data sharing and dissemination mandates, and limited use of available best technology for data governance. Various policy recommendations to address these problems are also considered. Efficient data governance can significantly benefit the general public as well; ensuring greater knowledge dissemination, awareness and their motivating inclusive participation in this massive river-basin rejuvenation drive.

Chapter 9 looks beyond data by employing the concept of knowledge management (KM) and several key related frameworks, and makes a convincing case that knowledge management is an essential tool in the Ganga Rejuvenation. It is also noted in the chapter that the tacit knowledge of people and communities is under appreciated and under utilized, implying that knowledge should be drawn from many sources, not just official ones. Through the analysis based on a KM framework of knowledge generation, validation, and dissemination, the chapter highlights some existing knowledge gaps in the Ganga Rejuvenation Project, providing targets for future efforts. In the area of pollution data, much more can be done to generate and validate useful knowledge, and to share that knowledge effectively. Through the application of data-information-knowledge-wisdom, the chapter demonstrates how existing data can be turned into more useful information, information into more practical knowledge, and finally knowledge into true wisdom in understanding Ganga river pollution and finding sustainable long-term solutions.

The last two chapters of the volume take international and comparative perspectives. Chapter 10 on regional collaboration suggests that better cooperation among riparian states of the Ganga basin can greatly benefit Ganga Rejuvenation. Water storage facilities built in Nepal can serve the interests of both countries by increasing river flow to India, particularly during the dry season, thereby augmenting the pollutant transporting and diluting capacities of the river while generating economic benefits for Nepal in terms of electricity generation. While the prospects of these projects has been dampened due to the mistrust and political chaos of Nepal, their potential importance to Ganga rejuvenation may provide strong impetus for India to make it a strategic priority. On the other hand, the Ganga clean up can also help improve the relationship between India and Bangladesh as conflicts over the quantity and quality of water reaching Bangladesh have strained the relationships between the two countries. The chapter argues that cooperation between the three countries on the Ganga River will not only help to clean and protect the river but would also be positive for security concerns. The chapter lays out a detailed set of recommendations on possible steps and actions that could be taken by the three countries.

Chapter 11 also emphasizes the importance of augmenting river flow as a way to reduce pollution, based on a systematic comparison between the Ganga and Yellow River in China. Lessons from the Yellow River set out in the chapter are: 1) set up environmental flows alongside pollutant abatement; 2) establish an umbrella water governance legal framework in the short run and explore the possibility of legislating a national Water Law in the long run; 3) and set an explicit amount control on environmental flows.

We hope that the ideas and recommendations presented in this book will assist in the development of more effective governance in the Ganga basin and thereby reduce pollution and return Ganga Ma to a healthier state. We, therefore, trust that there will be no need for another account in a decade or so that chronicles the failures of Namami Gange.

Chapter 2

HISTORICAL EVALUATION OF GANGA ACTION PLANS

Dawa Tshering, Denny Ariaputra, Soh Siew Luie, Tashi Tobgay and Wang Nian

Introduction

Indian Prime Minister Narendra Modi and his party, Bharatiya Janata Party (BJP), have committed to "ensure the cleanliness, purity and uninterrupted flow of the Ganga" in their 2014 election manifesto.[1] Recognizing the need for more resources to be dedicated to developing and regulating India's water resources, a new Ministry for Water Resources, River Development and Ganges Rejuvenation was established, and the existing programs to clean up the Ganga were shifted under the purview of this new Ministry. The Government of India also launched *Namami Ganga* in July 2014, a new Integrated Ganga Conservation Mission that has been set up to direct efforts in the conservation and improvement of the Ganga by "consolidating existing ongoing efforts and planning into a concrete action plan for the future".[2]

[1] Bharatiya Janata Party (2014).

[2] National Mission for Clean Ganga (2014).

This is not the first time that the Government of India has rolled out a program to clean up the Ganga. In 1985, the then Prime Minister Rajiv Gandhi launched the Ganga Action Plan (GAP) with the main objective of improving the water quality of Ganga to acceptable standards by preventing the pollution load from reaching the river.[3] The approach for GAP was to build a series of sewage and effluent treatment plants in major cities along the river, refurbish all existing treatment plans, expand existing sewer networks by connecting unserviced areas and construct electric crematoriums to address the issue of unburnt dead bodies in the river.

The Ganga Action Plan was implemented in two phases — GAP I covered 25 towns in three states and was expected to be completed in 1990.[4] However, the implementation of GAP I was extended by ten years to 2000 due to delays in meeting the targets. While GAP I was still in progress, GAP II was introduced in stages between 1993 and 1996, and covered 85 towns in seven states,[5] targeting not only the Ganga but also the three major tributaries of the Ganga, namely the Yamuna, Gomti and Damodar. Implementation of GAP II is still on-going.

Although a lot of resources has been spent on implementing GAP I and GAP II, there has been limited impact as evidenced by the fact that the biochemical oxygen demand (BOD) and the chemical oxygen demand (COD) of the river remain largely unchanged, while the coliform levels, although improved, are still high as compared to the target levels for Class B water quality (Table 1).

The Government of India also recognised that GAP had not been effective in achieving the objectives set for the program, and has tried a different approach to address the problem. The change was to adopt a more holistic approach, focusing on the Ganga Basin as a whole rather than on a few towns, which was seen to be an improvement over GAP. In line with this new strategy, the Government of India established the National Ganga River Basin Authority and Mission Clean Ganga in 2009.

---

[3] Ministry of Environment and Forest (2000).

[4] Ministry of Environment and Forest (2000).

[5] National Ganga River Basin Authority (2009).

Table 1: Water Quality Trend Data (2002–2010)[6]

No of Monitoring Locations	Year	Observed Range of Water Quality Parameters		
		BOD (mg/l)	COD (mg/l)	Total Coliform (MPN/100ml)
34	2002	0.5–16.8	1–30	300–25 × 10^5
34	2003	0.8–27	2–47.2	47–45 × 10^5
34	2004	0.7–14.4	—	11–45 × 10^5
39	2005	0.1–15.2	1–37.6	13–45 × 10^5
39	2006	0.1–16.4	4–38.4	1–25 × 10^5
39	2007	0–14	2.6–30	0–28 × 10^5
39	2008	0.5–21	1–34	0–101 × 10^5
57	2009	0.2–16	2–34.2	2–65 × 10^5
57	2010	0.2–15	—	3–14 × 10^5
CLASS B WATER QUALITY STANDARD[7]		**3 or less**	**5 or more**	**500 or less**

Nonetheless, the water quality has not shown significant improvements since the new strategy was implemented.[8]

As the new government under Prime Minister Modi works on developing a plan to restore the Ganga to its former glory, it is useful to take a step back and conduct a review of GAP I and GAP II so as to identify gaps in its planning and implementation that could be translated into learning points for *Namami Ganga*. There have already been several papers written that reviewed GAP.[9] This chapter presents a historical evaluation of GAP I

[6]Central Pollution Control Board (2013).

[7]The target standard for water quality under GAP was Class B i.e. suitable for outdoor bathing (Central Pollution Control Board, 2009).

[8]Central Pollution Control Board (2013).

[9]Hasan (2015); Narain (2014); Das and Tamminga (2012); Consortium of seven Indian Institutes of Technology (2011); World Bank (2011); Sankat Mochan Foundation (2000); Jaiswal, Ganga and GAP (2007).

and GAP II from a different perspective, by using the Stages Model of policy making to systematically review the policy processes of GAP I and GAP II. By taking a different perspective, this chapter seeks to enrich the understanding about GAP I and GAP II.

The chapter first provides an overview of the Stages Model. It then examines in greater detail the policy processes for GAP I and GAP II by referencing reports prepared by various government agencies and articles from academic journals as the main sources of information. The policy formulation and decision making stages are discussed together in this review as there is insufficient information available in the public domain to clearly differentiate between these two stages in the context of GAP. The study uses the Stages Model to identify the shortcomings in the policy process for GAP, and makes recommendations as to how similar shortcomings could be avoided for *Namami Ganga*.

Stages Model of Policy Making

The study of the policy process is part of the broader field of political studies or political science, with the emphasis being on the "outputs of politics"[10] and the process of policy making, including how problems are identified by the government, how government institutions come up with policies and how policies are implemented and evaluated as part of a continuous improvement process.[11] Studies of the policy process are usually case studies that make use of qualitative methods based on available information.[12]

Over the years, many different theoretical frameworks have been developed to explain the policy process, such as the Institutional Rational Choice, Multiple-Streams Framework and the Advocacy Coalition Framework.[13] We have decided to use the Stages Model, also known as the Stages Heuristic, for the purposes of evaluating GAP I and GAP II as it breaks down the complex policy making process into a set of distinct

[10] Hill (2004).
[11] Sabatier and Weible (1999).
[12] Hill (2004).
[13] Sabatier and Weible (1999).

stages which can be studied individually before putting everything together to look at the overall picture. However, it should be noted that although the Stages Model makes sense in theory, some stages may be skipped or the stages may not occur sequentially in practice.[14] Nonetheless, this model provides a simple and easy-to-understand structure for reviewing policies and helps to zoom in on specific stages of the process where there are areas for improvement.

First conceptualised by Lasswell, the initial Stages Model comprised seven stages which focused on the decision-making process within the government, namely intelligence, promotion, prescription, invocation, application, termination, and appraisal.[15] However, Lasswell's model was criticised as being overly simplistic and inaccurate as it did not consider the influence of extraneous factors on the decision-making process. In addition, Lasswell saw the policy process as being a sequential one, where each stage was distinct and followed from the previous one, and ignored the process as an entirety.[16]

Building on Lasswell's work, Gary Brewer improved upon the model by expanding the policy process to include the influence of other political actors beyond the government, and more importantly, introduced the concept that the policy process was an on-going cyclical process where policies are constantly being modified by policymakers. Brewer's work spurred the development of several other variants of the Stages Model in the 1970s and 1980s, the most well-known of which were set out in popular textbooks by Charles Jones and James Anderson.[17]

The Stages Model that is used in this study is the culmination of the work of these political scientists, which has been modified from Lasswell's initial seven-stage model into the five-stage model that public policymakers today are most familiar with. The five-stage model follows the logic of applied problem solving, with the five stages being (1) agenda setting, (2) policy formulation, (3) decision making, (4) policy implementation, and (5) policy evaluation. The model is typically depicted as shown in Figure 1.

---

[14] Howlett (2011).

[15] Howlett (2011); Lasswell (1951).

[16] Howlett (2011); Sabatier and Weible (1999).

[17] Howlett (2011).

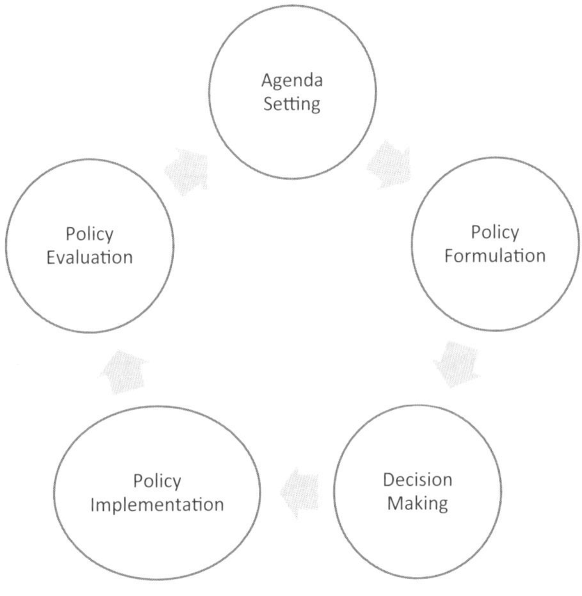

Figure 1: Stages Model of Policy Making

(1) *Agenda setting*

Agenda setting is the stage whereby an issue is recognised by the government as being a problem that they need to solve. Generally, an issue is brought to the attention of senior policymakers either through the initiative of mid-level government officials who have been studying the issue, or because of pressure from either the general public or organized political forces such as advocacy groups or non-governmental organisations. Other political actors who may influence whether the government recognises an issue as a problem include the media, public policy academics and international organisations such as World Bank.[18]

However, it is not sufficient for an issue to be brought to the attention of policy makers. All governments are faced with constraints and cannot be expected to put all issues that have been surfaced to them onto their agenda, which is "a list of issues or problem situations to which governmental officials and others in the policy community are paying

[18]Wu, *et al.*(2010).

some serious attention at any given time".[19] For agenda setting to be successful, the issue must be recognised as a problem and the need to quickly solve the problem must be regarded by the government as being a priority.[20]

There are a number of common challenges to the agenda setting process. First, because of public pressures and the short terms that politicians serve in the government, the government agenda is often dominated by responses to crisis and current issues. Issues that require a longer time to address tend to have lower priority on the agenda. Second, agenda setting is frequently taken over by politicians to gain political mileage, rather than about making real efforts to resolve the problems. Third, the government agenda may be hijacked by special interests groups which distort the relative importance of issues, such that the overall public interests are not being served by the government agenda. Last, poor agenda setting often leads to the implementation of programs with poor outcomes as it "leads to the dominance of ineffective and wasteful solutions".[21]

(2) *Policy formulation*

Policy formulation is the stage where the government begins to work out possible policy options that could be implemented as solutions to the problem. All the proposed policy options should be able to address the problem. In addition, they should be politically feasible, technically feasible and administratively feasible.[22] Policy options can be identified by researching best practices in other countries or in other sectors, collecting data to identify the root causes of the problem to hone in on what the solutions could be or by conducting public consultation to tap on the collective wisdom of the public or interested stakeholders. Besides ensuring that the proposed policy options are feasible, policymakers also need to identify the resources required to implement the solution effectively.[23]

---

[19]Wu, *et al.* (2010).

[20]Wu, *et al.* (2010).

[21]Wu, *et al.* (2010).

[22]Wu, *et al.* (2010).

[23]Wu, *et al.* (2010).

One of the common challenges that policymakers face during this stage is that the ultimate decision-makers i.e. the politicians and senior bureaucrats, do not have a clear idea about the intended outcomes that they would like to achieve. This leads to a lot of confusion during the policy formulation stage as policymakers try to come up with solutions that may solve the 'perceived' problem but without addressing the fundamental concerns of the decision-makers or achieving the desired outcomes.

Although the political challenge is often thought to be the biggest hurdles to policy formulation, technical challenges that arise because of the political challenges may sometimes be even more challenging. Most problems that policymakers have to deal with are wicked problems that are ill-defined and messy, and solutions for one aspect of the problem may have unintended consequences for another aspect of the problem.[24] In such situations, it becomes a challenge to even arrive at the root causes of the problem, let along understand what the proposed solutions should aim to achieve. The additional challenge of dealing with wicked problems is that there are never right-or-wrong solutions for such problems, only solutions which are better-or-worse. Policymakers are faced with the technical challenge of considering a diverse range of potential solutions, and then to identify solutions that would result in a net improvement to the situation.[25]

(3) *Decision making*

At this stage, the government makes a decision on the final policy option to be adopted. During the decision making stage, the best practice is for the policymakers to compare the options against one another based on objective criteria agreed upon by all the key stakeholders. Besides considering the benefits and costs of the various options, policy makers will also need to take into account any capability and capacity constraints faced by the government, as well as the political acceptability of the options from the perspective of all key stakeholders, and not just the politicians.

[24] Rittel and Webber (1973).

[25] Wu, *et al*. (2010).

One common challenge faced by policymakers during this stage is the lack of acceptance by stakeholders of the final solution that is selected.[26] As each stakeholder would have different concerns, it is never possible for a single solution to meet the needs of all stakeholders. Policymakers need to manage the communication process regarding the decision making process, especially the importance of explaining the reasons for choosing the policy option over another.

It is also important to project the possible outcomes and potential impact of implementing the policy options, particularly identifying any possible unintended conflicts with other government programs.[27]

(4) *Policy implementation*

This is the stage where the policy solution is put into effect. The key activity during this stage is to put in place the appropriate institutions and legislation to support implementation of the policy so that it is able to meet the objectives. Policymakers also need to ensure that sufficient and appropriate resources are made available to the implementers to support policy implementation.

The single biggest challenge commonly faced during this stage is that there was insufficient consideration of implementation issues during the policy formulation and decision making stages. This would lead to the formulation and selection of policy solutions that cannot be implemented in practice due to the presence of implementation barriers, or implementation of wasteful or ineffective solutions that are unable to make a significant impact to resolve the issue.[28]

One of the common implementation barriers faced by implementers is the lack of stakeholder support for the program, which is usually required to give legitimacy to the program.[29] Many stakeholders are involved in the policy process, and it is never possible to address the concerns of all stakeholders simultaneously. Although it is possible to manage

---

[26] Wu, *et al.* (2010).

[27] Bardach (2012).

[28] Wu, *et al.* (2010).

[29] Weimer and Vining (2004).

stakeholder expectations during the policy formulation and even the decision-making stages, it is not possible to ignore or gloss over the objections raised by stakeholders during the implementation stage, as these would prevent the finalization of plans and resource mobilization, resulting in delays. Political will to push forward may also become weak as mid-level bureaucrats are faced with difficult decisions about the tradeoffs to be made on the ground.[30]

Another common implementation barrier is the presence of institutional constraints that may make it difficult to implement the selected policy solution as designed. The first type of institutional constraints are the limits that are put upon the implementers because of existing provisions in the constitution and laws of the country, which may prevent implementers from adopting the most effective and efficient approaches during implementation. A second type of institutional constraint is the existence of multiple levels of government, which often increases the time needed to obtain approvals and allocation of required resources due to delays in the decision-making process. The third type of institutional constraint is the lack of operational capacity i.e. sufficiency of human and financial resources, and also access to manpower resources with the appropriate skills. Without overcoming the institutional barriers, it is generally not possible for programs to be implemented as planned.[31]

(5) *Policy evaluation*

Policy evaluation is the stage whereby an assessment is conducted to examine if the policy has been effective in achieving its objectives and in meeting its targets. The findings and recommendations arising from the policy evaluation process can then be used to aid the next round of policy design to see if the policy design or implementation could be improved, and to decide whether the program should be continued, terminated or modified.[32]

---

[30] Wu, *et al.* (2010).

[31] Wu, *et al.* (2010).

[32] Rossi, Lipsey and Freeman (2004).

There are two main activities in the policy evaluation process. The first is process evaluation, which measures how well a program is operating by assessing its operations and determining achievement of output which has been targeted. The second one is impact assessment, which measures the outcomes or the effects of implementing the program on the target participants.[33]

During this stage, data from the agenda setting stage until the implementation stage must be reviewed, and compared with the performance indicators that had been determined before implementation began. Such data is usually collected by oversight agencies or independent groups such as academics and non-governmental organisations. A good implementation plan will also suggest some ways in which ongoing data about the policy can be generated in the regular course of policy evaluation, for example, from reports, records, documents, feedback from stakeholders, diary entries, rating, tests, observation, and also physical evidence.[34]

A common challenge faced by evaluators is how to design the evaluation. When designing the evaluation, evaluators need to take into consideration the political context. However, as most policies are complex and involve a multitude of stakeholders, evaluators are often faced with trying to meet the information needs of various groups of stakeholders with conflicting interests, thus causing confusion to the evaluators about the perspective that they should take.[35] In addition, the sponsors of the evaluation may have a strong vested interest in demonstrating that the program had made a positive impact on the affected people, especially if they had invested heavily in the implementation of the policy. Hence, they may resist any evaluation results that might prove that the policies did not achieve the intended outcomes.[36]

Besides being sensitive to the political context for the evaluation, evaluators must also take care to ensure that the planned evaluations are both scientific and pragmatic. For the evaluation findings to be accepted

[33]Trochim (2009).

[34]Trochim (2009).

[35]Rossi, Lipsey and Freeman (2004).

[36]Rossi, Lipsey and Freeman (2004).

as being valid and credible, the evaluation design must be able to meet research standards in terms of its rigour. Appropriate and meaningful indicators must be chosen to ensure that the evaluation is not only measuring the outputs, but also the outcomes of the program. However, meeting such research standards generally require a lot of time and resources, and may not be practical. At the same time, it is important that policy evaluations must be provided in a timely manner and be meaningful for the consumers. Evaluators therefore need to find the appropriate balance between ensuring the validity of the evaluation and the usefulness of the evaluation findings.[37]

Review of the Policy Process For GAP I And GAP II

A. *Agenda setting stage*

(i) *Agenda setting for GAP I and GAP II*

Cleaning up the Ganga was a problem that was put on the Government of India's national agenda as a result of the personal intervention of the then Prime Minister Mrs. Indira Gandhi in the early 1980s. She directed the Central Board for the Prevention and Control of Water Pollution to perform a comprehensive study on the Ganga in 1979, which identified the high levels of pollution of the Ganga as a problem.[38] This had led to the development of the GAP, which was officially launched by her successor, ex-Prime Minister Rajeev Gandhi in 1985 after her death. The launch of GAP was also spurred on by the public interest suit filed by an environmental lawyer, M.C. Mehta in the Supreme Court of India, who had charged that the government and industries were responsible for the alarming rise in pollution in the Ganges.[39]

Outside of the government, it was mainly non-governmental organisations and environmental activists who were keenly interested in cleaning

[37]Rossi, Lipsey and Freeman (2004).

[38]Consortium of 7 Indian Institutes of Technology (2011).

[39]Das and Tamminga (2012).

up the Ganga. The general populace did not seem to think that the Ganga could be polluted,[40] as most saw the Ganga as a holy river with the ability to purify all things that come into it. According to the findings from a brief survey conducted by Das and Tamminga along the Ganges waterfront in Varanasi to assess people's perception of pollution in the Ganga,[41] most people perceived the pollution in the Ganga as being the result of "declining social and religious values", and that religious activities, meditation and worship were required to "invoke the Ganga's purifying power". Although some respondents recognized that the expanding city contributed to the growing pollution in the Ganga, there seemed to be a strong belief that the Ganga could never be impure as it was able to cleanse itself.[42]

(ii) *Shortcomings identified*

In today's society, agenda setting is no longer the sole responsibility of the government. Instead, the government needs to work with all the political actors to engage in consensus-oriented decision-making, where the people and the government agree on the issues to be addressed and the expected outcomes. There should be joint ownership and joint responsibility for the decisions, which then gives greater legitimacy to programs that are rolled out by the government.[43]

However, in the case of GAP I and GAP II, we think that there was insufficient public buy-in for the issue to be placed on the government agenda GAP at that time. This is evidenced by the fact that one of the aims of GAP was to promote public awareness about the need to clean the Ganga.[44] It has also been pointed out that the people's participation in the GAP was "symbolic rather than substantive", and the objective of promoting public awareness was mainly for political expediency.[45] For example,

[40] Das and Tamminga (2012).

[41] Das and Tamminga (2012).

[42] Das and Tamminga (2012).

[43] Ansell and Gash (2007).

[44] Ministry of Environment and Forest (2000).

[45] Ahmed (1994).

participation in Varanasi was limited to token forms such as sporadic ghat clean up campaigns and ghat-based street theatre and concerts which were run by volunteers without support from the local governments, which did not create a lasting sense of ownership of the programs.[46] Based on a scan of available literature, there was also no evidence of the public, the local community and other political actors being engaged during the agenda setting stage to seek their buy-in and support for the need to clean up the Ganga.

B. *Policy formulation and decision making stages*

(i) *Policy formulation and decision making for GAP I and GAP II*

"Centralised planning had been the norm in water resources management in India for several decades".[47] The policy for GAP was formulated by the National River Conservation Authority (NRCA), which had a hierarchical organizational structure comprised of five tiers, with the state government being involved only at the lowest tier. The local government and institutions were not included.[48] There was also no evidence from available literature that the NRCA had involved the public, non-governmental organisations and civil society organisations. Instead, it seemed that participation was only limited to high level politicians, bureaucrats and technocrats. This was pointed out in the Comptroller and Auditor General of India's report, "Audit did not find much evidence of any significant initiative on the part of the Ministry and the States to improve and promote public participation".[49]

Many stakeholders were not happy with the government's approach, and filed litigation and court cases against the government which eventually led to delays in the implementation of the program.[50] For example, in

---

[46] Das and Tamminga (2012).

[47] Das and Tamminga (2012).

[48] Consortium of 7 Indian Institutes of Technology (2011).

[49] Comptroller and Auditor General of India (2000).

[50] Narain (2014).

1987 when policing was activated along the river bank to monitor the polluters, there was a huge outcry from the public which led to a deterioration of the relationship between the government and the people. Another example was in 2001 when the National River Conservation Directorate sanctioned a budget of Rs. 416 crores for the construction of trunk sewers and interception drains, the public filed a case to the Supreme Court and the construction was halted.[51] A third example was when the government restricted the *dhobis* (washermen)[52] from washing along the river-front but allowing cattle wallowing, there was a huge outcry by the *dhobis* which resulted in a protest. Heeding to the outcry, the government built two large ghats downstream but the infrastructure did not serve the purpose as it was too far and also the design was not appropriate. After extensive discussions between the *dhobis* and the administration an alternate washing place was identified within the city by the *dhobis* which was accessible by the majority.

Besides involving only a limited group of people in the policy formulation and decision making stages, policy formulation and decision making seemed to have been based on the findings of the survey commissioned by then Prime Minister Mrs Indira Gandhi in 1979.[53] The survey had found that municipal wastewater generation from the 50 cities located along the river was estimated to be about 85% of the total pollution load,[54] with the remaining 15% of the pollution coming from industrial waste which was discharged into the river. As the eventual target of GAP was to prevent municipal waste from reaching the Ganga, it can be inferred that the decision was made to concentrate on the 85% of the pollutants that were contributing to pollution of the Ganga.

There was also no sharing of information about the GAP and its goals with the public, much less inviting active decision-making, creative problem-solving and shared prioritization.[55]

---

[51] Narain (2014).

[52] The *dhobis* use the riverfront for washing cloths (a hereditary occupation). Their livelihood depended on the Ganga.

[53] Consortium of 7 Indian Institutes of Technology (2011).

[54] Centre for Science and Environment (2014).

[55] Das and Tamminga (2012).

(ii) *Shortcomings identified*

There was no clear distinction between the policy formulation and decision making stages for GAP, nor was there evidence of the Government of India presenting a series of policy options for discussion with a broader group of stakeholders. Instead, it seemed that the government went directly from problem identification to formulating a single solution for implementation, rather than generating a series of policy options for consideration before going through the process of decision making. The lack of transparency about the discussions that took place within the Government of India during the policy formulation and decision making stages was a shortcoming as most people were unable to understand how the final decision for GAP was arrived at. For instance, no justifications were given as to why the plans for GAP only targeted municipal waste and was focused on finding technocratic solutions for sewage interception and building sewage treatment facilities, while not tackling other sources of pollution.

In actual fact, the industrial effluents attributed greatly to the Ganga pollution. Even though the private sector was one of the major contributor to pollution their involvement and participation was very minimum as the program was controlled and funded by the central government.[56] This could have been because the objective of GAP had been defined as improving water quality by focusing on preventing municipal waste from flowing into the river, and thus they were seen as being irrelevant. It could also have been because the government was concurrently trying to promote economic development through expansion of industrial activities. Whichever the reason, the lack of involvement of this important group of stakeholders resulted in the government only taking punitive measures against the industries, through prohibitive and disciplining tools against polluters.[57]

Besides the municipal and industrial waste flowing into the Ganga, the high water extraction level from the Ganga upstream for agriculture, hydroelectric power plants, industries and other uses had resulted in inadequate flow of water, causing parts of the riverbed to dry. Without sufficient water flow to dilute and assimilate waste, particularly during the

---

[56]Gupta (2005).

[57]Gupta (2005).

winter and summer months, waste would build up in the river and result in concentrated pollutants in the river. However, run-off from the agricultural fields which brought non-biodegradable pesticides into the Ganga received less attention and pollution from the rural areas was not addressed at all.[58]

There was also a lack of clarity about the envisioned outcomes, which made it difficult to come up with effective solutions that would address the problem. When GAP was first launched in 1985, there was only a broad target of improving the water quality to acceptable standards. It was not until two years later in 1987 that the term ‘acceptable standards’ was clarified to refer to Class B water quality i.e. suitable for outdoor bathing. There has also been criticism that setting Class B as the target standard was inappropriate given that many people use the water from Ganga directly for drinking, especially in the rural areas. Hence, the more appropriate target should have been Class A i.e. suitable as a direct source for drinking water. Furthermore, a technocratic approach towards abatement of pollution was imposed on a multifaceted community which has its own cultural and religious understandings of pollution. This resulted in an outcome that was different from what the program had envisioned.

Narrow participation of stakeholders in the policy formulation stage for GAP was another major shortcoming that has had bearing on GAP’s efficacy. By sticking to the conventional centralised planning approach without involving local stakeholders, there was a lack of coordination between the central, state and the local governments which led to implementation delays. For example, in West Bengal, the tender for construction of three pumping stations were awarded to a contractor before the state had acquired the land. This resulted in the need to re-issue the tender three times when the existing contractor could not complete the work without the period specified in the tender, at an extra cost of Rs. 41.85 lakhs.[59] There was also very little attempt to support the non-governmental organization’s activities nor were they provided any information on the GAP, its methodology and its impact on the quality of Ganga.

The narrow participation of stakeholders also resulted in insufficient consideration of the local context during the policy formulation and

[58]Jaiswal, Ganga Action Plan — A Critical Analysis (2007).

[59]Comptroller and Auditor General of India (2000).

decision making stages, which led to the implementation of solutions that were unable to address the problem. For instance, there was a gap between the total planned treatment capacity of the sewage treatment plants to be built under GAP and the actual sewage generated in the targeted cities. A study in 2009 showed that the sewage generated was 2,638 million litres per day (MLD) whereas the treatment capacity after building of the treatment plants under GAP was 1,174 MLD.[60] Further, the government did not take into consideration the expansion of cities and population growth in these cities in its plans. For example in Kanpur, the capacity of the sewage treatment plant that was built under GAP was 171 MLD. The capacity of this new plant was already lower than the estimated 200 MLD of sewage generated in 1985, and far below the 600 MLD of sewage that was generated in 2013.[61]

There was also little consideration of the state of sewage infrastructure development in the cities. Most of the cities along the Ganga basin did not have proper sewage systems. In Kanpur, unsewered area is 18,252 hectares compared to 7,558 hectares of sewered area.[62] In other cities like Allahabad and Varanasi, 70% to 85% of the sewage enter the river untreated due to the non-existence of an underground drainage system. Solid waste management was virtually non-existent and therefore solid waste eventually ended up in the open drains which lead to the Ganga. Although GAP's objective was to intercept, divert and treat the municipal waste before it is pumped into the river, discharges into the open drainages were not intercepted. In Varanasi, one third of the population lived in the slums with little or no access to any sanitation and sewage facilities resulting in open defecation. Thus, the sewage treatment plants could not help to effectively address the problem due to the lack of waste that reached the plant.

The choice of the technology used in the treatment plants also did not take into consideration the local circumstances. For instance, the Activated Sludge Process was one of the technologies adopted for the sewage treatment plants. However, it was found that in Kanpur, a sewage treatment

---

[60] Narain (2014).
[61] Narain (2014).
[62] Narain (2014).

plant which was supposed to have a capacity to treat 130 MLD of sewage could only treat 70 MLD due to a lack of sludge handling facilities. A pumping station with capacity of 162 MLD was installed but received less than 95 MLD due to lack of proper sewage system. In other cities, there was insufficient electricity to keep the treatment plants running.[63]

C. *Policy implementation stage*

(i) *Policy implementation for GAP I and GAP II*

Figure 2 provides an overview of the organizational structure of GAP. The Ministry of Environment and Forests (MoEF), which was overall-in-charge

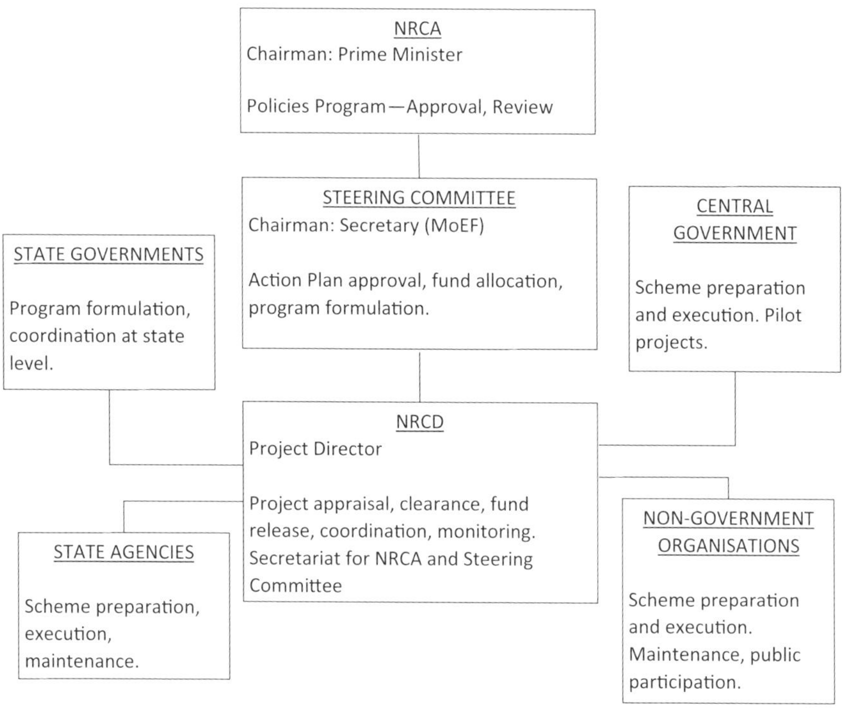

Figure 2: Organizational Structure for GAP

Source: Department of Environment, Government of India, (1985).

[63] Jaiswal, Ganga Action Plan — A Critical Analysis (2007).

for design and implementation of GAP, established two agencies to manage GAP as a government program. The NRCA was responsible for maintaining oversight of the policy framework for the program, while the National River Conservation Directorate (NRCD) was responsible for implementing the projects under GAP.

State governments and non-government organisations were responsible for scheme preparation, execution and maintenance. State River Conservation Authorities were established as the coordinating and monitoring agencies for GAP. Different parastatal agencies were also involved to implement the plans, such as setting up the treatment plants and carrying out the drainage interception and diversion work.[64] The local offices of the parastatal agencies were responsible for construction, operation and maintenance of the treatment plants in the individual towns. "The role of the municipal councils was limited to overseeing the implementation and operation".[65]

Although an organisational structure was established, Ganga pollution failed to retain its position of importance on the Government of India's agenda list. While Mrs Indira Gandhi and Mr Rajiv Gandhi were very concerned with the high levels of pollution of the Ganga and had kick-started efforts to clean up the river, successive prime ministers were not as interested in GAP, which resulted in a sharp drop in the level of political interest in the problem.[66] After their terms as Prime Minister had ended, neither their successors nor their political parties continued to consider cleaning the Ganga as an important project, which is evidenced by the fact that the NRCA which was chaired by the Prime Minister only met twice in 1994 and 1997, even though the meetings were supposed to be held annually.[67] This is very little when compared with the meeting frequency of other similar committees in other countries. For example, the National Productivity and Continuing Education Council established by the Singapore government in 2012 to oversee and drive efforts to uplift productivity would meet at least three years a year.

---

[64]Consortium of 7 Indian Institutes of Technology (2011).

[65]Consortium of 7 Indian Institutes of Technology (2011).

[66]Jaiswal, Ganga Action Plan — A Critical Analysis (2007).

[67]Consortium of 7 Indian Institutes of Technology (2011).

The Ganga Action Plan underwent several changes in its polices relating to the financial models for the treatment plants since it was launched. It initially started out in 1985 as being fully funded by the central government.[68] GAP II began in 1993 on a cost-sharing ratio of 50:50 between the central and state governments until 1997, when the funding model was changed to become fully funded by the central government. In March 2001, the funding model was changed again to a cost-sharing ratio of 70:30 between the central government and state government.[69] The decision by the NRCA in July 1997 that state governments shall defray the cost of operation and maintenance of the treatment plants built under GAP was never carried through.[70] There were also no plans about the longer term business model for the treatment plants, such as creating investment portfolios or incentive structures that would have ensured the project's long-term financial sustainability.[71]

As a result, the local governments did not have sufficient funds to operate and maintain the sewage and effluent treatment plants, which caused the sewage and effluent treatment plants to be left idle. For example, treatment plants in Bihar and Uttar Pradesh were not operational because the state and local governments could not provide funds.[72] Another example was that the non-governmental organisations were expected to adopt the ghats and maintain them but were not given neither the financial support nor any decision-making power.[73]

There was also a lack of operational capabilities at the local government level.[74] Local governments did not have trained personnel with the necessary technical expertise, skills, and experience to take charge of building, operating and maintaining the treatment plants in the future.[75] There were also no incentives for the local governments to become proficient at handling the sophisticated technology that was eventually

[68]Consortium of 7 Indian Institutes of Technology (2011).
[69]Jaiswal, Ganga Action Plan — A Critical Analysis (2007).
[70]Jaiswal, Ganga Action Plan — A Critical Analysis (2007).
[71]Consortium of 7 Indian Institutes of Technology (2011).
[72]Jaiswal, Ganga Action Plan — A Critical Analysis (2007).
[73]Das and Tamminga (2012).
[74]Jaiswal, Ganga Action Plan — A Critical Analysis (2007).
[75]Comptroller and Auditor General of India (2000).

installed. Local governments were not able to take over even after the judiciary had defined their duties and responsibilities vis-à-vis GAP and ruled that the responsibility for management, operation, and maintenance of GAP assets was theirs.[76]

This led to many instances of cost escalations, faulty designs and shoddy performance that were reported in association with the GAP.[77] One example happened in West Bengal, where the Kolkata Metropolitan Development Authority built two sewage treatment plants in 1994 and 1998 which pumped the treated sewage eastwards into a channel when the Ganga was to the west of the two treatment plants. As a result, these two sewage treatment plants did not contribute to helping to control the pollution flowing into the Ganga, and is an archetypal of the silo decision-making process commonly found in GAP. These problems with the treatment plants were in turn cited as one of the reasons why the local governments were reluctant to take over ownership of the plants after construction was completed.[78]

(ii) *Shortcomings identified*

It can be seen from Figure 2 that the local government institutions were not part of the organisational structure. Although the state governments were involved in the scheme preparation, the local governments were finally the ones tasked with building the treatment plants and subsequently overseeing the operation and maintenance of the plants. The shortcoming of this approach was that local governments had responsibility without accountability, which led to a lack of ownership of GAP by the local governments. The state of the treatment plants built under GAP in the states where GAP was executed can attest to the local governments' disinterest in GAP. For instance, in Uttar Pradesh, the Jal Nigam was in charge of planning and executing project activities under GAP like sewage management, and the local government was expected to take over after that.

[76]Jaiswal, Ganga Action Plan — A Critical Analysis (2007).

[77]Das and Tamminga (2012).

[78]Comptroller and Auditor General of India (2000).

However, the local government did not take over, citing its exclusion from the decision-making process right from the beginning as one of the reasons for its disinterest.[79] The statement that "GAP is no one's baby"[80] speaks volumes about how the people perceived GAP. This resulted in a lack of ownership of GAP by the local governments, which was further exacerbated by the fact that there were no incentives given to the local governments to encourage them to follow the instructions given by the NRCA and state governments.[81]

From the organisational structure of GAP, it can be inferred that decisions pertaining to the program were probably made at the central government level without inputs from the local governments. The consequences of this approach were implementation issues on the ground as decisions had been made without taking into consideration the financial, managerial, and infrastructural circumstances of the local governments. For instance, the central government decided on the type of technology to be installed without taking into account the local infrastructural, financial, and management capacity to operate and maintain the machinery.[82] The lack of a clear and viable financial model, and the lack of operational capabilities also further contributed to implementation problems and hampered the efficacy of GAP. This can be observed from the various instances of cost escalations, poor plant performance and treatment plants being left idle that were observed in some of the towns.[83]

The situation was further made worse by the lack of sustained political interest in GAP. This is evidenced by the fact that the NRCA only held two meetings since GAP was launched in 1985. This lack of sustained political interest was a barrier for GAP during the policy implementation stage.

---

[79] Jaiswal, Ganga Action Plan — A Critical Analysis (2007).

[80] Jaiswal, Ganga Action Plan — A Critical Analysis (2007).

[81] Jaiswal, Ganga Action Plan — A Critical Analysis (2007).

[82] Gupta (2005); Comptroller and Auditor General of India (2000).

[83] Das and Tamminga (2012); Comptroller and Auditor General of India (2000).

D. *Policy Evaluation Stage*

(i) *Policy evaluation for GAP I and GAP II*

When GAP was first launched in 1985, its objective was to improve the water quality of Ganga to acceptable standards by preventing the pollution load from reaching the river. In 1987, the objective of the plan was modified based on a recommendation from the Monitoring Committee of GAP. The modified objective was to restore the river water quality to the Designated Best Use class of Ganga, which is "Bathing Class" (Class B), as shown in Table 2.

For GAP I and GAP II, the Government of India adopted a multi-tier monitoring mechanism where all the action plans are monitored at both the Central as well as the State levels through a multi-tier monitoring mechanism (refer to Figure 2).[85] This included:

Central Level:

- Regular interaction and review by NRCD officials including frequent site visits. It also includes regular review by NRCD Project Director.
- Quarterly review of progress by the Steering Committee.
- Quarterly review of progress of scientific and technical aspects of the program as well as the impact of works on the river water quality by a Monitoring Committee headed by Member Environment, Planning Commissions.[86]

Table 2: Class B Water Quality Standard[84]

Indicator	Class B Water Quality Standard
pH	6.5 – 8.5
Biochemical Oxygen Demand	3 mg/l or less
Dissolved Oxygen	5 mg/l or more
Faecal Coliform	500 MPN/100ml or less

[84]The target standard for water quality under GAP was Class B i.e. suitable for outdoor bathing (Central Pollution Control Board, 2009).
[85]Jaiswal, Ganga Action Plan — A Critical Analysis (2007).
[86]Jaiswal, Ganga Action Plan — A Critical Analysis (2007).

- Quarterly review by a Standing Committee headed by the Union Minister of Environment & Forests.[87]
- Annual review of progress by the NRCA.

State Level[88]:

- The progress monitoring by a team of field engineers on day to day basis
- Monthly review of progress by the Chief Executive of the nodal implementing agency
- Citizen's Monitoring Committee in each town to review the progress and provide inputs for public participation and involvement
- Periodical review by the Divisional Project Monitoring Cells
- Periodical review by a High powered Committee under the Chairmanship of Chief Minister.

Alongside the official monitoring done by the government agencies, monitoring was also done by various academic and public institutions such as the Indian Institute of Technology in Kanpur and the University of Patna.

Throughout the implementation period of GAP I and GAP II, the official reports published by the Government India generally only explained the government's efforts to improve and maintain the level of water quality, and provided factual updates on the status of the building of sewage treatment plants and the utilization of central government budget, without including any qualitative assessments about the impact of GAP implementation on the people.[89] These reports showed that there was discernible improvement in terms of dissolved oxygen and biochemical oxygen demand levels. Government officials thus concluded that GAP has made a positive impact on the water quality of Ganga, as without the implementation of GAP, there would have been further deterioration in the

[87] Jaiswal, Ganga Action Plan — A Critical Analysis (2007).

[88] Jaiswal, Ganga Action Plan — A Critical Analysis (2007).

[89] This is based on a scan of all available reports published on the websites of the NMCG and NRCA.

Ganga water quality.[90] However, some critics have argued that GAP did not meet the expectations of the people.[91]

(ii) *Shortcomings identified*

It can be seen that the Government of India and the people had differing views on the success of GAP. This is an indication of an underlying disagreement about what GAP had been set up to achieve, the root cause of which is that there was no consensus between the government and public about the intended outcomes of GAP. Hence, the use of water quality indicators as indicators to support the evaluation of GAP, although reasonable from the perspective of government officials, was viewed by some people as being inappropriate. For example, Jaiswal has proposed that outcome indicators should have been used in addition to the output indicators.[92]

The Government of India adopted a multi-tier monitoring mechanism where all the action plans are monitored at both the Central as well as the State levels. Data collection was done by the government agencies, which was then used as inputs to evaluate the progress of the project implementation as well as the impact of the program. This was a strength of the monitoring system as it followed the principles of using first-hand data and doing peer reviews within the government institutions. The shortcoming of using this multi-tier monitoring mechanism was that it could only operate successfully if there was strong coordination between the different agencies and clarity in terms of the roles and responsibilities of each of the different monitoring agencies.

However, it seems that there was a lack of communication and coordination between the different monitoring agencies in India, which resulted in an ineffective monitoring system. This is evidenced by the fact that it had been recommended that "the reporting system and flow of

---

[90] National River Conservation Directorate (2009).

[91] Jaiswal, Ganga Action Plan — A Critical Analysis (2007).

[92] Jaiswal, Ganga Action Plan — A Critical Analysis (2007).

information from State Governments need to be activated and synergies built with implementing agencies and Central Government".[93] Although monitoring was also done by various academic and public institutions, there was insufficient sharing of the data between these non-government organisations and the government agencies to ensure that the data collected by these independent organizations were taken into consideration during the implementation reviews.[94]

Although the government agencies prepared many official reports regarding the status of GAP implementation, these reports were only published on government websites and not widely disseminated to the people, nor were the people engaged to provide their feedback about the program, as demonstrated by our inability to find any evidence of media coverage on the progress of GAP and any records of public engagement activities. It seemed as though the people were involved merely as spectators rather than as key stakeholders who had an interest in the program. This lack of effective public engagement was a shortcoming, as without the efforts of the Government of India to constantly nor effectively engage the public to keep them informed about the program and its progress, people tended to assume that the government's effort in cleaning the Ganga through GAP was a failure.[95]

E. *Summary of shortcomings*

A summary of the shortcomings identified in the policy process for GAP I and GAP II is presented in Table 3 below.

Recommendations

We have systematically reviewed GAP I and GAP II by breaking down the policy process into five distinct stages, and identified a total of 11

[93] Jaiswal, Ganga Action Plan — A Critical Analysis (2007).

[94] Jaiswal, Ganga Action Plan — A Critical Analysis (2007).

[95] Jaiswal, Ganga Action Plan — A Critical Analysis (2007); Sankat Mochan Foundation (2000).

Table 3: Summary of Shortcomings Identified

Stage(s)	Shortcomings Identified
Agenda Setting	(i) Insufficient public buy-in
Policy Formulation and Decision Making	(i) Lack of transparency about processes (ii) Lack of clarity about envisioned outcomes (iii) Narrow participation of stakeholders (iv) Insufficient consideration of local context
Policy Implementation	(i) Lack of ownership of GAP by local governments (ii) Insufficient consideration of circumstances of local governments (iii) Lack of sustained political interest in GAP
Policy Evaluation	(i) Use of inappropriate evaluation measures (ii) Lack of coordination between monitoring agencies (iii) Lack of effective public engagement

shortcomings across the five stages. However, besides reviewing each stage individually and identifying the specific shortcomings in each of the stages, it is also important to look at the overall picture by putting everything together. After all, some of the key criticisms of the Stages Model has been that the stages do not necessarily occur sequentially in practice and the model does not consider the interaction between the different stages.[96]

It can be seen that some of the shortcomings identified in one of the policy stages are very similar or related to shortcomings identified in another policy stage. For instance, insufficient consideration of the local context was identified as a shortcoming in the policy formulation and decision making stages, which is similar to the shortcoming of insufficient consideration of the circumstances of local governments during the policy implementation stage. Based on our analysis, we have come up with three recommendations on how these shortcomings could be avoided for Nanami Ganga.

[96] Sabatier and Weible (1999).

Recommendation 1: Define a clear vision for Namami Ganga

One of the recommended enhancements that could be made to the policy process for *Namami Ganga* is to define a clear vision of what the program intends to achieve in concrete terms and what people can expect to see at the end of program implementation. Currently, *Namami Ganga* is described as being an "Integrated Ganga Conservation Mission" that "approaches Ganga Rejuvenation by consolidating the existing ongoing efforts and planning for a concrete action plan for future".[97] All information currently available about *Namami Ganga* focuses only on the activities that have been planned, although there are mentions of plans to develop a long-term vision for the program.[98]

However, it is not sufficient to have a vision — the vision must be one which the people agree with. The people need to recognise that there are benefits to cleaning up the Ganga, which first requires them to be recognise that the current pollution levels of the Ganga is a problem. One of the ways that this could be done is by running awareness campaigns to make people aware of the problem, and framing the problem in more culturally acceptable ways such as talking about "raw sewage being poured on her body"[99] so as to garner the support of the people for *Namami Ganga*. It is important that the public are involved in defining the vision, as it would improving upon the shortcomings in GAP I and GAP II where there was insufficient public buy-in. Once the people become interested the program, this will naturally put pressure on the government to take greater interest in the program, thus overcoming the previous shortcoming of the lack of sustained political interest in GAP.

Once there is a clear vision that is supported by the people and the government, the shortcomings during the policy evaluation stage can also be improved upon. A clear vision makes it possible to conduct proper

[97] National Mission for Clean Ganga (2014).

[98] National Mission for Clean Ganga (2014).

[99] Das and Tamminga (2012).

impact assessment about Nanami Ganga, by facilitating the identification of appropriate monitoring indicators beyond measuring water quality.

Recommendation 2: Make Nanami Ganga a "people's program"

It is recommended that Nanami Ganga be made into a "People's Program", so that the people will take greater ownership of the program. This recommendation seeks to improve upon the shortcomings in GAP I and GAP II where there was narrow participation during the policy formulation stage and lack of effective public engagement during the policy evaluation stage. As cleaning up the Ganga is a wicked problem, there can be many differing views about what the problem is and also contradictory solutions that may solve the identified problem but cause problems in other areas. Since there is considerable uncertainty and ambiguity when addressing wicked problems, engaging the public more actively in the policy process would allow the Government of India to better understand all the different concerns of the stakeholders and work with them to formulate solutions that help to achieve public value outcomes that benefit users, citizens and the community.[100] It will also improve upon the previous shortcoming of the lack of transparency about what had occurred during the policy formulation and decision making stages.

People on the ground and working on the frontline are also the most knowledgeable about what is necessary for programs to be implemented successfully. By promoting open dialogue with these people and encouraging their active participation in all stages of the policy process, the people have the opportunity to express their own views and hear the views of other people, giving them a platform consensus building which will increase allow them to take greater ownership of the program. Making *Namami Ganga* a "People's Program" will also give greater legitimacy to the Government of India when implementing the action plans. This will increase the chances for success.

Government should also conduct surveys on the people so as to measure public awareness of *Namami Ganga* and to assess the impact of the

[100] Benington (2011).

program on the people. The findings from the surveys should be published to build a sense of crisis among the people about the importance of cleaning up the Ganga. By being transparent about what it is doing, this also helps to build trust between the people and the government, giving the government even greater legitimacy for its actions.

Recommendation 3: Involve state and local governments more closely in the policy process

One recommended enhancement to the policy process for *Namami Ganga* is to involve the state and local governments more closely in all stages of the policy process. This will address the shortcoming identified during the policy formulation stage where there was narrow participation of the state governments only in the lowest tier of the NRCA for GAP I and GAP II. In addition, this recommendation will also address other shortcomings, namely the insufficient consideration of the local context during the policy formulation stage, the lack of ownership of the program by the local government and insufficient consideration of the circumstances of local governments during the policy implementation stage, and the lack of coordination between monitoring agencies during the policy evaluation stage.

Specifically, the state and local governments should be consulted throughout the entire policy process, including during the stage of policy formulation. This would ensure that policy formulation occurs in an environment where the policymakers are fully apprised of the concerns and the constraints that the state and local governments may face during policy implementation. For instance, there should be greater communication upfront about the existing operational capabilities of the local governments, so that the implementation plans would take into consideration the time needed to build up local capabilities to operationalize the action plans and implementation timelines are then phased out accordingly.

The close involvement of the state and local governments in the policy formulation stage would also allow them to voice their concerns over matters such as the funding and financial model for implementation of the program, which was one of the shortcomings during the policy implementation stage for GAP I and GAP II. Indisputably, financing is the key driver of any undertaking, and particularly so for nationwide programs like Nanami Ganga that would impact many people. Therefore, it is imperative to have a viable financial model, which not only sets out the source of funding for the initial capital development stages, but also plans ahead as to how the treatment plants can be financially self-sustaining based on interest income from investments or by collecting fees from the people.

In addition, greater involvement of the state and local governments would mean that institutional barriers that may hinder policy implementation are highlighted during the policy formulation stage, thus ensuring that the decision-makers take into consideration the actions required to overcome these challenges when developing their implementation strategy. For instance, many of the states could not obtain the land required to build the sewage treatment plants during the implementation of GAP I and GAP II, thus leading to delays.[101] If this barrier had been identified earlier during the policy formulation stage, NRCA could have considered passing legislation to allow the state or local governments to acquire the land from the current owners.

An added advantage of having the state and local governments more involved in the policy process is that it builds up a feelings of trust and camaraderie between the central, state and local governments. Hence, state governments and local governments would be more open and willing to share their experiences with one another, and thus help each other to avoid having to solve the same problems multiple times. The improved communication and coordination between the different levels of government and also across states, would also make it possible for more frequent meetings to be held, thus positively impacting the monitoring and evaluation activities.

In addition to involving the state and local governments more closely in the policy process, there should also be more consultations with other ministries which may be impacted by the plans under *Namami Ganga* to

---

[101] Consortium of 7 Indian Institutes of Technology (2011).

clean up the river, or which may be overseeing other policies that could impact the implementation of the plans under *Namami Ganga*. Examples of such ministries include the Ministry of Agriculture, Ministry of Urban Development and the Ministry of New and Renewable Energy. This would ensure that a more integrated approach is taken for *Namami Ganga*, thus increasing the chances for success of the program.

Summary of recommendations

A summary of the recommendations and the shortcomings that they are intended to address is shown in Table 4 below.

Table 4: Summary of Recommendations

Recommendation	Shortcoming(s) that Recommendation Addresses
Define a clear vision for *Namami Ganga*	Policy formulation and decision making: — Lack of clarity about envisioned outcomes Policy evaluation: — Use of inappropriate evaluation measures
Make *Namami Ganga* a "People's Program"	Agenda setting: — Insufficient public buy-in — Insufficient political interest Policy formulation and decision making: — Lack of transparency about processes — Narrow participation of stakeholders — Lack of sustained political interest in GAP Policy evaluation: — Lack of effective public engagement
Involve state and local governments more closely in the policy process	Policy formulation and decision making: — Narrow participation of stakeholders — Insufficient consideration of local context Policy implementation: — Lack of ownership of GAP by local governments — Insufficient consideration of circumstances of local governments Policy evaluation: — Lack of coordination between monitoring agencies

Conclusion

Cleaning up the Ganga has been the vision of many Indian Prime Ministers since 1979. However, none of the programs that have been implemented since the problem was first put on the government's agenda in 1979 has been successful in improving the pollution levels of the Ganga. While Prime Minister Narendra Modi and his government works on developing a long-term vision for Nanami Ganga, and a strategy as to how they can achieve the vision, it is timely to conduct a review of the previous Ganga Action Plan (GAP) so as to see if there are learning points from the past so as to avoid making the same mistakes in the future.

A systematic review of GAP using the Stages Model of policy making has identified 11 shortcomings in the policy process for GAP in the stages of agenda setting, policy formulation, decision making, policy implementation and policy evaluation. Some of the shortcomings identified in one stage are very similar or related to shortcomings identified in another policy stage. After grouping the related shortcomings together, we have come up three recommendations as to how the policy process for Nanami Ganga could be enhanced. They are:

a) Define a clear vision for *Namami Ganga*;
b) Make *Namami Ganga* a "People's Program"
c) Involve state and local governments more closely in the policy process.

This study has made use of a different set of lenses to look at the past implementation of GAP. It is hoped that the findings from this study will be able to contribute to the wider debates about how the India can reduce the pollution levels of the Ganga and achieve the end goal of a 'Rejuvenated Ganga'.

References

Ahmed S. (1994). The rhetoric of participation re-examined: The state, NGOs, and the water users at Varanasi, Uttar Pradesh, India. *Environmentlist*, 14, 3–16.

Ansell, C, and A Gash. (2007). Collaborative Governance in Theory and Practice. *Journal of Public Administration Research & Theory*, 18, 543–571.

Bardach, E. (2012). *A Practical Guide for Policy Analysis: The Eightfold Path to More Effective Problem Solving*. Sage Publications.

Benington, J and M, Mark (2011). Public Value in Complex and Changing Times. *Public Value: Theory and Practice*, 1–30.

Bharatiya Janata Party (2014). BJP Manifesto 2014. http://bjpelectionmanifesto.com/.

Brewer, GD and P DeLeon (1983). *The Foundations of Policy Analysis*. Dorsey Press.

Central Pollution Control Board (2009). Ganga Water Quality Trend.

Central Pollution Control Board (2013). Pollution Assessment: River Ganga.

Centre for Science and Environment (2014). Ganga: The River, Its Pollution and What We Can Do to Clean It.

Centre for Science and Environment (2013). Status Paper for River Ganga: Past Failures and Current Challenges.

Comptroller and Auditor General of India (2000). Audit Report on Ganga Action Plan for Period Ended March 1999. Comptroller and Auditor General of India. http://www.saiindia.gov.in/english/home/public_folder/training/Structure%20Training%20Module/Enviornmental%20Auditing%20RTI,%20Mumbai/Day%203/GAP-CASE%20STUDY%20%5BCompatibility%20Mode%5D.pdf.

Comptroller and Auditor General of India (2000). CAG Report of 2000 Union Government: Environmental Auditing.

Consortium of 7 Indian Institutes of Technology (2011). SWOT Analysis of Ganga Action Plan.

Das, P and KR Tamminga (2012). The Ganges and the GAP: An Assessment of Efforts to Clean a Sacred River. *Sustainability*, 1647–1668.

Fischer, F, G Miller and Mara Sidney (2007). *Handbook of Public Policy Analysis: Theory, Politics, and Methods*. Boca Raton: CRC Press.

Gupta, KR (2005). *Encyclopaedia of Environment*. New Delhi, India: Atlantic Publishers & Distributors.

Hasan, S (2015). Water Quality of River Ganga — Pre and Post GAP: A Review. *International Journal of Advanced Research in Science, Engineering and Technology*, (2): 361–364.

Her Majesty's Treasury (2011). *The Magenta Book: Guidance for Evaluation*. http://www.hm-treasury.gov. uk/data_magentabook_index.htm.

Hill, M (2004). *The Public Policy Process*. Taylor & Francis.

Howlett, M (2011). Stages Model of Policy Making. In *International Encyclopedia of Political Science*, Bertrand, B, D Berg-Schlosser and L Morlino, pp. 2480–2482. Sage Publications.

Jaiswal, RK (2007). Ganga and GAP. *Eco Friends* — (2007). Ganga Action Plan — A Critical Analysis. *Kanpur: Eco Friends*, 28 May http://www.ecofriends.org/main/eganga/ganga_gap.htm.

Lasswell, H (1951). The Policy Orientation. In *The Policy Sciences: Recent Developments in Scope and Methods*, Lerner, D and HD Lasswell, pp. 3–15. Stanford University Press.

Ministry of Environment and Forest (2000). *National River Conservation Directorate.* http://www.moef.nic.in/sites/default/files/NRCD/obj.html (Accessed 2 June, 2015).

Narain, Sunita (2014) Ganga: The River, Its Pollution and What We Can Do to Clean It.

National Ganga River Basin Authority (2009). *Towns Covered in Ganga Action Plan Phase II.* http://www.moef.nic.in/sites/default/files/ngrba/towncovered2.html. (Accessed 3 June, 2015).

National Mission for Clean Ganga (2014). *National Mission for Clean Ganga.* https://nmcg.nic.in/NamamiGanga.aspx. (Accessed 4 February, 2015).

National River Conservation Directorate (2009). Status Report on River Ganga: State of Environment and River Quality.

Rittel, H and Webber M (1973). Dilemmas in a General Theory of Planning. *Policy Sciences* 4(2), 155–169.

Rossi, P, M Lipsey and Freeman H (2004). *Evaluation: A Systematic Approach.* Sage Publications.

Sabatier, PA and Christopher W (1999). *Theories of the Policy Process.* Westview Press.

Sankat Mochan Foundation (2000). *Failure of the Ganga Action Plan.* http://sankatmochan.tripod.com/GAPFailure.htm. (Accessed 8 June, 2015).

Trochim, WMK (2009). Evaluation Policy and Evaluation Practice. *New Directions for Evaluation*, 123, 13–32.

Weimer, D and A Vining (2004). *Policy Analysis: Concepts and Practice.* Pearson Prentice Hall.

World Bank (2011). India — National Ganga River Basin Project : Environmental and Social Analysis and Management Framework (ESMF) — Executive Summary.

Wu, X, M Ramesh, M Howlett and S Fritzen M (2010). *The Public Policy Primer: Managing the Policy Process.* Routledge.

Chapter 3

FORMING A BROAD UMBRELLA: FACILITATING INTER- AND INTRA-STATE COORDINATION

Hlaing Myat, Lay Sopagna, Sirilak Kakanopas, Soe Myint Aung, Visvanathakkurukkal Sokkanathan and Joost Buurman

1. Introduction

Even after millions of dollars were poured into clean-up efforts over more than two decades with high-level political commitments and rhetoric, the mighty and sacred mother Ganga of India is still severely polluted. The Ganga Action Plan (GAP-I, 1986 and GAP-II, 1993) did little to improve the water quality and restore aquatic resources, despite the Indian government's commitment to clean the river (CAG, 2000; Das and Tamminga, 2012). Freshwater is increasingly used for farms, industries and cities, and wastewater continues to be discharged, largely untreated, into the river (Das, 2011). The impacts from this pollution are visible: the health of the people who live along the riverbanks (Hamner *et al.*, 2006) as well as the health of aquatic life, including endangered species such as the Ganga dolphins is affected (NGRBA, 2010). The image of ancient holy cities like Varanasi is affected. Poor sanitation and visible pollution in the city is affecting tourism as well.[1]

[1] Source: Public lecture at Varanasi School of Management, December 2014.

One of the main reasons for the unsuccessful attempts to clean the river lies in the failure to establish effective institutional arrangements at national and state levels, and the inefficiency in planning, monitoring and implementing operations (IIT, 2011). One of the key characteristics of the GAP was concentration of power and decision-making in the national government or central-level authority. Overly centralised decision-making has led to many program delays, gaps and shortcomings when the GAP were implemented at various levels (Das and Tamminga, 2012). Institutional arrangements in the policy and regulatory framework of the GAP were ambiguous or duplicated other efforts. Moreover, the effort of Ganga clean-up has been complicated by conflicting interests and overlapping of functions between the national government and the states. GAP failed to achieve significant outcomes concerning the reduction of domestic sewerage and industrial pollution; actions outlined in GAP I and II lacked proper implementation (CAG, 2000). Coordination and cooperation between agencies, vertical as well as horizontal, was poor. There were cases where agencies implemented a particular program either in isolation or duplicated ongoing efforts, particularly with regards to programs that required cross-agency jurisdiction or multi-level jurisdiction.

Funding was not an issue in the GAP. Ever since the plan was launched, collaborative working between departments became rare and hampered by weak monitoring mechanisms and fragmented networking.[2] Many agencies have overlapping roles, while many others have little space to implement their plans. Stalemate and inaction are not uncommon. As a case in point, untreated wastewater continues to be discharged from paper, pulp and tannery industries.[3] The clean Ganga program suffered from divergent priorities of institutions due to differences in political affiliation and interest, social and administrative norms, and institutional arrangements. Instead of collaborating toward achieving common goals, many agencies engaged in perpetual disputes signalling conflict of interest and improper resource allocation. Central and state governments differed over GAP budget allocation, leaving states unable to formulate detailed project reports (CAG, 2000; IIT, 2011).

It is clear that establishing effective collaboration among various institutions and developing an alternative framework for interstate river basin

[2] Source: Central regulatory official at a public forum in New Delhi, December 2014.

[3] Source: Senior environmentalist at a public forum in New Delhi, December 2014.

management is essential to mitigate pollution (NGRBA, 2011a). To develop such a framework, a detailed understanding of the coordination and cooperation issues under GAP-I and GAP-II is required. This chapter examines three areas where better coordination at various levels may lead to better results of policies: coordination and cooperation between the national government and the states, between states (interstate) and within states (intrastate). This will help to understand how institutions can coordinate and cooperate with each other in cleaning the Ganga, and the factors that hinder smooth and effective coordination. Based on this, a balanced and effective model of coordination for the Ganga clean-up efforts will be suggested. This chapter will cover the period from the time when GAP-I was started, around 1986, until the recent developments under the management of the National Ganga River Basin Authority (NGRBA).

The theoretical framework for the analysis (Section 2) is based on Integrated Water Resources Management (IWRM). Our main focus is on system integration in IWRM, even though IWRM as a generic framework has much broader annotations and scope. In addition, we use the two interrelated concepts of 'delegation' (of power/authority) and 'incentive and regulatory structure' in the theoretical framework. Relevant literature on coordination and collaboration is also discussed. This chapter provides two types of analysis: diagnostic analysis (Section 3) is a close examination of what has impeded Ganga coordination. To build on our understanding of direct analysis of potential problem areas, we also provide a more suggestive discussion/synthesis (Section 4), with the intention of offering a broad set of alternative policy options for planners and policy makers in India. We conclude with policy recommendations for different coordination domains (Section 5).

2. Integrated Water Resources Management, Delegation, and Incentives

2.1. *Integrated basin approach*

Integrated Water Resources Management is a widely used framework for sustainable water use, and equitable and effective management of water resources in river basins around the globe. The earliest mention of IWRM goes back to 1933, with the establishment of the Tennessee Valley Authority in USA (Stålnacke and Gooch, 2010). It was first mainstreamed into the international water governance agenda in 1977 through the International Water Conference. In 1992, principles of IWRM were incorporated in the 'Dublin Principles' as a preparation to the United Nations Earth Summit (Van Beek and Lincklaen Arriens, 2014). In the European Commission's Water Framework Directive (EU, 2000), IWRM is a key analytical and operational framework. The adoption of Millennium Development Goals in 2000 and the Johannesburg Summit on Sustainable Development 2002 were other milestones for IWRM (Stålnacke and Gooch, 2010). In the past two decades there has been a growing consensus that the integration of ecosystem considerations into political, legal, economic, policy and institutional perspectives of trans-boundary river basin management is the most viable option for dealing with ever increasing pressures on river systems throughout the world. Over time, the IWRM framework has been gradually enhanced to adapt to various regional contexts and ever changing global challenges pertaining to water resources and river ecosystems.

The Global Water Partnership defines IWRM as "a process which promotes the coordinated development and management of water, land and related resources, in order to maximise the resultant economic and social welfare in an equitable manner without compromising the sustainability of vital eco-systems" (GWP, 2000). As opposed to the tradition sectoral, top-down planning and management for water resources, IWRM advocates an integrated approach. Integration should be pursued at different levels and in different domains: integration between human systems and natural systems, between the water sector and the economy, across levels of decision making, and across temporal and spatial scales (Medema *et al.*, 2008; Nesheim *et al.*, 2010; Van Beek and Lincklaen Arriens, 2014). When focussing on rivers, IWRM takes river basins or catchment areas as a unit of analysis rather than dealing with the river as an isolated stretch of water. Integrated Water Resources Management is a concept and a process, rather than an objective or goal, and as such is described in terms of principles and components. Practical implementation details are provided in the form of a toolbox of good practice (Medema *et al.*, 2008; http://www.gwp.org/ToolBox/).

Nesheim *et al.* (2010) discuss four principles derived from the IWRM concept, which are also relevant to our analysis of coordination and cooperation in the context of the Ganga clean-up. Box 1 briefly describes these principles. In addition, Medema *et al.* (2008) mention the component 'set of management instruments' as a necessary governance condition for implementation of IWRM, which we add as a fifth principle. Among the five principles, principle 3, effective governance, is most relevant in the context of the Ganga analysis. The other principles complement the analysis.

Box 1: Selected Principles of IWRM in the Ganga Context

1. IWRM as a mechanism for protection of the river catchments and the environment

This principle mainly encompasses the environmental aspects of terrestrial and aquatic ecosystems of the river basin. It also covers understanding of measurements and monitoring of both surface and ground water and ecosystem quality. The current practice in Ganga clean-up programs of focusing predominantly on some pollution indicators measured in the river is not sufficient for fulfilling the requirements of sustainable river basin management.

2. IWRM as a mechanism to ensure efficient and equitable use of water resources in the river basin

The premise of this principle is that water should be treated as both economic and social good. Water as an economic good means that adequate pricing mechanisms and incentives should be introduced to avoid wasteful and unsustainable use and ensure efficient allocation. For instance, for the Ganga ultimately polluters should pay the full cost of pollution abatement measures and water users should pay the economic cost of water. Water as a social good refers to the fact that it is a basic human right to have access to clean water. Economic and policy instruments should make sure that water use is equitable. Necessary institutional and legal arrangements should be in place and fully operational to fulfill this principle.

3. IWRM requires effective governance that ensures inter-sectoral integration and coordination

The institutional and inter-sector integration and coordination dimensions are the core of IWRM. Both formal and informal coordination mechanisms need to be established. Water resources planning and management should be based

(*Continued*)

Box 1: (*Continued*)

on a participatory approach involving stakeholders and policy-makers at all levels. IWRM promotes decentralisation to the lowest possible decision-making level to increase participation, including participation by vulnerable groups. Effective governance, coordination and cooperation are severely lacking in the Ganga basin.

4. IWRM requires institutional capacity building

This principle relates to all other principles because institutional capacity is essential for effective implementation. In this regard, capacity building means that different parties involved should possess not only sufficient information and knowledge, but also expertise and incentives. Capacity building should take place at all levels. In the case of Ganga, this principle is useful not only as operational guidance, but also as an evaluative tool.

5. IWRM as an effective information and decision management system

Successful implementation of IWRM requires a set of management instruments that can provide the data and information to correctly assess water resources, and make appropriate water management decisions. To develop an integrated basin management model, the analysis usually starts with identification of water users, their demands and expectations. Then the model compiles basin-related information into an integrated information system, which can be used in later stages for decision support. After decisions are implemented, management systems should help in assessment of the performance of the measures and the impacts of the programs.

Based on Nesheim et al. (2010), Medema et al. (2008) and www.gwp.org.

So far we have focussed on interagency cooperation and coordination in the Ganga basin. In fact, principle 3 states that IWRM requires more than interagency coordination. It requires that all relevant sectors for river basin management should be integrated into formal and overarching structures. At the same time, informal networks and exchange mechanisms should be established, because both formal and informal channels are needed for effective integration and coordination. An important aspect in this is the participation of all stakeholders, which should raise awareness of the importance of a clean and ecologically sustainable river. Participation also concerns having access to information to participate in decision-making, which links to principle 5. Principle 3 covers the core of the present study. Integrated governing bodies such as NGRBA, which comprise various ministries and representatives from sub-national governments, have been formed for the Ganga clean-up, both at national and state level. However, there are also some important institutions working outside of NGRBA's jurisdiction or not fully integrated into the NGRBA structure. Here, the IWRM framework provides us with three useful measures, namely integration, coordination and stakeholder participation, to analyse and evaluate the structure and functioning of Ganga clean-up institutions.

Integrated Water Resources Management has not remained without criticism. The major drawback of IWRM as a framework is its lack of clear methodologies (Stålnacke and Gooch, 2010). A number of commenters argued that IWRM is quite vague as a framework and difficult to implement (Nesheim *et al.*, 2010). Others pointed out that IWRM is a universal concept and is questionable if it always can be applied in a heterogeneous world, and particularly in developing countries (Biswas, 2008). Some highlight that IWRM should be linked to a country's overall sustainable development strategies and public administration framework; otherwise, it will not work effectively or will be very difficult to implement (Lenton and Muller, 2012, p13). To address some of the limitations of the IWRM concept, we propose two additional concepts for a more comprehensive analysis: delegation of power and authority, and incentive and regulatory structures.

2.2. *Delegation of power and authority*

The concept of delegation of power and authority is derived from decentralisation, which is encouraged by IWRM (principle 3). Decentralisation is a widely used term that has many interpretations and implementations. It can be described as the transfer of power and reallocation of resources from higher to lower level governments in a political-administrative and territorial hierarchy (Agrawal and Ribot, 1999). It requires devolution of functions and all administrative, political and economic attributes (Faguet, 2014). A distinction can be made between administrative decentralisation (or de-concentration), whereby power and authority is delegated to lower-level authorities who still remain accountable to higher-level authorities, and political decentralisation, whereby the lower-level authorities will become downwardly accountable to the local population (Faguet, 2014). According to the World Bank's interpretation (World Bank, 2001), decentralisation improves complex bureaucratic procedure and decision-making by drawing a clear line between policy concentration at the top levels and routine tasks at the local levels. In this manner, the central government can reach more people with services and take into account the local context and needs. Delegation of power and authority can also address issues of overlapping functions among different government actors.

Decentralisation or delegation of power alone is not sufficient in ensuring effective management and coordination. In many cases, delegation will end up as the monopolistic actions of local elites, especially when oversight and regulatory capacity is weak. In other cases, delegation does not produce sustainable results because power is decentralised without sufficient financial resources. Ribot (2002) emphasised that decentralisation can lead to conflict. "Transferring power without accountable representation is dangerous. Establishing accountable representation without powers is empty." (Ribot, 2002, p1). In practice, many decentralisation reforms have been unsuccessful as delegation was incomplete (Agrawal and Ribot, 1999; Ribot, 2002). Delegation should include power, accountability and resources. This links to the third analytical concept of incentives and regulation.

2.3. *Incentives and regulatory structures*

Integrated Water Resources Management and delegation will not work as intended without adequate incentives and regulatory structures in place. Incentives motivate people and organisations to do (or not do) something and can come in several forms: e.g. monetary rewards, promotion, higher esteem, etc. Many disciplines, economics in particular, study incentives from several perspectives. For instance, incentives to pollute or protect water quality by different actors are extensively studied in environmental economics. The present chapter focuses mainly on incentive structures between different government levels, within the bureaucratic and political system. Dixit (2002) mentions in this respect that public sector agencies have special characteristics that require specific attention in the analysis of incentives. These include a multiplicity of stakeholders, often conflicting interests about ends and means, and many tiers of management. Incentives can be formalised in the regulatory structure, but they can also be external. Regulation refers to the imposition of rules, which can be backed by fines or penalties.

Incentives and regulation can be analysed using the Principal Agent Theory. This theory emerged as a modelling technique in economics to explain asymmetries in information and authority, and has been further developed in the policy sciences to provide insights into the role of information asymmetry and incentives in political and bureaucratic relationships (Dixit, 2002; Miller, 2005). In the Principal Agent Theory, the principal has power and authority to devise incentives for agents in order to induce actions in the interest of principal. The principal is informationally disadvantaged, as he or she cannot directly observe the effort the agent makes to take an action, though the outcome of the action is measurable. However, the outcome of the action is also dependent on some other random variables. This leads to the moral hazard problem, whereby the agent's actions affect the principal's payoff (Dixit, 2002). For instance, the state government may delegate sewage treatment to local governments in aiming to meet certain standards. However, the amount of sewage treated may depend on available funding for the construction of sewage treatment plants, the duration of power outages, and economic growth affecting sewage growth, in addition to the local government's efforts to increase sewage treatment. It would be very expensive for the state government to monitor in detail all efforts taken by all local governments, hence they have an information disadvantage. Not meeting sewage treatment standards may, for instance, affect funding provided to the state by the central government (the payoff).

Alternatives to the use of incentives are strict monitoring and regulation of the actions of the agents. However, this would involve very large transaction costs (monitoring and enforcement costs) and some actions are not possible to monitor (information asymmetry). Hence, the principle will never fully regulate, but use a mix of incentives and regulation. Specifically on pollution control, several authors have proposed innovative incentive structures, in addition to traditional means of enforcement such as fines and penalties. For example, drawing lessons from emission controls in

Canada, Foulon *et al.* (2000) suggested that public disclosure of non-abiding organisations could effectively serve as an additional and strong incentive for pollution control.

2.4. *Coordination and collaboration*

An overarching theme in analysis of the Ganga clean-up is the issue of coordination and collaboration. Analytical frameworks for coordination and collaboration have a large overlap with the topics discussed above: IWRM, delegation of power and authority, and incentives and regulatory structures. Yet, in the governance literature some useful additional insights can be found on the coordination and collaboration between government institutions to achieve policy outcomes.

According to Bryson *et al.* (2006) the value of collaborative governance lies in the establishment and sustainment of a regime of mutual gain that taps on people's common desires and interest for building a better world. That regime is defined as "sets of implicit or explicit principles, rules, norms, and decision-making procedures around which actors' expectations converge in a given area" (Krasner, 1983 as cited in Bryson *et al.*, 2006). However, there are times that governments struggle to create a regime that brings multiple agencies together to achieve a given policy goal. Government operating agencies can fail because they lack an external orientation, are not properly organised internally or are indifferent to fairness of allocation of income or wealth (Blandl, 1998) as cited in Bryson and Crosby, 2005). This could be the result of monopoly practices, distracted monitoring, inappropriate pricing mechanism, bounded reality or imperfect information. Bryson and Crosby (2005) further distil from the literature some reasons why several types of government agencies can fail to function or take appropriate actions. These include:

- Shortcomings in higher-level government direction-setting due to legislative or bureaucratic gridlock, lack of visionary leadership, or faulty understanding of the problem issue or needs;
- Compromising of accountability due to dysfunctional oversight bodies that may be underfunded, wrongly organised, wrongly incentivised or have inadequate authority;

- Inadequate support from support agencies due to poor organisational design, insufficient resources, or perverse incentives;
- Failure of regulatory agencies due to poor organisational design, insufficient resources, perverse incentives, inadequate authority, capture by the regulated, poor information, or corruption.

2.5. *Theoretical framework*

Figure 1 summarises the theoretical framework for this chapter. Integrated Water Resources Management, delegation of power, and incentives and regulatory structures will be used respectively to analyse a specific layer of institutional relationship. The IWRM framework will be used to discuss the central level institutions. Elements of IWRM can also be relevant for coordination and collaboration issues at other levels, hence it may also be applied in those discussions. Similarly, the framework of delegation of power is specifically useful in the discussion of power and resources sharing between central and sub-national institutions, and how the sharing arrangements influence outcomes, though some elements are also relevant

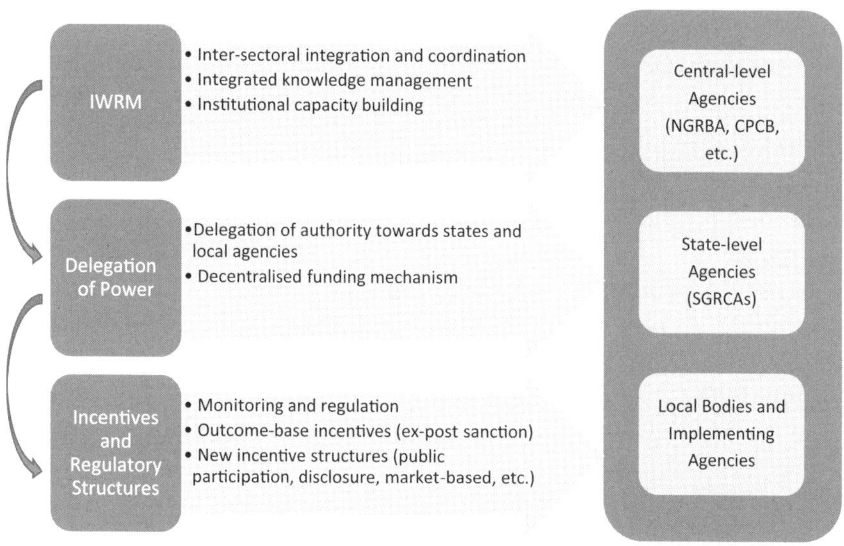

Figure 1: Theoretical Framework

for the discussion of other layers of institutional relationship. The framework of incentives and regulatory structures is most suited for the discussion of local-level implementation, but can also play a role at other levels.

3. Major Coordination Issues Facing Ganga Clean-Up Programmes: Inter- and Intra-State Coordination

3.1. *Central-level and interstate coordination*

The following sections closely examine what has impeded coordination and collaboration in the efforts to clean-up the Ganga River. Although the terms coordination and collaboration do not mean the same, we will use them interchangeably for short and be only specific when required. We start our discussion by looking at the organisation, legal and regulatory framework, and institutional arrangements, before analysing the major issues in coordination.

3.1.1. *Organisation overview*

In matters related to Ganga clean-up, the Government of India functions in a predominantly hierarchical order. The chain of command is established by the central government, mainly through NGRBA and concerned union ministries including the Union Ministry of Environment & Forest (MOEF), and Union Ministry of Water Resources, River Development & Ganga Rejuvenation. The institutional arrangement for Ganga clean-up programmes follows the three government layers in India: central, state and local.

Figure 2 depicts the key public agencies working on Ganga projects at different layers of the government. At the central level, the NGRBA was established in 2009 and is overseen by the Parliament, Supreme Court and Auditor General. The National Green Tribunal plays a role at the central level to handle environmental disputes involving multidisciplinary issues.[4]

[4]The websites (and sources of information) are: MOEF: http://www.moef.nic.in/; NGRBA: http://www.moef.nic.in/sites/default/files/ngrba/index.html; Green Tribunal: www.green tribunal.gov.in/.

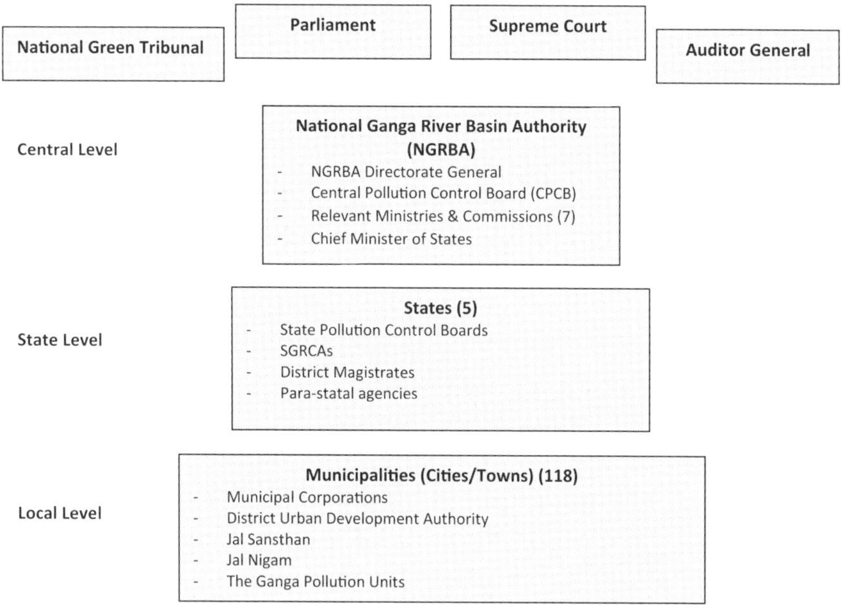

Figure 2: Major Institutions at Various Government Levels in Ganga Clean-up Programmes

The Ganga basin covers five states and 118 municipalities. At the local level Jal Sansthan and Jal Nigam are organisations responsible for development and management of water and sanitation in Uttarakhand and Uttar Pradesh respectively. Although their area of operation is state-wide, they effectively operate at the local level.

3.1.2. *Legal and regulatory framework*

Legislative power is exercised by the parliament in the democratic political system of India. The parliament has promulgated various laws giving mandates and responsibilities for Ganga pollution abatement and control to the central government, state governments, and pollution control boards. India has a long tradition of setting a regulatory and legal framework for water resources regulation and management; some of the earliest formal provisions date back to British colonial era (Cullet and Gupta, 2009). At present, there is a broad and diverse set of legal and regulatory regimes dealing with water and water resources management, including regulation for interstate river basin such as the Ganga. Directly related to Ganga river basin management are amongst others the Water Prevention and Control of Pollution Act (1974), the Environmental Protection Act (1986), and legally binding framework documents such as the National Water Policy (2012).

Although many of these laws and regulations have been in place for some time, implementation and regulation remain problematic. Cullet and Gupta (2009) mention that the water law in India is fragmented and patchy, partly as it is both a central and state level issue, and partly as water law is also included in environmental and health laws. Furthermore, the inadequacy of agency-to-agency collaboration hampers successful formulation of a legal framework to effectively oversee projects and deliver outcomes.

In addition to the Parliament, the Supreme Court of India can provide special jurisdiction — especially in case of serious pollution problems and critical failure of compliance and performance — authorising the central government, union ministries, or agencies, to take remedial actions. For example, the Supreme Court had asked the National Green Tribunal to deal with highly polluting industries, which discharged toxic effluents into the river Ganga on a large scale.[5] The Supreme Court has also conducted hearings on Public Interest Litigation (PIL) cases related to Ganga pollution since the late 1970s (Singhal, 2012). Under a PIL, any concerned individual can file a lawsuit against any party, including government agencies, to settle matters related to serious environmental damage or failure to protect the natural environment. Furthermore, the Comptroller and Auditor General of India (CAG) promotes accountability by auditing use of public funds and outcomes produced by the government agencies.[6] For example, for Ganga Action Plan I and II, the Auditor General checked all official reports and documents of the government agencies involved in using budgets for projects implemented in four states (CAG, 2000).

---

[5] Media report from DNA Online Newspaper, "*Take action against units polluting Ganga, Supreme Court tells National Green Tribunal*". Available at http://www.dnaindia.com/india/report-take-action-against-units-polluting-ganga-supreme-court-tells-national-green-tribunal-2030342. Accessed on 31 October 2014.

[6] Comptroller and Auditor General of India. Available at http://www.saiindia.gov.in/english/index.html. Accessed on 31 October 2017.

3.1.3. *Institutional arrangements at the central level*

In 2008, the Prime Minister declared the Ganga River to be the "National River" of India. In 2009, the Government of India established NGRBA at the national level as a coordinating authority for Ganga river clean-up programmes and projects. The NGRBA is chaired by the prime minister and comprises seven central-level ministries: Ministry of Environment and Forest; Ministry of Finance; Ministry of Urban Development; Ministry of Water Resources, River Development and Ganga Rejuvenation; Ministry of Power; Ministry of Science and Technology; and Planning Commission. The Chief Ministers of the five riparian states — Uttarakhand, Uttar Pradesh, Bihar, Jharkhand and West Bengal — are also members of NGRBA. As the highest-level authority, the primary mandate of the NGRBA is to address the pressing concerns of pollution control.[7] It handles financing, planning, coordination and monitoring for river conservation activities, and ensures that all required resources are in place for execution of projects by state agencies and urban local bodies (NGRBA, 2011b).

Monitoring of water quality is primarily the task of the Central Board Control Pollution (CBCP), which is an independent agency under the Ministry of Environment and Forests. CBCP gives advice to the central government on matters concerning water pollution abatement, water quality enhancement and sewage management.[8] Moreover, CBCP inspects wastewater treatment installations, air-pollution control equipment, industrial plants and manufacturing processes. At the state level, CPCB has its counterparts in State Pollution Control Boards (SPCB).

3.1.4. *Major issues in central-level and interstate coordination*

The issues in central level and interstate coordination can be placed under the headings institutional weakness, imbalance of power and conflicts of

---

[7] Bharat Sharma (2014) *Cleaning the Ganga, Step by Step*. Available at http://www.thehindu.com/opinion/op-ed/cleaning-the-ganga-step-by-step/article6105068.ece. Accessed on 27 August 2014.

[8] Central Pollution Control Board, Ministry of Environment and Forests, India. Available at http://www.cpcb.nic.in/Functions.php.

interest. These issues play at the central level, between the central and state level, and between the states.

Institutional Weakness

According to IIT (2011) inherent weaknesses in the design of GAP created ambiguity in policy and legal frameworks. As a result, there was a lack of clarity of roles for the agencies involved in GAP implementation. However, this is not an uncommon situation: wherever multiple stakeholders engage in execution, it can lead to fragmentation, isolation, overlap and duplication of tasks and responsibilities. The major issues in GAP are, however, ambiguities and loopholes that deterred agencies in exercising clear roles in implementation. From IWRM perspective this is an integration problem. Integrated Water Resources Management requires effective governance to ensure integration and coordination, both for existing institutions as well for newly created agencies and departments, at and across different levels.

Das and Tamminga (2012) pointed out the creation of a massive bureaucracy comprised of various governmental units and committee structures for the Ganga clean-up programs. Despite, or perhaps because of, these many bureaucratic layers, CAG (2000) reported that GAP failed to establish collaboration among the five beneficiary states to address pollution from both point and non-point sources. It is almost impossible to clean a river with a length of 2,500 km without the cooperation and coordination of state agencies. Three problems in institutional design prevented effective cooperation and coordination. The first problem is the rigid funding system or inadequate delegation of fiscal resources. The second problem is political affiliation: if a state is governed by a different political party from the one in power in Delhi the potential for conflicts is higher. The third problem is delays in authorisation of projects due to many bureaucratic layers and inefficient procedures. An extreme example is the Master Plan 2030 project from the Japan International Cooperation Agency, which took nearly a decade to get final approval.[9] Institutional weakness at the top level was also clear: the apex coordinating agency

---

[9]Source: Personal communication with a representative of the JICA office, India, 2 December 2014.

NRCA, which is the predecessor of NGRBA, only met twice before the year 2000, in 1994 and 1997 (CAG, 2000). Ultimately, all of these institutional limitations could be used as pretext for non-actions, failure to achieve outcomes, complacency and finger-pointing.

Imbalance of Power

The relation between the states and the central government is complex, and differs from state to state. On paper, power is focussed at the central government: critical decisions regarding planning, implementation and financing rest with the central government. However, the wealth of states determines their negotiation position: wealthy, large states such as Uttar Pradesh are in a better position to challenge decisions from the central government than smaller states that are more dependent on funding from the central government.

The GAP had a centralised, top-down approach, with minimal participation from the bottom (Das and Tamminga, 2012; Das, 2011).[10] In the Ganga programmes, power imbalance manifests itself in two ways. Firstly, the planning and financial power is vested heavily in the central authority, leaving little room for state level agencies to exercise control over the implementation. Secondly, since state and local level agencies have the executive power and do almost all the implementation, the central authority lacks a comprehensive and consistent mechanism to monitor the effectiveness of the projects and the outcomes in terms of improved quality of water. A result of this is, for instance, that the role of the Central Pollution Control Board has been reduced to a mere technical support and statistical unit which lacks binding power over the State Pollution Control Boards. This is a classical principal-agent problem in decentralisation that needs to be solved with the right incentives and regulation.

Conflicts of Interest

Conflicts of interest are present throughout the GAP at different levels and among different stakeholders. Different political affiliations is one reason

---

[10] See also: Ganga Rejuvenation: Old Water in a New Bottle? Available at http://www.lilainteractions.in/ganga-rejuvenation. Accessed 14 May 2015.

that can lead to conflicts regarding the Ganga clean-up programmes. In a hearing on the failures to clean the Ganga, the Supreme Court observed that most states were governed by opposition political parties.[11] But there are many other reasons. A plethora of stakeholders, including diverse government agencies at different levels, parastatal organizations,[12] local politicians, and civil society are involved in the Ganga clean-up projects. Different agencies pursue different agendas, reflecting different mandates and interests. Communication between agencies on these issues is often lacking. Without intervention from the central government, a clean Ganga will never be an automatic goal for all.

Under the NGRBA environmental and social management framework two categories of stakeholders are identified (NGRBA, 2011a). Primary stakeholders consist of groups that benefit from and are affected by the projects, e.g. dhobi, fisherman, crematoria societies, and industries. Secondary stakeholders are government organisations at central, state and local level which implement projects, such as municipalities, executive agencies and project management groups, and NGOs and research institutes which have an interest in development, community participation and community welfare. Conflicting interests occur naturally among the stakeholders, and is compounded by ineffective communication and sharing of knowledge on water resources and related issues. One prominent Ganga activist expressed the view that "[t]he first step is to make the institutions connected with rivers accountable and transparent".[13]

Conflicts also occur among riparian states, which have different portions of the Ganga River within their territory. The longest stretch of Ganga is located in Uttar Pradesh (1,140 kilometres). The majority of polluting industries are located in this state, hence it contributes most to

---

[11] See: Need states' cooperation for a clean Ganga: Centre to SC. *The Mathrubhumi*, 24 September 2014. Available at http://www.mathrubhumi.com/english/news/latest-news/need-states-cooperation-for-a-clean-ganga-centre-to-sc-152246.html. Accessed 7 June 2015.

[12] In India context, they are semi-state owned organizations having some political authority and serving the state directly. Usually, they are Public Works organisations under the direct control of State governments.

[13] See: Make Ganga users pay for clean-up: Activists. *The Times of India*, 7 July 2014. Available at http://timesofindia.indiatimes.com/india/Make-Ganga-users-pay-for-clean-up-Activists/articleshow/37930619.cms. Accessed 4 December 2015.

the pollution of the Ganga. Out of the total 764 industries along the Ganga, 687 industries or 90 percent reside in Uttar Pradesh (CPCB, 2012). IWRM teaches us that river basins are vital sources for water sharing, development and management. However, sharing within a basin can be a source of conflicts, and the situation may exacerbate when the parliament fails to use its legislative oversight power — granted by Entry 56 of the Union List under the Constitution — to deal with interstate conflicts. The proposed River Board, authorised by the same constitutional provision, has never been established (Richards and Singh, 2001). Interstate conflicts over the Ganga are not well documented, though we can use the case of the Yamuna, a major tributary of the Ganga, as an example of conflict over shared ecological resources, investments and basin management. The Yamuna flows through the states of Uttar Pradesh, Himachal Pradesh, Haryana, Rajasthan and Delhi. A major stretch of Yamuna is effectively controlled by Uttar Pradesh and Haryana. Following the ever-increasing demand of the rapidly urbanising State of Delhi, conflicts arose about the sharing of water of the Yamuna. The case landed in Supreme Court for a Public Interest Litigation. Since then, conflicts over sharing of Yamuna water see no sign of diminishing (Mohan and Routray, 2011). Conflicts of similar nature exist among the five riparian states, which share the rights to use water resources and the responsibilities for keeping the water clean. At the same time, the states also pollute the waters to different degrees.

3.2. *Intrastate coordination*

3.2.1. *Institutional arrangements at state level*

Under the NGRBA, State Ganga River Conservation Authorities (SGRCA) have been formed in each of the riparian states to serve as the highest policy and decision-making structure (NGRBA, 2011b). Under the SGRCA, State Executive Committees (SEC) and State Programme Management Groups (SPMG) are formed as registered societies of the respective state governments that mainly take on the responsibilities of planning and implementation, with the exception of Jharkhand where the stretch of Ganga is very short. The membership of SGRCAs is very broad, including the chief minister, state ministers from environment, forest, finance, urban development, irrigation, public works, and housing, and mayors from major towns (GoI, 2009). By composition and by definition, SGRCAs would be perfect vehicles for intrastate coordination. However, there are several factors that limit the effectiveness of SGRCAs as apex bodies for intrastate coordination.

The State Pollution Control Boards (SPCB) — the Central Pollution Control Board's counterparts — are responsible for monitoring of water quality, and for enforcement of laws, rules and regulations. Progress on cleaning the Ganga is supposedly monitored by SPCB engineering teams on daily basis and the chief executive of the implementing agency should review results on a monthly basis. In addition, Citizens' Monitoring Committees are formed in each municipality to independently review the progress. However, the effectiveness or the level of genuine participation is in question. For example, various reports highlighted that resources for citizen participation were underutilised and Citizen's Monitoring Committees were either symbolic, ineffective, or non-existence under GAPs (CAG, 2000; Das and Tamminga, 2012).

3.2.2. *Major issues in intrastate coordination*

The major issues in intrastate coordination can be placed under the headings two-tiered system, ground level issues and incentives and enforcement.

Two-tiered System and Lack of Integration

In the design of the NGRBA an integrated river basin approach has been adopted, and mechanisms at both central and state level are designed to support and ensure multi-sectoral integration and coordination. In line with this, the SGRCAs have broad mandates and broad representations from different sectors. Within the states only two tiers of institutions are present: state-level institutions such as SGRCA and project level institutions, such as Jal Nigams or Urban Local Bodies.

At the project level, local level agencies implement either the extension projects of GAP-II or new programs under NGRBA. GAPs were widely criticised — also by NGRBA itself — for not being framed along an integrated basin approach. Projects are over-focused on one sector: control of pollution through sewers and sewage treatment plants (STPs). However, current project activities suffer from the same 'non-integrated' focus on construction or upgrading of sewer networks and STPs. For example, in the latest progress report of the Uttar Pradesh State Ganga River Conservation Agency (July 2014)[14] the projects under NGRBA are mainly sewers and STPs.

SGRCAs have broad mandates that include components of integrated river basin management, but there is no such 'integration' at the level of project implementation. Multi-sectoral integration is a component that is lost between state and local, project-level institutions. For example, the sanitation plan of Varanasi has no single reference to the Uttar Pradesh State Ganga River Conservation Agency (VMC, 2011). IIT (2011) also highlighted the failure to deal with non-pollution factors such as unsustainable upstream water extraction and diversion, and non-point-source pollution such as agricultural run-off. In disregard of their broad mandate and composition, SGRCAs seem to have no role in the original design of the river basin management plan; rather NGRBA develops the plan at the central level. As a result the scope for coordination within respective states — in line with the integrated river basin management framework they are supposed to oversee — is effectively limited.

Project-level Issues

A similar array of problems with integration can be observed at the local level implementation and monitoring agencies. Common issues mentioned in various reports include institutional complexity and fragmentation, inconsistent financial models, lack of an integrated data management system, shortage of manpower, and lack of institutional capacity (Alley, 2014; Das and Tamminga, 2012; VMC, 2011), which all contribute to poor integration. Even though two decades of experience was obtained in GAPs, SGRCAs have not been equipped with the necessary resources and mechanism to deal with these issues.

Integration of stakeholders at project level has also been poor. Even though successive GAPs and the NGRBA tried to encourage public

[14]The report is a spreadsheet file that can be found at http://sgrca.org/static/AboutUpsgrca.aspx.

participation on paper, the roles of communities and civil society have been minor in practice. When communities do take part, it is not more than in symbolic actions such as *ad hoc* cleaning and awareness campaigns. Meaningful participation in programme design, implementation and monitoring is lacking (Das and Tamminga, 2012).

Therefore, it is quite reasonable to conclude that even though there is a broad institutional structure for coordination and cooperation within states, in line with an integrated basin approach, not much comes out of it in practice. The subsequent point probably offers an explanation.

Incentives and Enforcement

A possible explanation for failed coordination at the project implementation level is the poor incentive and enforcement structures. Normally, functioning incentive structures require two conditions. First, agencies should have enough discretionary authority, information and resources to produce the results required by the principal (in this case the central government). Second, the principal should devise adequate provision for monitoring of the implementation process and for penalising underperformance and non-compliance of the agents (in this case state or local implementation agencies) (Foulon *et al.*, 2000). However, this was not the case in the GAP. India's federal structure and diverse politics make any punitive action difficult to uphold. Instead of penalising giving rewards, setting competitive targets of introducing public oversight may work better.

GAP I and II failed to include and consult local-level agencies in the planning and design process. In addition, readiness and institutional capacity of local-level agencies have never been adequately assessed prior to implementation (Alley, 2014; Das and Tamminga, 2012). Even though agencies such as State Pollution Control Boards and State Project Steering Committees have regulatory and enforcement mandates, there are no effective sanctions in place for underperformance, non-compliance and potential malpractices. Moreover, the practice of tendering construction projects exclusively to parastatal agencies under state governments limits competition and complicates developing effective incentives. Ganga clean-up programmes lack both *ex ante* incentives and *ex post* sanctions. As a result, performance and efficiency suffer.

4. Moving Forward

The analysis of major coordination issues in Ganga clean-up programmes in the previous sections used IWRM as a main analytical framework supplemented by the concepts of delegation of power, and incentives and regulatory arrangements. This section continues the discussion along the lines of these three frameworks, though the focus is shifting from diagnostic analysis to suggestive discussion. This is a synthesis section that sets the ground for subsequent section on policy recommendations.

4.1. *Recent developments in Ganga Rejuvenation*

Since 2011 the World Bank has been supporting the government of India through the National Ganga River Basin Project. The total support amounts to US$ 1 billion and the primary focus of the project is institutional capacity building, infrastructure investment in five riparian states, industrial pollution abatement, water quality monitoring, and setting up a Ganga Knowledge Center.[15] The World Bank believes that high-level political will is of utmost priority in sustaining the Ganga clean-up in the long run, and stakeholder engagement and mobilisation are of crucial importance in cleaning the river.[16] Stakeholders include local inhabitants, religious leaders, elected representatives, prominent civil societies and NGOs.

In July 2014, Prime Minister Narendra Modi declared the objective to rejuvenate the Ganga. Two months later Ganga rejuvenation was allocated US$ 334 million under the Ministry of Water Resources, River Development and Ganga Rejuvenation. The objective is to clean the Ganga in three years, with no untreated sewage or industrial effluent discharge by 2020.[17]

---

[15]The World Bank, US$1 billion Support from World Bank for Ganga Clean-up, 31 May 2011. Available at http://web.worldbank.org/external/default/main?pagePK=34370&piPK=34424&theSitePK=4607&menuPK=34463&contentMDK=22928173. Accessed 12 May 2014.

[16]The World Bank, World Bank Five Points Plan to Clean Ganga, 19 June 2014. Available at http://www.worldbank.org/en/news/opinion/2014/06/19/five-point-plan-to-clean-the-ganga. Accessed on 21 May 2015.

[17]*The Diplomat*, "Cleaning up the Ganges," 16 September 2014. Available at http://thediplomat.com/2014/09/cleaning-up-the-ganges. Accessed on 16 September 2014.

It now seems that political will at the highest leadership level and all necessary financial resources and technical support is in place. The question is if the current institutional framework and coordination arrangements lead to the desired and much needed outcome of a clean Ganga.

4.2. *The need for integration*

The sources of pollution in the Ganga basin are widespread and complex, and cannot be addressed by just technical solutions. Ganga cleaning requires an integrated, multi-stakeholder approach and institutions that facilitate collaboration. Some sources of pollution, such as agricultural runoff and dead bodies from traditional crematory sites are particularly complex and difficult to address due to the socio-political, economic and religious context (Das and Tamminga, 2012). The interlinkages along the 2,500 kilometre long river require integrated institutional arrangements. Two related issues are standardisation of pollution and water quality parameters and criteria, and standardised operation of comprehensive monitoring.

CAG (2000) reports that as of the year 2000 GAP had met only 39 percent of its primary target of sewage treatment. Standardisation of pollution and water quality monitoring was lacking, since the Ministry of Environment and the central level agencies (including CPCB) left technical details to the state governments. State governments did not come up with uniform criteria. As a result, every aspect of clean-up projects, including project prioritisation, estimation and evaluation, became problematic. Monitoring has been chaotic, because there was no stringent and integrated mechanism. Reporting from the states to the central government has never been consistent. Despite attempts from the Supreme Court to intervene from time to time through PIL, the GAPs were not fully and meaningfully implemented.

Being aware of institutional problems of the GAPs, the government of India, with the support of the World Bank, now commits to work towards holistic and integrated institutional arrangements supporting collaboration. It is most likely that the government of India and the World Bank will move forward with the adoption of IWRM as the new paradigm for Ganga river basin management. However, as mentioned in Section 2, IWRM is not free from criticism; one of the major criticisms is the alleged failure of IWRM to address institutional diversity. Since institutions differ from one another in terms of size, interests and level of consolidation, any integration effort may inadvertently lead to centralisation and thus reduce responsiveness, which is contrary to the presupposed objective of this paradigm (Biswas, 2008).

To counter the tendency of IWRM towards centralisation rather than institutional integration, effective delegation of power and authority is required alongside strong incentive and regulatory structures. Together these are the three components of the theoretical framework (Figure 1), which are used below to develop recommendations.

4.3. *The need for delegation*

4.3.1. *Delegation and decentralisation of decision making*

In 1985 the Central Ganga Authority (CGA) was established to oversee the implementation of the GAP. The chairman of the CGA was the Prime Minister and its members were Union Ministers and Chief Ministers, senior bureaucrats and technical experts. They represented a high degree of political power. GAP I and II were predominantly planned and placed under the central government. However, to achieve efficiency and meet the desired outcomes, local governments, as the representatives of local citizens, should be allowed to participate in the formulation, decision-making and implementation of policies.[18] Interests from all stakeholders, including the less powerful ones, would need to be taken into account for effective collaboration (Merrill-Sands and Sheridan, 1996).

As we have seen in the previous section, the framework of the GAPs lacked clear roles and responsibilities, especially for financing mechanisms. State and local governments had no clear-cut roles in implementation and monitoring, and the GAPs created overlapping and conflicting jurisdictions among many government agencies. Hence, to improve this situation it is necessary that the central government, in consultation with respective regional governments, clearly defines the level of authority,

[18] http://www1.worldbank.org/publicsector/decentralization/political.htm.

transparency, accountability and participation for state and local government units in order to effectively implement any future clean-up project. At the same time, accountability of government agencies must be upheld through checks and balances by legislatures and independent third party organisations.

The idea of coupling institutional integration (IWRM) and delegation (decentralised governance) may seem uncomfortable at first glance. However, there are several examples of successful integration of the two, both in and outside India. Box 3 explains the successful case of National Rural Health Mission (NRHM) in India, where traditional forms of local governance institutions are effectively integrated into a national framework to provide universal health care for the most vulnerable population.

Box 3: The Case of NRHM: Integrated and Delegated Governance that Works for Better Outcomes

In 2005, the NRHM was set up in India. The primary objective was to provide universal health care to the most vulnerable of the rural population, making it one the world's largest public health care systems. The NRHM is a sub-mission under the overarching National Health Mission (NHM). In India, health is governed by the states, like in the case of water resources. However, NRHM goes beyond the state level by integrating agencies and organisations at all levels in the entire program design, including local problem prioritisation, local planning, national program design, financing and management, and monitoring and accountability mechanisms.

The extent of institutional integration under the NRHM regime is quite profound and innovative. The NRHM takes Panchayati Raj Institutions (PRI) as a basic program unit. Panchayati Raj is a traditional local governance unit common in South Asia comprised of elderly and respectable persons from the village. In India, the Panchayati Raj system exists in parallel with modern governance arrangements such as municipalities, village and district authorities. In the case of NRHM, PRI is both at a starting point and end point of the program circle.

NRHM can be viewed as highly decentralised and highly integrated authority at the same time. Planning starts at the village level, jointly done by

(*Continued*)

Box 3: (*Continued*)

the Village Health and Sanitation Committee and the PRI, in a very bottom-up fashion. Village-level plans are reconciled into district plans, which are reconciled into state plans and finally national plans. Apart from this critical planning role, the PRI as a basic governance unit, also have a larger role in implementation and monitoring of programs at district and state levels. (However, it should be noted that not all PRI enjoy the same level of empowerment and autonomy across states).

Integration arrangements under NRHM are quite intensive. The Mission Steering Committee (MSC) is the highest authority chaired by the Union Minister of Health and Family Welfare, and the membership comprises of other central level ministers, including the Minister of Panchayati Raj, Minister of Rural Development, and Minister of Human Resource Development, representatives from the Planning Commission, and other senior bureaucrats and health professionals from central and state governments. The MSC has an executive arm called the Empowered Program Committee, which is also a good example of multi-sector and multi-layered integration. All the concerned departments on community health, women and family affairs are integrated in this program committee and at the state level all state and non-state actors working on community health and hospital managements are integrated as well. Another innovative aspect is the integration of conventional western medical knowledge and traditional medical knowledge and practices (such as yoga and homoeopathy). Program wise, NHRM integrates all the parallel services provisions such as drinking water, sanitation, and disease control, and pools financial and human resource under its broad umbrella.

Although evidence of success by the NRHM is debated, NRHM is widely recognised by its innovative design, especially regarding the successful merger of decentralised and integrated health care systems.

Sources: *GoI (2002, 2011); Pandey (2009); Sharma (2009).*

The case of institutional arrangements under the NRHM shows how the central government can empower state governments to work for the interests of their constituencies in delivery of essential services through integration and delegation in an extensive program. Even though the attempt at integration and delegation does not always work uniformly across jurisdictions and states, there seemed to be a significant improvement in trust and commitment of both health professionals and local communities (Lahariya *et al.*, 2006). In general, despite poor financial commitment and allocation towards health provision — less than five percent of GDP in 2010 — India has a reasonably sound public health care system and NRHM has been one of the contributions behind recent developments in this sector.[19]

There is a significant difference between health and water, though: it is quite unlikely that states will have major conflicts over public health-related issues, while, interstate natural resources like transboundary rivers have always been sources of conflict. An example of a successful institutional arrangement for federal-state and interstate relationships can be found in the USA. The system in the USA works mainly by balancing conflicting interests of the federal government and the states. The Federal-State Partnership and Federal-Interstate Compacts are set up to strengthen collaboration and dispute resolution (Mandarano *et al.*, 2008). Interstate Water Compacts usually resolve issues among states, sometimes based on Supreme Court litigations. It also administers clear rights and responsibilities for all concerned states, amid ever-changing needs, politics and social concerns. It works out solutions for on-going collaboration and negotiation and adjustments through various 'commissions' and 'associations'. For instance, under interstate compacts informal structures for communication were created, such as interstate associations, providing a forum for collaboration (e.g. the Upper Mississippi River Basin Association since 1982). There are some limitations, though, for the US interstate compacts such as the informal role of the federal government and the power of the congress that can effectively compromise deliberation and decisions in compacts. However, they serve as reliable and efficient mechanisms for resolution of conflicts between the federal government and states, and also among states (Mandarano *et al.*, 2008).

4.3.2. *Delegation of financial resources*

Involving practitioners (local-level bureaucrats) in policy and project design and the budget formulation process is an effective way to encourage

[19]Mainstream Weekly, Report of a Survey on National Rural Health Mission (NRHM) and Panchayats. Available at http://www.mainstreamweekly.net/article4000.html. Accessed on 21 June 2015.

their commitments in implementation (Bryson and Crosby, 2005). It would be less substantial to delegate or decentralise without transferring sufficient fiscal resources to lower order jurisdictions. State and local governments have been struggling to fulfil responsibilities of the Ganga clean-up programs, despite their implementation and monitoring roles through decentralised arrangements (Das and Tamminga, 2012). Experience in the GAPs show that unsustainable or intermittent funding has been a major reason for failure. In fact, the design of the funding system had never been unanimously agreed since the beginning of GAP I. The payment for capital investments and operation and maintenance remained an area of constant dispute between the central and state governments.[20] In the early 1990s, half the funds had to be borne by the states. The system was later changed to the central government funding hundred percent of the costs. In 2001, the funding structure changed again to a 70:30 ratio between the central and state governments.[21] Still, projects were delayed due to complicated fund clearance requirements by different ministries.

The Minister of State for Environment convincingly argued that there is no shortage of funds for the Ganga clean-up: the total expenditure for GAP I & II as of 31 March 2014 amounted to Rs. 986.34 crore (about USD 150 million) over 25 years.[22] The current funding commitment towards Ganga is even more impressive. In addition to the USD 1 billion funding support from the World Bank, approximately USD 115 million and USD 430 million were pledged by the central and the five state governments respectively.[23] Furthermore, India did introduce measures for

---

[20] Centre for Science and Environment "Status Paper for River Ganga: Past failures and current challenges". http://www.cseindia.org/userfiles/Status%20Paper%20Ganga%20 2013.pdf. Accessed on 13 December 2015.

[21] EcoFriends, "Ganga Action Plan-A critical analysis". http://www.ecofriends.org/main/eganga/images/Critical%20analysis%20of%20GAP.pdf. Accessed on 13 December 2015.

[22] Minister of State for Environment, Forest and Climate Change Prakash Javadekar replies to a query on 'Ganga Action Plan', Indian Parliament Q&A (July 16, 2014). Actually, it was the total expenditure released by the Central government. State governments may have some amount borne by their own funds.

[23] World Bank Press Release No: 2011/518/SAR. http://web.worldbank.org/WBSITE/EXTERNAL/NEWS/0,,contentMDK:22928173~pagePK:34370~piPK:34424~theSitePK:4607,00.html. Accessed on 27 August 2014.

funding sustainability at the local level. In 2006, the government introduced a new tax system to collect revenue from STPs through by-products from sewage treatment operations like bio-gas for power generation and sludge as bio fertiliser. The idea was to minimise the operation and maintenance cost burden of local governments.

It is most likely that the main source of the problem is not the funding pool for Ganga clean-up projects, but the design of financial arrangements. Bottlenecks in financial arrangements were a common feature of past initiatives. Lessons from the NRHM may be useful for the Ganga clean-up projects and the NGRBA. Many vertical health programs run and governed by different government organisations at multiple levels are horizontally integrated under the NRHM umbrella. The funding arrangement under the NRHM is also unified, i.e. the central government provides all the funds. At the same time it is decentralised enough in the sense that States and local level implementing units enjoy reasonably degree of certainty and autonomy over the financial flow, since the budgeting was done in a bottom-up fashion. The funding structure for Ganga clean-up should be integrated enough so that consistent and sustained fund flow is ensured, while at the same time it is decentralised enough for budget planning and allocation to address local realities and needs. Delegation should improve efficiency by reducing transaction costs while creating correct incentive structures, if other important conditions (next sub-section) are also met.

4.4. *The need for incentives and enforcement*

An institutional problem that is evident from several studies and reports reviewing the previous Ganga clean-up efforts is lack of resources, capacity, or commitment, or a combination of these, from the local-level implementation bodies. A central-level effort requiring integration (IWRM) and delegation of authority will unlikely result in the desired outcomes without adequate incentives and regulatory mechanisms in place. Our analysis of local level implementation and monitoring in the previous section revealed that the Ganga programmes lack both ex ante incentives and *ex post* sanctions. Here, we briefly discuss what types of incentives and regulatory mechanisms would work in the Ganga context. However, this discussion should be viewed as suggestive, not prescriptive, since our purpose is to provide policy makers with more options and set the stage for more detailed discussions.

4.4.1. *Shared purposes and goals*

Stakeholders who hold different, contesting or contradictory values and beliefs would end up jeopardizing the Ganga programmes. Differences in approach and objectives are inherent and natural. A common and shared purpose needs emphasis (Parkinson, 2006). Differences and disagreements could result from cultural, religious, political factors or other factors. Our primary focus here is on the different political affiliations and bureaucratic interests among the central, state and local-level agencies. Focusing on the strengths and skills of participating agencies enables good coordination. At the same time, religious and cultural stereotyping should be barred by every participant (Merrill-Sands and Sheridan, 1996). Rejuvenation of Ganga is clearly a national cause with significant political support and commitment. However, for the individual states, the priority would be utilisation, not necessarily rejuvenation. This seemingly natural difference in orientation becomes more complicated if the contested and partisan nature of Indian politics is taken into account.

Any effective coordination requires a broad purpose, mandates, commitment and sharing of resources. Clear description of roles, designation of formal leadership, clear collaboration structures, and built-in flexibility that adjusts to local conditions and changes are also necessary conditions (Arino and De La Torre, 1998; Bryson and Crosby, 2005; Page, 2004). In practice, this is not easy to achieve. An example is that Prime Minister Narendra Modi has been campaigning for and promoting the Ganga as a national river, but partisan and riparian politics between national and state governments have prevented the river from becoming truly national.[24] The fact that different political parties lead central, state and local governments

[24] From personal communication with people in New Delhi and Varanasi the authors understand that sub-national governments do not always have the level of inspiration and determination that are consistent with the national level agenda.

tends to hamper decision-making, funding flow and implementation, though this is a 'natural' consequence of the large and diverse political system in India.

To achieve successful coordination and outcomes, all parties should strive to achieve a common ground by pursuing similar goals, purposes and priorities. It may not be possible to meet all interests or to reconcile all differences, but efforts need to be made to identify commonalities that can be a basis for action. The first step would be mapping all conflicts, existing and potential or implicit and explicit. Many believe that conflict is inevitable but recognising and respecting differences is an integral part of good collaboration. Skilful facilitators could be helpful in negotiation and conflict resolution (Bryson *et al.*, 2006). In fact, conflicts could be good places to start collaboration processes as they are not always negative; conflicts can represent signs of change that are being intensified, and challenging issues that rise to the surface (Winer and Ray, 1994). In the current institutional framework, it seems that there is no room for conflict or, in other words, conflict is not explicitly recognised.

Political leadership is also important to achieve successful coordination and outcomes. To get all public agencies to stand for common values and goals for Ganga clean-up, political will and leadership that could harness collective support toward shared understanding of the mission is required. Parkinson (2006) indicated that key successes in realising collaboration and shared purpose depend heavily on open and frequent dialogues, mutual respect, understanding and trust by putting collaboration above self-interest and having clear roles and policy guidelines. Past GAPs and current NGRBA activities seem to focus heavily on hardware, i.e. the construction of treatment plants, sewers, cannels, etc. For successful programmes the focus should shift to software, i.e. shared understanding and shared goals among multiple state agencies. The critical component of this software architecture is explained in the following paragraphs.

4.4.2. *Shared information*

Open and frequent, but not overwhelming, communication contributes to collaboration success. In this communication, collaborating parties should expect distortion and misinterpretation of messages, and have an opportunity for correction and clarification (Merrill-Sands and Sheridan, 1996). Communication is integral to building rapport, trust, and establishing a working relationship among all actors. In addition to the formal channels, informal channels should be established to promote people to people connections across sectors and different government tiers to enable a better, more informed and cohesive working environment (Parkinson, 2006).

The IWRM approach to water governance suggests a platform for trust building and information sharing among key stakeholders at both national and sub-national levels. Within the current institutional configuration of Ganga river basin conservation, a number of formal cross-sectoral and multi-level platforms are in place, such as the NGRBA, CPCB and the Ganga Knowledge Centre. However, in the previous sections we have described the 'institutional fragmentation' and how this exists at all levels. There is a need for specific measures to build trust and knowledge sharing platforms. Edelenbos and van Meerkerk (2015) suggest that informal networking and boundary spanning leadership are necessary conditions for developing trust and shared knowledge among stakeholders at various levels and domains. Although institutions such as NGRBA and SGRBAs are still key players, broadening participation across disciplines and stakeholders beyond public institutions at both central level and program level may result in the wealth of diverse knowledge and alternatives.

The IWRM framework is widely regarded as being ambiguous in term of definition and demands a very high capacity from existing agencies and organisations (Medema *et al.*, 2008). Moreover, it needs a high level of institutional integration and knowledge exchange through both formal and informal channels. Since India has already developed many functioning formal institutional arrangements, there may now be a case for a complementary and informal effort to build trust and focus on knowledge integration mechanisms along with formal structures like NGRBA and the Ganga Knowledge Centre. In this regard, informality is highly recommended because it provides opportunities for public office holders to speak with more openness and flexibility, without worrying too much about official status. In informal forums, officials may be more convincing if allowed to step outside official constraints. Academia and civil society representatives can also more effectively contribute to this safe, inclusive and non-attributive environment. Eventually, it will serve as a building block for supporting more informed and integrated institutional arrangements at official and policy levels. A successful example of interagency information and knowledge exchange is the Delaware Interstate Water Management in the USA. In this interstate management framework, federal, state and local authorities have established a channel of communication and regularly meet each other to share data and problems, while informal mechanisms, like interstate associations, were very helpful in promoting trust and mutual understanding (Medema *et al.*, 2008).

This idea mainly derives from Track II diplomacy in international affairs, where informal forums serve the purpose of exchange and trust building, which prevent information gaps, misinterpretation and miscalculation among key stakeholders. Both the NGRBA and the Ganga Knowledge Centre can take a lead in broadening participation and initiating public outreach, but none have so far substantially invested into it. At the moment, the World Bank supports the institutionalisation of the Ganga Knowledge Centre under the NGRBA program.[25] For more details, please refer to the 'Knowledge Management' chapter in this book.

4.4.3. *Regulation and sanction*

In the previous paragraphs, we discussed sharing goals and information as effective *ex ante* incentives for local implementation and monitoring units, in combination with delegated authority and resources. Now we will discuss how *ex post* sanctions were ineffective or simply lacking in the Ganga context. With regards to regulation, two specific issues stand out prominently. First is regulation over projects, i.e., construction of infrastructures, operation and maintenance. Second, regulation over water quality, including both municipal and industrial aspects of the problem.

---

[25] One of the Ganga Knowledge Centre's mandate is to facilitate stakeholder dialogues, encourage and broaden public participation and build partnership with universities, central, state and local government agencies, NGOs, and the private sector. Source: "Appraisal Note for Proposal on Ganga Knowledge Centre under the Institutional Development Component of NGRBA Programme Assisted by the World Bank", http://www.moef.nic.in/sites/default/files/ngrba/Appraisal%20Note%20GKC.pdf. Accessed on 4 June, 2015.

Under the NGRBA framework, main bodies responsible for oversight of all infrastructure projects are Program Management Groups (PMG) at the central level and State Program Management Groups (SPMGs) at the state level. However, since the state governments also contribute funds, SPMGs have to manage two funding sources and are accountable to two authorities (NGRBA, 2011b). Typically this creates loopholes in regulatory frameworks, and evidence from previous GAPs is abundant. In many cases, SPMGs tend to emphasise only the accomplishments with regard to planned infrastructures, without evaluating quality and appropriateness. Under the GAPs, many infrastructure projects, although constructed as planned, ended up quite irrelevant and non-functioning. Alley (2014) provides a number of examples of such cases. In one, an STP only operates at only eight percent of its design capacity. In another, the sewage cannel, sewer and treatment plant were not connected, thus completely non-functioning.

Foulon *et al.* (2000) provide two different but complementary mechanisms for regulation: 'generic regulatory tools' like fines and penalties and 'market or public-based instruments'. The authors present a simple mathematical formulation for pollution that states that pollution is a function of regulation, traditional and public enforcement, and control variables X. Although the primary analysis of their research is compliance of polluting industries, the key lesson of their study can be extended to since executing agencies under Ganga clean-up programmes.

In order to create effective *ex post* sanction to get the incentives right for local-level execution agencies, current regulatory and enforcement mechanism should be integrated and harmonised. Currently, links are lost between PMG/SPMGs and CPCB/SPCBs. Even within CPCB jurisdiction, there is no effective control of the central-level board over state boards. These lost linkages should be corrected. In addition to conventional enforcement mechanisms, public-based enforcement mechanisms should be introduced through public disclosure of bidding for all infrastructures and public disclosure of performance and compliance by executing agencies. The current role of Citizen's Monitoring Committees can also be considered as a good starting point to create system of public disclosure.

5. Policy Recommendations

Despite concerted efforts, time and resources invested in the clean Ganga missions, there are large areas for improvement, specifically in terms of coordinated aims to cultivate rapport among key public institutions. To ensure desirable outcomes, good planning should be accompanied by matching coordination and implementation. This chapter seeks to offer modest recommendations, which might stand a greater prospect of achieving results than more radical plans.

It is recommended that:

5.1. *Central-state/local level coordination*

- **State and local governments should be given more responsibilities in Ganga clean-up projects**, including in planning, design, budgeting, and operating physical facilities and plants (i.e. planning and setting up of STPs and implementation), in a bid to devise holistic and integrated plans and reduce contradiction and conflict between master plans. NGRBA should continue to serve in an oversight role to check and balance and monitor the performance of the states and local governments. The execution should be left to state and local governments whose feedback on technical aspects and funding allocations may also be relevant.
- **States and local governments should be empowered to take part in the development of action plans and budgeting based on their contextual situations**. It would incentivise them to take responsible actions during the implementation without much delay as experienced in the past. For example, engage local government in the type and location of STPs. This would allow them to be aware of the background of the project from the beginning and for the effort to gain from the actual situation on the ground.
- **Coordinated and integrated governance based on the balance of power (central-states/local) should be established within the Ganga clean-up framework**. The power imbalance should be addressed as early as possible. This requires political commitment for delegating some political, administrative and funding decisions and

works through a new political and administrative arrangement that promotes more understanding, collaboration and dispute resolution. The lack of power balance between central and state/local levels has harmed the Ganga clean-up program over the past 25 years. The success of NRHM in India in coordinating multiple levels of governance under the mission's overarching umbrella should be a recommended indicator for the Ganga river basin pollution abatement mission.

- **Regulations and policies concerning resource allocation should be aligned towards more delegated administrative and fiscal policies**. The review aims at enabling improved management, improving readiness and timely intervention by state and local government in controlling and abating pollution. Though the funding can remain fully central as is the case, it might be in the interest of the program to allow the states more leeway to encourage participation and give them a stake. Situations on the ground demand skilful local leadership that can create a sense of shared responsibility and cooperation among groups with different affiliations. It also facilitates information exchange and the decision-making process to strengthen collaborative network at the local level.

Interstate Coordination

- To promote interstate coordination and networking, NGRBA should ensure that all voices of member states are heard and taken into account, particularly serving as a platform for deepening coordination, information sharing, dispute resolution between states (e.g. addressing nonpoint source pollution and collective actions).
- **In addition to the broad NGRBA umbrella, formal central-interstate coordination structures (e.g. Ganga river basin compact, adapted from US interstate river basin compact) should be developed with the membership of all five riparian states**. This compact creates a platform for regular and joint exercise of jurisdiction over Ganga's shared water resource and pollution control as well as exchange of information and problem solving (i.e. nonpoint source pollution). This Compact could also serve as enforcement, monitoring and conflict resolution body rather than seeking litigation.

Intrastate Coordination

- Together with delegated administrative power and fiscal resources, capacity building and training for local governments should be provided to ensure that they not only have the resources necessary to get the job done but also the means to fulfill requirements and achieve targets.
- The incentives for their accomplishments should be in put in place accompanied by enforcement and sanction.

General Recommendations (applied to all coordination areas)

- **Informal institutions (e.g. associations) for central-state, interstate and intrastate collaboration should be established in compliment to the formal channel of communication**. This informal channel facilitates coordination, sharing information and concerns. The members of associations are the representatives of the central, state and municipal governments, with the establishment of personal connections that ultimately work collectively toward a more informed and integrated group working on common ground and priorities.
- **An effort should be made to ensure that public agencies associated with the Ganga clean-up are insulated from political change**. This can be done by political consultations and making a clean Ganga a national objective so that political parties will be wary of doing anything that deviates from it, amid constant changes in government and state leaderships.
- **Regular meetings, dialogues and information sharing between central-state, interstate and state-local government through both formal and informal platforms should be strengthened for better collaboration, trust-building and mutual respect**.
- **Information-gathering and sharing systems such as the Ganga Knowledge Center should be strengthened to be a more integrated and dedicated information-gathering system**. It could serve as a reporting mechanism and provide constant pollution and other Ganga-related feedbacks and information in a standardised manner, so as to gauge the impact of the program across all five states. Both NGRBA and GKC should also periodically organise a national level forum aimed for Ganga conservation with broad participation. This forum

will serve as informal networking, trust-building and knowledge sharing channel among key state actors and non-state actors like industry representatives and civil society, as well as a platform for academia to present research findings and assessment reports.

6. Conclusion

The Ganga clean-up is complex both in scope and scale. It involves multiple stakeholders from central to state and local governments, from NGOs to religious societies. Lessons learnt from Ganga clean-up program over the past years tell policy makers that they could not simply continue on a similar path dependency. Change in governance is needed or the Ganga will remain polluted. The Supreme Court of India warned in 2014 that this unchanged pollution control mechanism was a testimony of 'failure, frustration and disaster' and that it will not improve the Ganga's condition 'even 200 years later'.

First of all, the traditional way of governance, as witnessed by a top-down approach, should be revisited. It is necessary to maintain a balance between top-down and bottom-up approaches of governance. It needs a concerted effort and commitment from the people in power to delegate more political authorities and financial resources to the state and local governments. Delegation of power and financial resources must be accompanied with clear rules and responsibilities for all levels of government, and oversight by stringent monitoring and evaluation mechanism. At the same time, incentives and sanctions must be introduced to create a sense of ownership, atmosphere of competitiveness, and threat for ill-performance and malpractices. The central government should lead the process of reviewing and addressing conflicting, ambiguous and dysfunctional policies and regulatory frameworks. In this process, all tiers of government should be involved to help frame clear-cut policies for Ganga rejuvenation.

Lessons learnt from India and other parts of the world have shown that improving collaboration and participation among government layers can be successful through incentives and delegation of administrative authority and funds. However, resistance is always there from all sides. All governments, regardless of whether they are central, state or local, are quite reluctant to accept change if they think change will compromise their traditional power and authority. It is the primary role of Government of India to create an environment in which various governmental relationships can improve, with shared knowledge, trust and respect, to realise common values and goals. This is, of course, a difficult undertaking but it is worth the effort and necessary to achieve the clean-up of the Ganga. The example of the inter-state water resources management model in the USA manifests the need for collaborative governance in handling complex issues that cross inter-state frontiers and administrative jurisdictions.

Indeed, the complicated Ganga clean-up programme demands strong leadership and political determination so that effective coordination between national and sub-national actors can be realised amid real and potential conflicts. Conflicts and dispute can be counterproductive while they may also be catalysts for the democratic system of governance. Successful coordination and problem solving require a platform for on-going dialogues and/or a third-party mechanism for conflict resolution. In the current situation, the NGRBA is in the best position to offer oversight and guidance, not execution. It can be a catalyst in integrating all concerned ministries, central-level agencies, states and local executing agencies, under one broad umbrella, apart from its traditional roles of enforcement and mediating.

Multiple agencies are involved in Ganga clean-up with differing mandates, rules and procedures. Delegation of functions can improve coordination between the central government and states, and between states. At the same time, the regulatory authority should consult stakeholders at multiple levels. One of the important stakeholder groups in the governance process is local residents. Consultation should not be limited to powerful stakeholder groups. Rather, it should encompass views from academia, civil society and environmentalists, river user communities, residents, farmers and businesses. In fact, a participatory process means rebalancing between top-down and bottom-up decision-making. This action links integrated water resource management with the aim to ensure that interstate institution coordination efforts result into a common strategic direction that benefits the nation as a whole. With everything linked to IWRM, aspects of interstate and intrastate institutional coordination in Ganga clean-up needs a common strategic direction, which is indeed possible in the Indian context.

Though they are framed in different policy contexts, the Ganga clean-up can draw some important lessons from the NRHM case.

Given the complexity of the Ganga river basin and its management for pollution abatement, achieving results quickly seems a difficult task but modest improvements through collaboration among public institutions grounded on common values and goals seems doable. By increasing consultation and divesting more decision making to states and local governments both in execution and fund, followed by more rigorous enforcement and monitoring mechanism, the fate of the Ganga could change for the better.

References

Agrawal, A and J Ribot (1999). Accountability in Decentralization: A Framework with South Asian and West African Cases. *The Journal of Developing Areas*, 33(4), 473–502.

Alley, KD (2014). Varanasi's Ganga Wastewater Management: Why has it remained such an Intractable Problem? *South Asia Network on Dams, Rivers and People*.

Arino, A and J De La Torre (1998). Learning From Failure: Towards an Evolutionary Model of Collaborative Ventures. *Organization Science*, 9(3), 306–325.

Biswas, AK (2008). Integrated Water Resources Management: Is It Working? *International Journal of Water Resources Development*, 24(1), 5–22.

Blandl, J (1998). *Money and Good Intentions are not Enough*. Washington DC: Brookings Institute.

Bryson, JM and BC Crosby (2005). Leadership and the Creation of Cross-Sector Regimes of Mutual Gain. 8th National Public Management Research Conference, University of Southern California, Los Angeles.

Bryson, JM, BC Crosby and MM Stone (2006). The Design and Implementation of Cross-Sector Collaborations: Propositions from the Literature. *Public Administration Review*, 66, 44–55.

CAG (2000). Audit Report on Ganga Action Plan. New Delhi: Comptroller and Auditor General of India.

CPCB (2012). Status of Water Quality in India. Delhi: Ministry of Environment and Forests, Central Pollution Control Board.

Cullet, P and J Gupta (2009). India: Evolution of Water Law and Policy. In *The Evolution of the Law and Politics of Water*, Dellapenna, J and J Gupta (eds.), Springer Netherlands 157–173.

Das, P and KR Tamminga (2012). The Ganges and the GAP: An Assessment of Efforts to Clean a Sacred River. *Sustainability*, 4(8), 1647.

Das, S (2011). Cleaning of the Ganga. *Journal of the Geological Society of India*, 78(2), 124–130.

Dixit, A (2002). Incentives and Organizations in the Public Sector: An Interpretative Review. *The Journal of Human Resources*, 37(4), 696–727.

Draft Sanitation Plan — Main Document. Varanasi Municipal Corporation and GIZ ASEM (2011). Available online at www.nnuns.org/data/dpr/csp%20 Ex%20summary-UNN.pdf. Accessed 15 May 2015.

Edelenbos, J and I van Meerkerk (2015). Connective Capacity in Water Governance Practices: The Meaning of Trust and Boundary Spanning for Integrated Performance. *Current Opinion in Environmental Sustainability*, 12, 25–29.

EU (2000). Directive 2000/60/EC of the European Parliament and of the Council of the European Union of 23 October 2000 establishing a framework for Community action in the field of water policy. European Union, ed. *OJ L 327, 22.12.2000.*

Faguet, J-P (2014). Decentralization and Governance. *World Development*, 53, 2–13.

Foulon, J, P Lanoie and B Laplante (2000). Incentives for Pollution Control — Regulation and Public Disclosure. *Research Working Paper*. Washington DC: World Bank.

GoI (2002). Tenth Five Year Plan 2002–2007, Volume II: Sectoral Policies and Programmes. Delhi: Government of India, Planning Commission.

GoI (2009). Notification of Ministry of Environment and Forest. India, G. O., ed.1570. Delhi.

GoI (2011). Evaluation Study of National Rural Health Mission (NRHM) in 7 States New Delhi: Planning Commission, Government of India.

GWP (2000). Towards Water Security: A Framework for Action. Stockholm: Global Water Partnership.

Hamner, S, A Tripathi, RK Mishra, N Bouskill, SC Broadaway, BH Pyle and TE Ford (2006). The Role of Water Use Patterns and Sewage Pollution in Incidence of Water-borne/Enteric Diseases along the Ganges River in Varanasi, India. *International Journal of Environmental Health Research*, 16(2), 113–132.

IIT (2011). SWOT Analysis of Ganga Action Plan. Indian Institutes of Technology. GRB EMP: Ganga River Basin Environment Management Plan. www.nmcg.nic.in/writereaddata/fileupload/50_006GEN.pdf. Accessed 15 May 2015.

Krasner, SD (1983). Structural Causes and Regime Consequences: Regimes as Intervening Variables. In *International Regimes*, Krasner, SD (ed.), Ithaca, NY: Cornell University Press 1–21.

Lahariya, C, H Khandekar, J Prasuna and Meenakshi (2006). A Critical Review of National Rural Health Mission in India. *The Internet Journal of Health*, 6(1).

Lenton, R and M Muller (eds.) (2012). *Integrated Water Resources Management in Practice: Better Water Management for Development*. London: Earthscan.

Mandarano, LA, JP Featherstone and K Paulsen (2008). Institutions for Interstate Water Resources Management. *JAWRA Journal of the American Water Resources Association*, 44(1), 136–147.

Medema, W, BS McIntosh and PJ Jeffrey (2008). From Premise to Practice: A Critical Assessment of Integrated Water Resources Management and Adaptive Management Approaches in the Water Sector. *Ecology and Society*, 13(2).

Merrill-Sands, D and B Sheridan (1996). Developing and Managing Collaborative Alliances: Lessons from a Review of the Literature. *Organizational Change Briefing Note*. Boston: Simmons Institute for Leadership and Change.

Miller, GJ (2005). The Political Evolution of Principal-agent Models. *Annual Review of Political Science*, 8(1), 203–225.

Mohan, NS and S Routray (2011). Resolving Inter-state Water Sharing Disputes. *Seminar*(626).

Nesheim, I, D McNeill, KJ Joy, S Manasi, D Nhung, M Portela and S Paranjape (2010). The Challenge and Status of IWRM in Four River Basins in Europe and Asia. *Irrigation and Drainage Systems*, 24(3–4), 205–221.

NGRBA (2010). The Conservation Action Plan for the Gangetic Dolphin 2010–2020. Delhi: National Ganga River Basin Authority, Ministry of Environment & Forests, Government of India.

NGRBA (2011a). Environment and Social Management Framework (ESMF) for World Bank Assisted National Ganga River Basin Project. National Ganga River Basin Authority, Ministry of Environment and Forests, Government of India.

NGRBA (2011b). NGRBA Programme Framework. National Ganga River Basin Authority, Ministry of Environment and Forests, Government of India.

Page, S (2004). Measuring Accountability for Results in Interagency Collaboratives. *Public Administration Review*, 64(5), 591–606.

Pandey, SK (2009). National Rural Health Mission: A Multi-Level Governance Approach to Rural Health Care in India. *XXI IPSA World Congress of Political Science*. Santiago.

Parkinson, C (2006). Building Successful Collaborations: A Guide to Collaboration Among Non-profit Agencies and Between Non-profit Agencies and Businesses. Cambridge & North Dumfries Community Foundation.

Ribot, J (2002). Democratic Decentralization of Natural Resources: Institutionalizing Popular Participation. Washington: World Resources Institute.

Richards, A and N Singh (2001). Inter State Water Disputes in India: Institutions and Policies. UCSC Department of Economics Working Paper: University of California, Santa Cruz, Department of Economics.

Sharma, AK (2009). National Rural Health Mission: Time to Take Stock. *Indian Journal of Community Medicine: Official Publication of Indian Association of Preventive & Social Medicine*, 34(3), 175–182.

Singhal, AK (2012). Legal Cases on Ganga Pollution and Judicial Activism. *Journal of Applied Sciences in Environmental Sanitation*, 7(1), 61–64.

Stålnacke, P and G Gooch (2010). Integrated Water Resources Management. *Irrigation and Drainage Systems*, 24(3–4), 155–159.

Van Beek, E and W Lincklaen Arriens (2014). Water Security: Putting the Concept into Practice. *TEC Background Papers*. Stockholm: Global Water Partnership Technical Committee.

Winer, M and K Ray (1994). *Collaboration Handbook: Creating, Sustaining, and Enjoying the Journey*: ERIC.

World Bank (2001). Decentralization & Subnational Regional Economics: What Why, and Where. www.worldbank.org/publicsector/dezentralization/what.htm. Accessed on 5 May 2015.

Chapter 4

EDUCATING THE YOUTH: FACILITATING A SUSTAINABLE TRANSFORMATION OF GANGA

Low Lay Chin, Mark McKiernan, May Yu Lwin, Siow Boon Teik, Erwin, Thamil Selvan Govintharaju

1. Introduction

Rising concerns over the exponentially-increasing pollution levels in the waters of the River Ganga have made it imperative for policy-makers to look beyond the immediate technical answers towards a long-term sustainable solution. Environmental education (EE) is often considered by educators around the world as an important means of promoting academic and social growth of young people and encouraging conservation of the natural environment.

With half of India's population below the age of 27 (Worldometers, 2015), the pivotal role of this group of youth in nationwide efforts must be acknowledged and their enthusiasm leveraged upon. Strengthening the contribution of these future leaders of India through EE is thus essential to ensure that the world's second most populous country is adequately equipped for tomorrow's challenges to have a long-lasting positive impact on the river.

Education as a means of influencing people towards greater environmental proficiency will tackle the issue at its source by stamping out activities that contribute to the severity of the issue at hand. The ultimate goal then, of deploying EE as a solution, is to ensure that the youth develop into individuals with the ambition, commitment and skills needed to protect and improve the health of River Ganga. They can act as key drivers for change in the attitudes and behaviours of the community-at-large, creating a future society well equipped to live more harmoniously with the Ganga.

There are numerous research publications and articles that look into EE and the importance of starting EE in young children. Our contribution to this vast field of study is through primary research in Varanasi via focus group discussions amongst teachers and online surveys of teachers and school children. Further, in addition to our analytical review of the current state of EE in India, we explore recommendations to substantially elevate the status of EE and improve the quality of EE teaching, using examples of good practices from within India and other parts of the world. By starting small and focusing on Varanasi as a central point, we hope that the findings and recommendations of this exploratory research can be used to support a larger scale transformation in India.

Starting with the background of our approach to this study and a historical overview of EE in India, the rest of the chapter presents an analysis of the current state of EE and proposes recommendations to elevate the status of EE and improve the quality of EE teaching. The primary findings on the current state of EE reveal that (i) there is insufficient emphasis and priority accorded to EE as a whole, (ii) the EE curriculum, though adequate, is too broad and not customised to the local context, (iii) the pedagogical approaches used in EE are not very effective and there is a lack of quality in the training provided to teachers and (iv) extra-curricular activities have not achieved the desired impact.

To address these findings, the recommendations proposed include, (i) inducting the Minister responsible for school education into the National Ganga River Basin Authority (NGRBA) committee, (ii) establishing a youth division within the National Mission for Clean Ganga (NMCG), (iii) creating a customised formal EE curriculum for Varanasi, (iv) establishing a collaborative EE teacher development platform for teachers to learn, develop and share effective practices, and (v) instituting student-led Eco-Clubs aimed at informing, inspiring and mobilising the society as a whole.

2. Background Information

2.1. *Approach of study*

The national agenda of the River Ganga clean-up has focused mainly on the technical aspects such as waste management systems, water flow quantity, industrial effluents and agricultural run-offs. However, as authors of this study who share a common conviction, we believe that while addressing the above aspects may alleviate the symptoms of pollution, for the outcomes to be sustainable, EE must play a key role in helping people come to a more acute awareness of the ramifications of their actions. In terms of our professional backgrounds, we hail from different countries, namely the United Kingdom (UK), Singapore and Myanmar and bring with us extensive work experience in the educational field, both from a policy-making and practical standpoint. One of our authors has numerous years of experience in developing and delivering education policies for the UK government, while two have been involved in strategic public policy-making in the educational sector in Singapore. Additionally, another author is an educator who has taught undergraduate students in Singapore for more than a decade. Finally, we also have a senior civil servant from Myanmar in our team. It is our hope that this chapter will serve as an impetus for education to take on a more sizeable role in the River Ganga clean-up.

For this study, we have chosen to target high school students in Varanasi. Targeting this group is crucial as the future lies in their hands, and it will be easier to shape their attitudes and mindsets. With a higher level of education than their parents, these children could also spread the knowledge within their families and communities and to future generations. They could actively participate in efforts to clean up the River Ganga and rally others to do the same.

This study made use of information and data from both primary and secondary research. Besides reviewing literature on the theories and practices of using EE as a policy intervention for sustainable environmental protection, we also looked into the National Framework for EE curriculum for India and how different school boards have adapted this suggested curriculum.

On a study trip to understand issues relating to the pollution woes of the River Ganga in early December 2014, focus group discussions were arranged with 14 high school teachers from W.H. Smith Memorial School and the Sunbeam group of schools from Varanasi, India. The interviews were informative and they have reaffirmed the importance of education as a policy intervention and many of our findings from the literature review.

As a follow-up to the study trip, two sets of online survey questionnaires were also sent out between January and February 2015 to three different groups of schools in Varanasi, namely the W.H. Smith Memorial School, the Sunbeam group of schools, and Delhi Public School, Varanasi (see Annex 1 for a sample of the questionnaires). Of the 52 teachers who responded to the online survey, most of them were from Delhi Public School, Varanasi, with a couple from other schools. Fifty seven students from Delhi Public School responded. The students were from class levels 7 to 12, with 24 students from level 8 and 19 students from level 12. No students from the other two groups of schools responded.

The purposes of the surveys were (a) to assess the impact of EE in raising awareness, and in changing habits and attitudes among the school children, and (b) to assess the adequacy of EE from the teachers' viewpoint as currently practiced in Varanasi.

2.2. *History of environmental education in India*

To appreciate the saliency of EE to India's environmental preservation efforts, it is necessary to understand the historical context and development of EE over the years. The 1977 Tbilisi Declaration (GDRC, 2015) defined the goals of EE as follows:

1. to foster clear awareness of, and concern about, economic, social, political, and ecological interdependence in urban and rural areas;
2. to provide every person with opportunities to acquire the knowledge, values, attitudes, commitment, and skills needed to protect and improve the environment;
3. to create new patterns of behaviour of individuals, groups and society as a whole towards the environment.

EE provides "an important response" to modern human activities which, left unchecked, can endanger the habitats and ecosystems of different species (Cutter-Mackenzie, 2014, p. 25). At a United Nations Conference on the Human Environment in 1972, EE was reported as "one of the most critical elements of an all-out attack on the world's environmental crisis" (UNESCO-UNEP, 1976). It achieves this through helping the younger generation understand the nature and complexity of environmental challenges and developing their ability to take necessary action (Seema *et al.*, 2014, p. 271).

The concept of environmental protection is not new to India, having traditionally been interwoven with many facets of Indian life. The protection of the environment and its intricate links with daily life has traditionally been core to the fabric of Indian society (Ravindranath, 2007, p. 192). Coverage of the environment in Indian education can be traced back to the legendary leader of the Indian independence movement, Mahatma Gandhi, with his "Nai Taleem of Basic Education" in 1937. This aimed to foster "reflective learners...who were integrated into community life through engagement in productive work and who desired to undertake service for humanity" (Haigh, 2008, p. 244). Environmental education was seen as the crucial link between education and environmental protection, with former Indian Prime Minister Indira Gandhi viewing it as critical to developing understanding amongst the Indian people that environmental destruction was detrimental to both the individual and the community (Halder, 2012, p. 2223). It was formally introduced into the Indian school system between 1964 and 1966 in response to recommendations made by the Kothari Commission.

Another major leap forward came with the setting up of the Centre for Environment Education (CEE) in 1984. The CEE helped to champion the incorporation of EE into the school system at various levels. This led to the first National Policy on Education in 1986, and subsequently the Environment Orientation to School Education Scheme in 1988 (Almeida and Cutter-Mackenzie, 2011, p. 125), which directed the assimilation of EE into the school curriculum. In December 2003, the Supreme Court directed that EE be made a mandatory subject to be taught in all schools in every state (Almeida and Cutter-Mackenzie, 2011, p. 126), paving the way for an elevated status accorded to EE. This built on an earlier court judgement in 1991 that prescribed the need for courses on the environment in higher education institutes. In June 2005, most state boards rolled out the subject *Environmental Science* in their schools. As such, EE in schools has the potential to play an essential role in reconnecting young people with the once strong tradition by imbuing them with the right knowledge, values and attitudes towards the environment and in nurturing what Seema *et al.* (2014, 272) calls "environmentally-active and responsible citizens" who can be influencers to their families and communities, thereby effecting long-term change.

3. The State of Environmental Education in India

Despite the significance of EE in India, continued environmental degradation, such as that faced by the River Ganga, puts in doubt its efficacy in encouraging environmental sustainability. While pockets of effective practice in EE exist in India, it is not consistently achieving the desired impact. For example, a two-year-long study conducted by the Pune-based Bharati Vidyapeeth Institute for Environmental Education and Research (BVIEER) found that EE has not led to greater concern for the environment (India Together, 2013). To better understand the state of EE in schools in India and the reasons for the subdued impact, this section examines (1) the emphasis and priority accorded to EE, (2) EE in the school curriculum, (3) the quality of EE teaching and (4) extra-curricular learning related to EE.

3.1. *Emphasis and priority accorded to EE*

Notably, our primary research shows strong support for the paramount role that EE plays in bringing about a cleaner River Ganga. Eighty-three percent of the Varanasi teachers surveyed for this study agreed that education had the potential to play a large or reasonably large role in a sustainable clean-up of the River Ganga. Focus group discussions with teachers in Varanasi echoed this finding. Some of the teachers interviewed opined that "…children are impressionable so (the) impact is long lasting…" and that "…education is the long term sustainable solution!" There was overwhelming consensus among the other participants that education was a viable means of achieving a sustainable co-existence with River Ganga, though they acknowledged that this had to happen in tandem with other measures such as waste management and urban development to address infrastructural deficiencies.

The focus group participants felt that equipping children with a sound EE had the potential to achieve enduring impact: in the short term, children could influence their communities and in future, they would become the next generation of workers and business owners with the capacity to effect massive changes to environmental practices. As shown in Tables 1 and 2, students surveyed for this study strongly appreciated the importance of having a clean River Ganga (96% rated this important (4) or very important (5)) and were interested in contributing to cleaning the river (95% of them indicated high levels of interest). Eighty-nine percent of them felt that learning about environment protection would enable them to help keep the River Ganga clean (see Table 3).

Table 1: How Important is it to You in Having a Clean River Ganga?

Scale	5 (Very Important)	4	3	2	1 (Not Important At All)
No. of respondents	49	6	1	1	0
% of respondents	86%	10%	2%	2%	0%

Table 2: Would You be Interested in Doing More to Clean up the River Ganga?

Scale	5 (Very Interested)	4	3	2	1 (Not Interested At All)
No. of respondents	33	21	3	0	0
% of respondents	58%	37%	5%	0%	0%

Table 3: Do You Think that Learning About Environmental Protection will Help You Keep River Ganga clean?

Scale	5 (To A Large Extent)	4	3	2	1 (Not At All)
No. of respondents	29	22	5	1	0
% of respondents	51%	38%	9%	2%	0%

But while these students were aware of the importance of EE and were generally interested in doing more, less than half of teachers surveyed felt that the students were actively practicing what they were taught (though the survey results showed that more students were interested in cleaning up the River Ganga than what their teachers had thought). Both teachers and students were asked to indicate the top three reasons for this. As Figure 1 shows, the primary reason cited by both groups was the need to focus on other subjects. One of the teachers during the focus group discussion attributed this to the parents' sentiments of time allocated to EE being regarded as "...some useful periods wasted."

This finding echoes other research such as a study conducted by Halder (2012, p. 2228) that surveyed 120 schools in North Bengal, which showed that the majority of schools (67.5%) provided only one class per week. Twenty three percent provided only two classes per week and the remaining tenth of schools did not provide a regular EE class at all. The students in Halder's study were typically assessed on their knowledge of the environment through tests — 65% of schools arranged two evaluation tests in a single academic year. Most students, however, did not consider

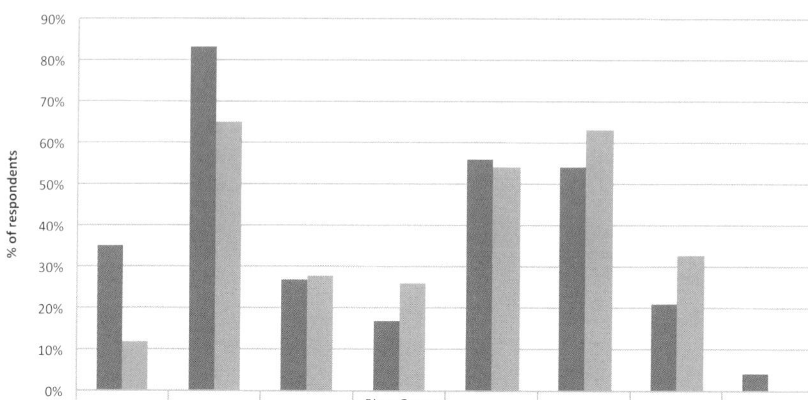

	Not interested in cleaning up the River Ganga	Need to focus on other school subjects and examinations	What is taught in school is not relevant to River Ganga	River Ganga can protect and clean itself, there is no need for me to do anything more	It is someone else's responsibility to clean and not school children	As school students, their efforts make no difference and they feel powerless	It is too troublesome to do what the school teaches	Other
■ Teachers	35%	83%	27%	17%	56%	54%	21%	4%
■ Students	12%	65%	28%	26%	54%	63%	33%	0

Figure 1: Reasons that Prevent Students from Practicing What They Have Been Taught

the evaluation tests to be as important as the other subjects they were studying and apportioned less effort to them (Halder, 2012, p. 2230).

The next most common reasons cited for students failing to put what they had been taught into practice were: the perception that, as school students, they were powerless, a feeling that their efforts made no difference; and the belief that it was someone else's responsibility to clean the River Ganga. Summing up the feeling of helplessness when faced with a problem that is perceived to have no solution, one of the teachers suggested that students might be asking themselves a simple question, "Why bother?"

Focus group participants identified the lack of focus on local issues in EE, limited effectiveness of EE on the students, and the non-prioritising of EE by students and parents as the key challenges to teaching EE. As shown in Figure 2, these findings were reflected in the results from the survey of teachers with the most commonly-cited challenges being: the lack of state-specific curriculum; parents and students prioritising other subjects before EE; and disinterested students.

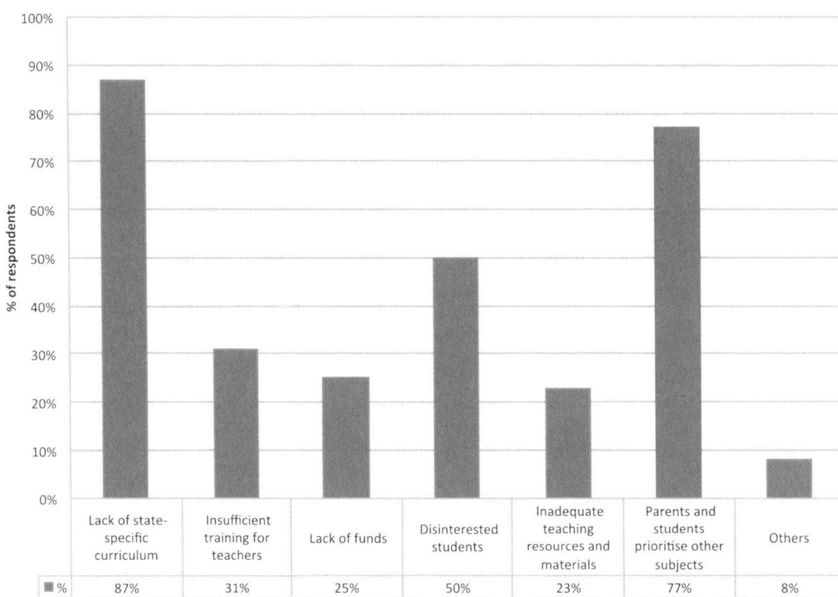

Figure 2: Main Challenges Teachers Face in Teaching EE

3.2. *Environmental education in the school curriculum*

Aside from issues relating to the insufficient priority and emphasis accorded to EE, we also examined EE in the school curriculum.

The December 2003 directive passed by the Supreme Court of India mandating that EE be taught as a subject in all schools is not in itself useful without the organizational capacity to implement it. While the National Commission for Education, Research and Training (NCERT) was asked to develop an appropriate EE curriculum for all grades (see Annex 2 for an extract of the NCERT curriculum), the directive meant additional burden on an "already over-stretched education system" (Almeida and Cutter-Mackenzie, 2011, p. 126).

Moreover, there has been unevenness in the implementation of EE. For instance, there has yet to be a general agreement on how to deliver EE in normal school curriculum (Sonowal, 2009, p. 16). In 2005, the National Curriculum Framework (NCF) was drafted, representing a big step toward improving the consistency of implementation in EE. From grades one to five, the NCF recommends the use of activities and a standalone subject known as Environmental Studies (EVS) to deliver EE. At the secondary levels, the NCF suggests the use of an "infusion-based model[1] and project-based studies" (Almeida and Cutter-Mackenzie, 2011, p. 127). Although their directives helped to propel EE forward, the Supreme Court's decision to transfer significant responsibility to the state education boards "to adapt the syllabi from NCERT to their respective education systems" without stipulating the means to achieve such an end has left the interpretation of the policy "open to numerous interpretations and loopholes" (Almeida and Cutter-Mackenzie, 2011, p. 127). With different states and cities being allowed their own interpretations of the broad curriculum and thus producing a variety of textbooks, it is not surprising that implementation of EE in schools across different states in India has been largely inconsistent.

[1]An infusion-based model is one in which topics are not taught as part of a separate subject but are integrated into various subjects like Geography, Social Studies, Science etc.

Table 4: How Adequate is the Current Curriculum Coverage on Environment Protection of the River Ganga?

Scale	5 (Very Adequate)	4	3	2	1 (Extremely Inadequate)
No. of respondents	2	20	14	12	4
% of respondents	4%	38%	27%	23%	8%

While the Varanasi teachers from the focus group conducted for this study agreed that the curriculum content was generally adequate, some observed that the curriculum for EE was very broad and insufficiently localised. They noted that it contained elements which were of little relevance to Varanasi, such as those relating to earthquakes and tsunamis. For EE to have a more personal connection with the students, they suggested that the curriculum needed greater customisation. One of the focus group participants described the main problem as such, "...curriculum should focus on the problems for the city rather than for the whole nation."

In addition, the teachers at the focus group discussion also highlighted the lack of a mechanism for teachers, as users of the curriculum, to provide their views to the curriculum designers. This is an important point for policy makers to note as, although 83% of the teachers surveyed felt that education could play a large role in cleaning up the River Ganga, only 42% rated the current coverage of environmental protection of the River Ganga in the curriculum as adequate or very adequate (see Table 4).

3.3. *The quality of environmental education teaching in India*

Although the preceding analysis points to the possibility of an improvement in the EE curriculum for Varanasi, it should also be noted that no matter how robust, relevant and engaging a curriculum, as the oft-quoted aphorism goes, "the quality of an education system cannot exceed the quality of its teachers" (Barber and Mourshed, 2007, p. 16). This section examines the pedagogical approaches currently used in EE and the initial and in-service training of teachers in the subject to assess the quality of teaching in this field.

3.3.1. *The difference a good teacher can make*

Research has found that the quality of the teacher is the most important determinant of the quality of education (Azam and Kingdon, 2015). The results of Azam and Kingdon's study of grade 10 to 12 pupils in linked private schools in Uttar Pradesh imply that teacher quality can have a significant impact on pupil attainment. For example, their research to assess teacher value added found that on average an improvement in teacher quality of one standard deviation would move an average pupil from the 50th to the 65th percentile (Azam and Kingdon, 2015).

While the quality of teachers can make a significant difference, it is widely acknowledged that there is room for improvement in the quality of teaching in Indian schools. UNESCO's 2013/4 Education for All report, for instance, confirms the endurance of a number of long standing issues which affect teacher quality in India, including high levels of absenteeism, low motivation, weak accountability and inadequate training (UNESCO, 2014). The Indian government has launched a number of ambitious initiatives to improve the quality of teaching in recent years. Overall, during the period of the 12th five-year plan (2012–2017), the government has sanctioned 6300 crore rupees (approximately US$991 million) for improving teacher education (*India Education Review*, 2013). It will, however, take time for this activity to affect all schools and for the results to filter through to improving outcomes for pupils.

3.3.2. *Environmental education pedagogy*

The effectiveness of approaches to teaching and learning (pedagogy) adopted in the classroom can have a significant influence on the quality of EE. Research into teaching practices in developing countries found that effective pedagogical practices were "student-centred and teacher-led but informed by social constructivist approaches that see knowledge as socially constructed and learning as a social process" (Department for International Development, 2014). Unfortunately, much of the criticism of the quality of teaching in India is linked to the prevalent lecture-style approach. This is summed up well in a 2012 position paper from the National Focus group on Teacher education for curriculum renewal which refers to the problem thus:

> *"The concern for curriculum renewal emerges from the tenacious character of an education system that continues to view teachers as 'dispensers of information' and children as 'passive recipients' of an 'education' that is sought to be 'delivered' in four-walled class-rooms.*" (National Focus Group on Teacher Education for Curriculum Renewal, 2012).

With research suggesting that it is ineffective merely to tell young people about environmental issues (Gayford, 2009), the endurance of this approach is problematic for the quality of EE teaching.

Much evidence now exists to support the effectiveness of more active, participatory and collaborative approaches to teaching and learning (Barratt and Barratt Hacking, 2011). Through these approaches pupils can achieve better outcomes both academically and in terms of practical impact. For example, being involved in active teaching and learning for EE can improve pupils' retention of knowledge, motivation, high-order learning skills, and develop their practical skills (Alexandar and Poyyamoli, 2014). In addition, pupils are more likely to transfer their learning to daily life and develop effective ways of contributing to local improvements if they have been taught using these approaches (Gayford, 2009 and Percy-Smith *et al.*, 2009). This was a view echoed by some of the Varanasi teachers in the focus group discussions who felt that rote learning of EE in a classroom had limited effectiveness. As one teacher put it, "Pupils learn better by doing."

But what do these alternative approaches look like in practice? Based on his analysis of 32 examples of schools recognised for undertaking effective learning for sustainability in the UK, Dr Chris Gayford identifies 19 key pedagogical approaches. By way of example, these include:

- pupils taking responsibility for researching aspects of a topic and presenting their findings;
- the use of group work, collaborative tasks and peer learning;
- consideration of how pupils can individually or collectively take action to address sustainability issues; and
- involvement of parents or their community in activities (Gayford, 2010).

In addition, improved educational outcomes and motivation was achieved when pupils helped to plan or evaluate the effectiveness of their activities (Gayford, 2009, p. 17).

Other research has found that field based activities are effective but are 'limited or negligible' in the environmental sciences at school level in India (Alexandar and Poyyamoli, 2014). Engaging pupils through these approaches is particularly important as many of the environmental challenges blighting the River Ganga are a result of economic development and, without clear and transparent debate about the socio-political and economic consequences and an appreciation of the trade-offs, pupils may feel that what is taught in EE contradicts the harsh realities of many of their lives (Sonowal, 2009, p. 17).

This is not to say there are no examples of effective practice in India. On the contrary, the CEE, for example, has done much to promote effective practice in EE through school-based projects, capacity building work and teaching resources. In Uttar Pradesh, for instance, CEE has worked on the *Pani Parikshan* project which involves school children in monitoring water quality of the Gomti river. The project produced a kit and a guide for teachers to facilitate the educational activities.

There is a lack of evidence, however, that such projects have had a wide reach or impacted school practice in a sustained way. Many of the schools in the North Bengal survey mentioned earlier, for example, followed an improved method to teaching EE by using teaching and learning materials to engage pupils as opposed to lecture-style teaching. But there was a wide variation in the use of these methods between school districts (ranging between 40% and 70%) (Halder, 2012, p. 2230). This suggests that there is a lack of consistency with which effective pedagogies are applied. This is not helped by the lack of emphasis on improving the teaching of EE in the government initiatives referred to above.

3.3.3. *Environmental education in teacher education and continuing professional development*

Clearly, with many teachers unfamiliar with the more interactive pedagogical approaches discussed above, there is a major need for good quality training and development opportunities. In fact, Gayford notes that the key pedagogical approaches he identifies were only successful where, amongst other things, "...teachers' subject knowledge was secure and extensive enough to support pupils' enquiry, independent thinking and debate, and there was good professional development within the school." (Gayford, 2010, p. 4).

The surveys of teachers and pupils undertaken for this study suggest that both groups have a high level of confidence in the quality of EE teaching. Sixty nine percent of teacher respondents felt that they were very well or well equipped/supported to teach EE of the River Ganga (see Table 5). As shown in Table 6, 80% of pupil respondents felt that their teachers were excellent or good in teaching about environmental protection.

During the focus group discussions, however, the Varanasi teachers expressed concern about the scarcity of EE resources available to them, whether in terms of materials or funding, exacerbated by the lack of training for teachers in the subject area. This is echoed by approximately one third of teachers who responded to the survey who also felt that training was insufficient; more than half of these (9/16) being Environmental Science teachers. So, while there is confidence in the existing teaching, there is also an acknowledgement among many that further training is needed.

Table 5: How Well Equipped and/or Supported are You to Teach EE of River Ganga?

Scale	5 (Very Well)	4	3	2	1 (Not Able To)
No. of respondents	13	23	11	3	2
% of respondents	25%	44%	21%	6%	4%

Table 6: How Good are Your Teachers at Teaching You About the Environment Protection of the River Ganga?

Scale	5 (Excellent Teaching)	4	3	2	1 (Not Good At All)
No. of respondents	23	23	9	0	2
% of respondents	40%	40%	16%	0%	4%

3.3.3.1. Preparation of teachers entering the profession

The NCTE applauded the increased recognition of EE from the Supreme Court in 2003 but acknowledged that it had yet to attain "its rightful place", even in teacher education. Unfortunately, the NCTE's acknowledgement of the importance of EE does not appear to be fully reflected in the curriculum framework itself; their 2009 update of the framework gives scant attention to EE (National Council for Teacher Education, 2009).

Teacher education providers have autonomy to design their own curricula. Following wide-ranging consultation, the NCTE periodically sets out expectations of providers' curricula in the curriculum framework. Though not providing clear direction, the current curriculum framework for two-year Bachelor of Education programmes, for instance, provides encouragement for EE to be included in teacher education programmes. It is, therefore, a matter for individual teacher education providers, informed by the NCTE's framework, to decide whether and the extent to which EE forms part of teacher education.

It should be noted that, in response to the 2003 Supreme Court directive, the University Grants Commission (UGC) developed a six-month compulsory course in environmental studies for all undergraduates. It has been the subject of controversy, though, as some institutions have yet to implement the module leading the National Green Tribunal recently to ask the UGC to monitor implementation of the core module. Where it is being delivered, the module could contribute some way toward developing a student-teacher's knowledge but without further subject knowledge development and accompanying training in pedagogical methods, this will be insufficient.

Some university-based teacher education institutions have gone beyond providing the module mandated by the UGC and are responding to the NCTE guidelines to provide additional training. For example, Banaras Hindu University (BHU), a major teacher educator in Varanasi, provides all student-teachers, irrespective of subject specialism, with a course in environmental studies involving working with the community, as well as an additional EE component for those specialising in science.[2]

[2] Information from discussion with Prof. H.C.S Rathore, Dean of Faculty of Education, BHU on 14 May 2015.

3.3.3.2. Professional development for existing teachers

In response to concerns about the quality of teacher education and training, in 1987, a programme was launched to establish a new infrastructure for teacher training in India. A network of Institutes of Advanced Study in Education (IASE), Colleges of Teacher Education (CTEs) and a District Institute of Education and Training (DIET) were established to provide in-service training for secondary teachers.

In-service training, however, has been criticised more recently for not being sufficiently well linked to pre-service training or sufficiently focused on teachers' needs (Srivastava, 1999) (and ironically for being lecture-based itself even when focused on practical aspects such as activity-based teaching) (National Focus Group on Teacher Education for Curriculum Renewal, 2012, p. 5). Organizations, such as the CEE, produce resource materials and run some projects but there is little evidence that their work has been taken to scale. A promising exception to this is CEE's work to train master trainers in the Uttar Pradesh DIET.

This analysis shows that, in line with the views of the C.P.R. Environmental Education Centre (a Centre of Excellence of the Ministry of Environment, Forests and Climate Change), EE could be much more effective if the pedagogical skills of teachers were enhanced to deliver lessons in different instructional formats (Roberts, 2009, p. 447). With adequate teacher training and by focusing more on practical experience related to the environment rather than predominantly classroom-based learning, (Roberts, 2009, p. 447) pupils will be better able to understand how theories and concepts learnt in books are played out in real life. While there are examples of excellent practice in EE in Indian education, thus far there has neither been a major drive to boost the competencies and capabilities of teachers to teach EE, nor a paradigm shift from the use of conventional lecture-style teaching methods to adopt dynamic approaches to teaching EE (Gupta *et al.*, 2011, p. 31). It is far too often the case that the methods employed in classrooms are inadequate, and that teacher education and in-service training are not equipping the teachers sufficiently well.

3.4. *Extra-curricular learning*

Beyond the EE curriculum and teaching quality, it is also useful to explore how we can further imbue the correct values of environmental protection into school children such that it will become a life-long habit.

In 1999, Louise Chawla conducted an interview-based research among 30 environmentalists from Kentucky, USA and 26 environmentalists from Norway. The environmentalists were asked what motivated them to protect the environment. The respondents attributed their motivation and actions to a similar set of sources, among which were (1) spending extended time outdoors in the natural environment, often during childhood, (2) the influence of parents or family members and (3) teachers or classes (Chawla, 1999, p. 15). The out-of-school experiences feature prominently in the environmentalists' memories. Chawla's research points to a need to have EE included not only in classroom instructions and field studies but also in (a) preservation of natural areas, (b) outreach to parents to enable them to serve as role models, and (c) support for community organizations where there are adult role models (Chawla, 1999, p. 25).

In India, the Ministry of Environment, Forests and Climate Change (MoEFCC) set up the National Green Corps (NGC) programme in 2001 to "encourage and mobilise participation of school children in various environment education activities in their localities." (Roberts, 2009, p. 448). The goal of NGC is being realised through the establishment of Eco-Clubs in every district across India, with an emphasis on action-oriented environmental programmes (Roberts, 2009, p. 448).

An evaluation of the Eco-Clubs initiatives shows that the concept and outreach of the environmental programmes is generally sound and well organised, and it has led to positive impacts within the boundaries of educational institutions. However, further scrutiny has revealed that these benefits do not extend to the wider community. This is mainly attributed to the lack of importance accorded to those programmes and the flaws in managing and executing them (Roberts, 2009, p. 455). One obvious weakness lies in the fact that, although the state governments are free to establish as many Eco-Clubs as they want, financial assistance is only made available by the MoEFCC to a certain number of clubs within a particular district. This often results in many of the Eco-Clubs not being able to execute activities beyond a certain number due to the limited amount of funding. Another finding has revealed that the current organisational structure of the NGC Eco-Club Management is too top-heavy, too cumbersome and is plagued by a serious lack of accountability and breakdown in communication. The leaders within the current system do not offer the required level of time and commitment for the Eco-Clubs to succeed on a consistent basis (Roberts, 2009, p. 455).

The evaluation also identified some barriers to student participation in the programme. These included the fear that adults will not listen to them, non-receipt of resource materials from the assigned supervising agency, insufficient Eco-Club meetings to assist with carrying out individual projects and insufficient training of their teachers (Roberts, 2009, p. 457). Notwithstanding the problems faced by the Eco-Clubs programme, however, it has provided an excellent platform for students to participate directly in activities aimed at preserving the environment, such as cleaning up beaches and rivers. Such action-oriented learning is instrumental in reinforcing curricular learning, instilling strong values and promoting environmentally-conscious behaviour. There is also a clear imperative for parents and the community-at-large to be involved to boost publicity and elevate the status of Eco-Clubs at the national level (Roberts, 2009, p. 458).

The findings of the online surveys undertaken for this study provide further support the view that the NGC is not achieving the desired impact. Although the students' and teachers' desires to participate further in clean-up projects and awareness-creation activities among families and friends (see Figure 3) indicate some enthusiasm and interest, there are also strong indications that impediments exist which prevent this enthusiasm from being translated into action.

As highlighted earlier in Figure 1, more than 70% of both students and teachers surveyed indicated the need to focus on core school subjects and examinations as the primary obstacle to their engagement in environment related activities. Seventy percent of student respondents also indicated the lack of power they are able to wield in influencing or mobilising the public on environmental awareness issues.

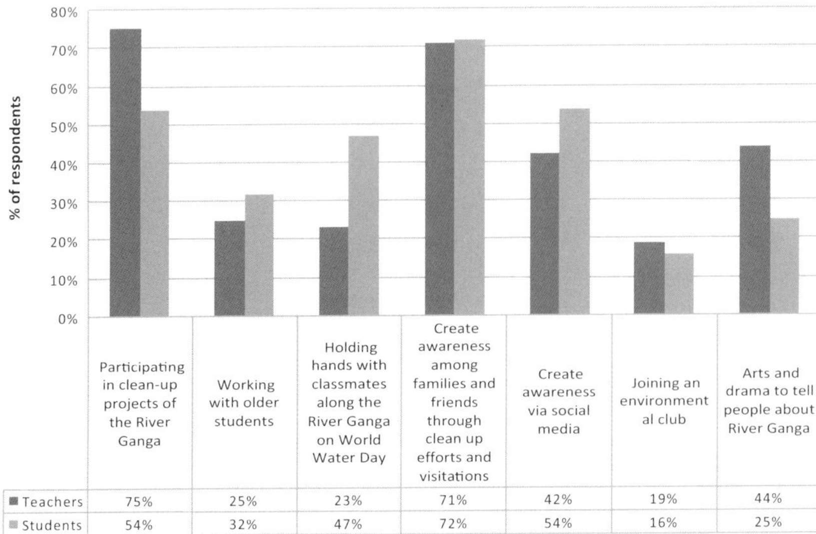

Figure 3: Interesting Activities that Students Would Like to Participate In

In summary, the preceding sections, which looked at the state of EE in India, have underlined the frailties of EE in the Indian education system as follows:

- There is insufficient emphasis placed on EE as compared to other subjects;
- Coverage of environmental issues in EE tends to be overly-broad and not locale-specific;
- There is room for improvement in the quality of EE teaching, particularly in the areas of teacher training and pedagogy; and
- EE activities beyond the classroom have only had a muted impact in the wider community due to management and implementation gaps.

In spite of the challenges surrounding EE in India, these are by no means insurmountable. Concrete steps will have to be taken to ensure that education achieves its full potential in bringing about a sustainable change in behaviours and attitudes towards the cleanliness of the River Ganga.

4. Recommendations for Enhancing Environmental Education

As demonstrated in the analysis, while a sound legislative framework for EE in schools is in place and much activity is in train, barriers exist which reduce its effectiveness. Maximising the impact of EE requires a combination of government intervention and grassroots action. This is reflected in the recommendations set out in this section.

Recommendations 1 and 2 are designed to ensure that EE receives appropriate status, and becomes an integrated element of the wider efforts to clean River Ganga, releasing central and state resources to drive improvements. Recommendations 3 to 5 are designed to harness local expertise, knowledge and passion to improve the impact that EE has on the River Ganga.

As the Figure 4 below shows, taken together these recommendations create a self-improving model by placing responsibility with communities and providing them with the support and motivation to make a significant contribution to the sustainable clean-up of River Ganga.

The recommendations may be implemented for communities throughout the Ganga Basin, however it is proposed that they are initially applied as a pilot project to secondary education in Varanasi. The impact of this project should be monitored and lessons learnt before rolling out more widely.

(i) Include the Minister with responsibility for school education on the National Ganga River Basin Authority (NGRBA) committee

The analysis has indicated the potentially significant positive effect that EE can have in the clean-up of River Ganga. Though EE can be, and is often crowded out by competition from other subjects, it is still held in high regard by the Indian people.

It is notable that the body with responsibility for the implementation of Indian Prime Minister Narendra Modi's priority of cleaning up the River Ganga does not include an Education Representative. As the education of all citizens is a key way to effect a sustainable approach to a clean River Ganga, the absence of a Ministry of Human Resource Development (MHRD) representative is a significant omission. If EE were perceived

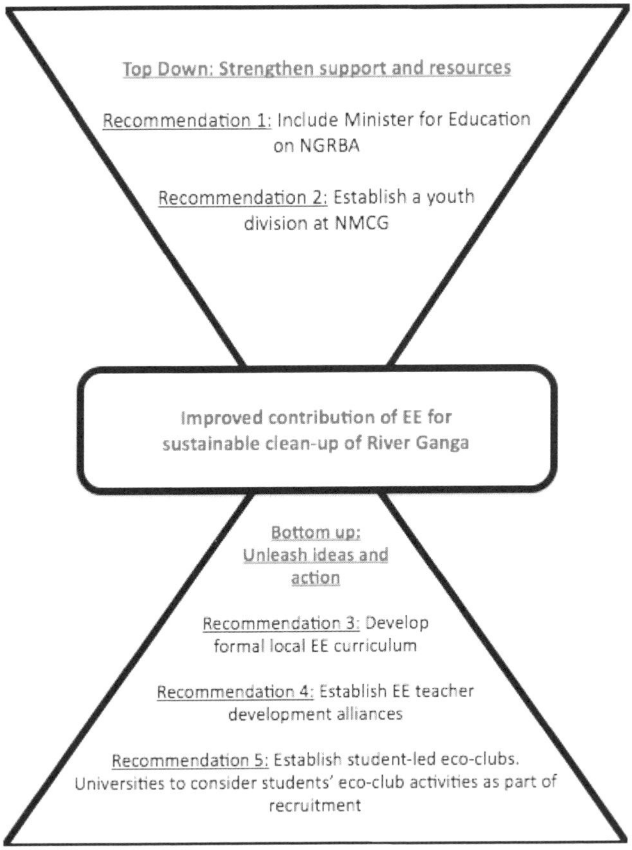

Figure 4: A Model of Recommendation for Environmental Education

more strongly to be part of the solution to this national priority for India, the relevant Minister would be making both a crucial contribution to solving the Ganga problem, while at the same time pursuing educational priorities to develop teachers. The inclusion of the Minister in charge of MHRD on the NGRBA committee will be symbolic in signifying the attention that should be placed on EE, both in Varanasi and across India. Moreover, it has the potential to ensure that EE is backed by sufficient resources and ministerial interest to improve the chances of it having a meaningful impact.

The representation of the Ministry of Education in high level anti-pollution committees has been practiced in other nations. In August 2012, Jamaican Prime Minister Portia Simpson Miller set up an Inter-Ministerial Committee to clean up the environment for the island nation. The Ministry of Education was also represented in the high level committee (*Caribbean Journal*, 2012). Another example is that of Singapore. Singapore's first national campaign as an independent nation was the 'Keep Singapore Clean' campaign. A national committee was formed in August of 1968, comprising representatives from various government agencies including the Ministry of Education. It was launched on 1 October 1968 by the late founding Prime Minister Lee Kuan Yew. Mr Lee explained that "cleaner communities would lead to a more pleasant life and keep morale high and sickness rate low, thus creating the necessary social conditions for higher economic growth through industry and tourism." Although this campaign was aimed at everyone, children campaign to ensure that good habits were cultivated from a young age. The success of this inaugural campaign has enabled it to be carried out yearly till 1990 when it was merged with another Garden City campaign to form the "Clean and Green Week" campaign, which still exists presently (National Library Board, Singapore Infopedia, 2012).

(ii) Establish a youth division within the National Mission for Clean Ganga (NMCG)

In India, there is formal EE in schools and there are also non-formal EE programmes administered by individual school teachers, in collaboration with various non-governmental organizations (NGOs) and the CEE under the MoEFCC. In addition, most of the EE initiatives and themes in the formal education system have been contributed by sectoral industries such as the environment, fisheries, agriculture, forestry and natural resources. The specific content developed by them is mostly geared towards increasing knowledge and skills but may not be comprehensive enough to achieve sustainability. Gupta *et al.* (2011, p. 31) expressed in their study that "it is not possible to get the necessary full commitment from the government towards EE activities unless it is addressed in totality." Although the CEE works closely with local schools, local NGOs, and

State Departments of Education, it does not have the authority to amend the state curriculum, the basic structure of textbooks or examinations (CEE, 2007, p. 19).

The lack of coordination for various EE efforts is exemplified by the fact that there are a number of groups which students can volunteer to join which have activities that create awareness of the environment and they are not under one umbrella. For example, under the Indian military, the National Cadet Corps has adopted community development activities with the aim of imbuing amongst cadets selfless service to the community, dignity of labour, importance of self-help, the need to protect the environment and to assist weaker sections of the society in their upliftment (National Cadet Corps, 2015). Under the Ministry of Youth Affairs and Sports, the National Service Scheme primarily focuses on the development of students' personality through community service (National Service Scheme, 2015). Under the MoEFCC, the *Paryavaran Vahinis* aims to create environmental awareness "through active participation and reporting of illegal acts pertaining to forests, wildlife, pollution, environmental degradation and cruelty to animals."(MoEFCC, 2015). Recently, the NGRBA has passed an order to set up three district *Ganga Vahinis*, each headed by a district magistrate, in Kanpur, Allahabad and Varanasi to keep the river pollution-free. The *Ganga Vahini* will be set up in the pattern of Red Cross, which would comprise youth, students, ex-servicemen and others. It would protect the river and spread the message of its rejuvenation across India (*The Times of India*, 2015). These groups and CEE programmes are not coordinated by one central body.

Without a centralised department to coordinate such efforts, it can lead to a parochial view of issues that may lack the comprehensiveness and breadth needed for sustainability. The effectiveness of EE requires a dedicated, whole-of-government approach to be more holistic and impactful.

For the purpose of this study on a sustainable clean-up of River Ganga, and in line with the first recommendation of including the Minister of Education in the NGRBA, we recommend that the NMCG set up a Youth Division at both the national and state levels, to provide resources and hold communities accountable for improving the impact of EE on the rejuvenation of River Ganga. The roles of this Youth Division (coordinated on the national level and executed by state level units) would include:

- Liaising with relevant civil servants and EE experts to ensure that the NGRBA are serviced with the best advice on how EE can most effectively contribute to the clean-up of the River Ganga;
- Ensuring that decisions related to EE made by the NGRBA are implemented efficiently; and
- Co-ordinating all EE programmes for schools, both formal and non-formal EE, that can contribute to the clean-up of River Ganga. This will ensure that synergies between EE and other NMCG efforts relating to the clean up of River Ganga can be identified and exploited.

The responsibilities of this Youth Division would relate to EE insofar as it is directly focused on the clean-up of River Ganga. On the ground level, it may not always be possible to disaggregate this from the wider EE activities and programmes, for example, eco-clubs that undertake wider environmental activities as well as Ganga-related activities. Although the work of this Youth Division may help improve EE in general with high-level, ministerial coordination among various environmental groups and programmes, and as a result, elevating the status of EE, it should remain focused on maximising the impact on the clean-up of River Ganga.

We suggest that the Youth Division runs a pilot programme in Varanasi. Initially, it could work with local schools and NGOS in Varanasi to:

- Develop clear annual objectives for activities related to EE on the clean-up of River Ganga;
- Develop their capacity to implement the objectives by providing the necessary expertise and financial resources;
- Develop evaluation methods to assess the impact of their activities; and
- Hold the local groups accountable for the impact of their activities on the clean-up of River Ganga.

Such a dedicated unit was also formed in Singapore within the Ministry of Environment and Water Resources — the National

Environment Agency's (NEA) 3P (Private, Public, People) Division. The partnership approach towards EE among the 3Ps has been successful in Singapore. There is a Youth & Schools Department under the 3P Division that specifically works with schools and youths. It aligns its approach with other stakeholders, for example, with the Ministry of Education's (MOE) "Values-in-Action" Character and Citizenship Education curriculum so that MOE, principals and teachers see that NEA's programmes fit their objectives too.[3]

The NEA strongly believes that it is crucial to shape students' attitudes and behaviours, rather than changing bad habits. In partnering schools, NEA would usually work with the school teachers in charge of environment education (or known as the Environmental Education Advisors [EEAs]) in developing programmes and activities for students rather than working with students directly. In doing so, it is easier to incorporate the programmes into the schools' curriculum. Schools have been empowered to organise their own environmental programmes and NEA provides some platforms for the schools to showcase their projects. One such platform is the annual Youth for the Environment Day, where schools organise environmental programmes to celebrate Earth Day.[4]

There are numerous other targeted programmes that NEA runs with schools and other partners annually, such as "The Learning Journey Programme", spearheaded by the MOE, which enable students to visit national institutions and public installations to widen their knowledge. These include tours of the NEWater Visitor Centre, incineration plants and talks on waste management. The "Environmental Education Advisor (EEA) Network" runs annual workshops, seminars and retreats by NEA to update EEAs from all schools and for them to provide their feedback on current environment trends and issues. It also facilitates a network of core EEAs who will initiate programmes for students. The "Clean Singapore Learning Trail (Beaches) Programme" encourages students to take ownership of the beaches and to clean the beaches for all to enjoy (NEA, 2015).

---

[3] Information from email interview with NEA, coordinated by Mr Darren Tan, Senior Executive, Schools & Youth Department, 3P Network Division, NEA on 6 February 2015.

[4] Information from email interview with NEA, coordinated by Mr Darren Tan, Senior Executive, Schools & Youth Department, 3P Network Division, NEA on 6 February 2015.

Not only does NEA's Youth & Schools Department work with schools to come up with initiatives, it also has an Environment Club Fund for all schools to apply for use in their activities (NEA, 2015).

In conclusion, a strong centralised coordinating body will play an instrumental role in elevating the priority and emphasis of EE.

(iii) Produce a formal Varanasi EE curriculum to be infused into the various subjects for use across all schools

As mentioned earlier, there are formal and non-formal EE programmes being used to teach EE to students. The formal EE curriculum is more general and covers a whole plethora of issues relating to EE. The non-formal EE programmes are generally more contextualised to the local issues. Since every student has to cover the formal EE programmes while not everyone will take part in the non-formal EE programmes, the impact of EE on the problem of River Ganga and pollution in general in Varanasi has not been seen as effective.

Moreover, presently, different schools adopt different curricula from different school boards and hence there are differences in emphasis, infusion, and teaching of EE to the students. In order to make a concerted effort in keeping River Ganga and Varanasi clean, it is imperative to have a local formal EE curriculum specially tailored to Varanasi, for infusion into the various subjects from class 6 to 12.

Although it is important for students to learn about the many EE issues that exist in the world, the fact is that it takes up a lot of classroom time and many students lose interest in topics that they do not find relevant. Due to time constraints and emphasis placed on examinable subjects, parents and students do not prioritise EE into their learning, particularly for the contextualised non-formal EE activities.

Targeting specific environmental issues in Varanasi can yield many benefits. Both teachers and students will be able to identify with the challenges of environment protection and will be able to apply what they have studied in their very own backyard. The impact of such a locale-specific curriculum should not be underestimated. When the topics taught are close to the hearts of the teachers, they are likely to be more passionate about imparting the knowledge to the students. If EE is tailored to Varanasi and the students can learn specific ways to help keep Varanasi and River Ganga clean and understand how their quality of life will be improved, they are more likely to feel that they can play an important role in the effort.

An empirical support for having contextualised curriculum to enable effective learning can be found in a large scale three year study conducted by the Centre for the Use of Research and Evidence in Education (CUREE) in UK. This study was done among hundreds of schools, teachers and pupils between 2007 and 2010 to explore the successes and obstacles in curriculum development. One of the key findings shows that "…creating opportunities for learners to make connections between the school curriculum and the 'real world' promotes effective learning. Schools could do more...to create curriculum experiences in which learners draw on outside experiences…" (CUREE, 2015). The 'real world' experiences for Varanasi students have to be what they can see and experience outside of school, hence the importance of a locale-specific curriculum.

The Youth Division for EE that is proposed in recommendation 2 could lead in the process of coming up with a localised formal Varanasi EE curriculum to be infused into the various subjects. It could get inputs from teachers, the various school boards, NGOs, CEE North, universities and Uttar Pradesh Pollution Control Board experts.

It is also important to have an evaluation survey among teachers and students after a year of implementation to assess the relevance of the infused localised EE curriculum. Regular reviews of the curriculum must also be scheduled for long-term efficacy.

(iv) Establish EE teacher development alliances

As demonstrated in the earlier analysis, participatory approaches to teaching EE can make a significant difference to its impact. It is essential, therefore, that teachers are trained to use these approaches and develop the confidence and capability to deploy them when designing and delivering EE lessons. In line with the view of the National Focus Group on Teacher Education for Curriculum Renewal, teachers should also be given the scope to innovate, evaluate and share effective practice. While activities are in train to ensure teachers have a grounding in these approaches, sustaining their use once new teachers enter the classroom, and changing the transmission-based approaches used by many existing teachers, remains the more significant challenge. To address this in Varanasi, it is proposed that EE teacher development alliances, made up of groups of secondary schools, be established for teachers to learn together and develop their own effective practice.

Strong evidence exists to support the effectiveness of a collaborative approach to teachers' continuing professional development (CPD). The *Evidence for Policy Practice and Information and Co-ordinating Centre* (EPPI Centre) in the UK has defined collaborative CPD as programmes with "...specific plans to encourage and enable shared learning and support between *at least* two teacher colleagues on a sustained basis" (sustained being at least 12 weeks or one term) (EPPI Centre, 2015). Based on research in the UK, such collaborative programmes have been shown to have many positive impacts on teachers, amongst them increased confidence and belief in self-efficacy and increased commitment to changing practice and willingness to try new things. In turn, evidence suggests that pupils demonstrated increased motivation, confidence and participation, and improvements in learning (EPPI Centre, 2015).

Following local dialogue to determine an appropriate alliance size, and any natural allegiances (e.g. religious, examination board), it is proposed that schools in Varanasi be grouped into alliances. Drawing on the experience of the Teaching School model operating in England — itself informed by research on effective CPD — each alliance would be headed by a lead secondary school with an excellent track record of delivering impactful EE. The lead school would receive a grant to operate as coordinator for the alliance for an initial period of one year. It would have responsibilities for managing resources on behalf of the alliance and identifying, developing and coordinating expertise (both from teachers within the alliance and other sources, e.g. BHU, CEE and relevant NGOs) in order to provide development opportunities for teachers from across the alliance. In addition, it would serve as the main point of contact for the alliance with the state and national youth divisions at NMCG.

The development opportunities provided should be focused on improving pupil learning in EE, and improving the immediate positive impact that schools are having on the River Ganga. The lead schools would also operate as the main agents for sharing learning between alliances. The alliance, through the lead school, would have to evaluate the impact of the initiative and report back to NMCG through the Youth Division.

It would be important for the learning opportunities offered to teachers through the alliances to be informed by research on effective CPD practice. A review of teacher impact studies conducted by the EPPI Centre identified the positive benefits of CPD with the following characteristics:

- use of peer support;
- explicit use of specialist expertise;
- explicit mention of involving the teachers in applying and refining new knowledge and skills and experimenting with ways of integrating them in their day-to-day practice;
- involving consultation with the teachers about their own starting points, the focus of the CPD, the pace of the CPD or the scope of the CPD;
- involving teachers observing one another as an integral part of the CPD; and
- involving specialists in observation and reflection (Cordingley *et al.*, 2005).

The lead school would be selected through open competition based on the assessment of applications against criteria. It is recommended that this be developed by the Youth Division (see recommendation 2) in collaboration with the CEE and that consideration be given to including the following expectations within the criteria:

- evidence of sustained effective teaching and learning which is in line with the activity-based approaches suggested by research;
- consistent high levels of pupil performance in EE;
- experience of engaging with relevant environmental NGOs; and
- experience of developing and delivering impactful learning opportunities for teachers as part of a sustained approach to CPD.

In designing the detail of this approach, models and potential synergies from other initiatives to improve school education should be identified. It will be important, for example, to consider the experience of the government's current efforts to improve secondary education through developing and sharing professional development materials by 'key resource persons' and 'master trainers' (National Council for Education Research and Training, 2015). Policy makers should also consider whether there may be any synergies with the Schools and Teachers Innovating for Results' (STIR's) innovation networks which are being developed to support teachers to implement and evaluate 'micro-innovations' in schools.

(v) Have Student-Led Eco-Clubs: To Inform, Inspire and Mobilise

As student-led learning takes on more significant roles in schooling experiences around the world, a restructured Eco-Club model, in which a core group of enthusiastic students take the lead, will propel interests to greater heights. While the teachers can continue to provide overarching guidance on administrative aspects, the control of the learning experience will be shifted from the teachers to the students.

These students will not only focus on raising awareness on issues affecting River Ganga, they will also use the Eco-Club movements as a platform to *inform, inspire and mobilise* the community-at-large around conscious living and sustainable practices pertaining to River Ganga. The students will be fully empowered to develop their own Ganga-related community projects, from conceptualisation right up to execution, and in doing so, will act as a catalyst in inspiring individuals and organisations to take actions that ensure a sustainable future for River Ganga.

"Give a man a fish and you feed him for a day; teach a man to fish and you feed him for a lifetime." This is an age-old adage that has been heard and used countless times but its relevance and applicability could not be more evident than when allowing students to direct their own learning. Moreover, students who take part in peer-led learning will also develop a greater sense of belonging and those who undertake the role of "peer leader" will reap the added benefits of deepening their own subject learning and inter-cultural awareness.

Eco-Zones and Eco-Forum

In order to have a more systematic and structured implementation of the environmental programmes, the Eco-Clubs in Varanasi, which will be made up of and presided over by a team of students, can be clustered into four distinct eco-zones, namely, the North, South, East and West. Each eco-zone will assume responsibility for the execution of a particular set of Ganga-related initiatives for a specific period of time within their respective zones.

Though the number of environmental-related programmes that India has initiated over the past decade is probably unparalleled by even the most developed of nations, adopting a three-phased approach in executing those initiatives can possibly increase their effectiveness. Phase-1 activities such as awareness programmes can be used to inform the public, phase-2 activities such as environmental camps can be used to inspire and connect people emotionally and finally, phase-3 activities such as mass trash-removal movements can be used to mobilise the community. Each phase will have a clear start point, well-defined tasks and a definite end point and, by systematically coordinating the various activities within zones, impact measurement of the various activities and review of the different phases can be done much more efficiently.

Each zone will be led by a zonal head who will be responsible for conducting a quarterly or half-yearly forum within Varanasi, where the presidents and, if necessary, the other student members of the various Eco-Clubs, come together to share their respective Eco-Club's experiences and to plan the implementation of future outreach programmes on a collective basis. These forums will not only serve as a vibrant space for the intellectual engagement of the youths of Varanasi, it can also be utilised as a platform for them to engage parents, business leaders and the community-at-large.

As the future leaders of India, it is important that the students are empowered to lead from a young age and by discussing their ideas with their peers and members of the public, they are presented with an opportunity to think critically about the social, economic and the political impact of the pollution issues pertaining to River Ganga.

Approval of programmes will also take place within the confines of these forums and the mutually agreed initiatives will then be brought back to be implemented in their respective eco-zones. Proper records and documentations of the contents of the forums and the approved programmes have to be maintained, and disseminated where necessary, by the respective presidents of the Eco-Clubs.

Accord priority to Eco-Club members for entry into Institutions of Higher Learning

Although our primary research has shown that there is a strong belief amongst teachers and students that only education-related solutions to the Ganga problems will have sustainable effects in the long term, the current levels of education-related impact are less than desirable. The main reason cited for this is the emphasis accorded, by both parents and students, to other primary subjects such as Mathematics and Science. This is not a surprising finding, as will be the case in any other part of the world, the natural tendency of any pragmatic parent would be to render priority to those "core" subjects that will improve the prospects of their child in gaining entry to institutions of higher learning. In view of this, a possible way to circumvent this "problem" and elevate the importance given to EE as a whole will be to adopt Singapore's Ministry of Education's (MOE) "Values In Action" (VIA) model. Values in Action projects are learning experiences that support students' development as socially responsible citizens through the learning and application of values, knowledge and skills. The VIA projects provide students with the opportunity to participate in meaningful activities which contribute to the community, such as environmental conservation and care for the elderly and disabled, and it aims to help students understand that everyone can play a part in bettering the community. With teachers providing the overarching guidance and students initiating their own projects, the VIA initiative serves as an important first step in a student's life to encourage long-term involvement in community service (Ministry of Education, Singapore, 2015).

Besides the primary objective of entrenching social responsibility into the youths, involvement in VIA projects also serves to facilitate a student's entry into institutions of higher learning. A qualitative approach has been adopted by the city-state, where students' involvement and achievements in VIA projects are assessed for evidence of desired attributes such as leadership, creative flair, passion and commitment and used as one of the criteria for entry into its universities. The universities will not only consider outstanding achievement and leadership but will also take into account the level of participation and representation of the student in the VIA projects. The National University of Singapore (NUS), for example, sets aside 10% of its university places for consideration of exceptional candidates. These are students who, besides possessing good academic grades, have attained exceptional achievements in other areas such as, (i) winning a medal at the International Olympiads (Biology, Chemistry, Informatics, Mathematics and Physics), (ii) winning the National Science & Talent Search, (iii) representing Singapore in Arts and/or Sports, (iv) have actively participated in community service and volunteer programmes, (v) possessing work experience relevant to the course applied for (supporting document/s is/are required), or are (vi) holding key leadership positions in community organizations, sports and athletic clubs (outside school) (National University of Singapore, 2015). The adoption of such a model, that is, to accord priority for students involved in the Eco-Clubs, into the Varanasi education system might just prove to be the catalyst to elevate the reputation of EE amongst parents and students alike.

At a 2013 convocation ceremony of the Indian Institute of Technology (IIT) — Banaras Hindu University (BHU), Mr Rajendra Singh, the 2015 Stockholm Water Prize Laureate, declared that "There is a need for an education system that encourages development without damaging the natural resources." (*The Times of India*, 2013). He further asserted, and rightfully so, that there are no laws that can totally control the destruction of nature and only good character building through education can bring about sustainable development which will also be favourable for the nature and its dependents. The VIA model, in complementing the classroom teaching, does just that for Singapore, a small city-state with people as the only resource. It is increasingly recognised as not merely a 'nice-to-have' or a peripheral programme, but as an integral part of the socio-economic development processes required to ensure a better quality of life for all.

5. Conclusion

This study has demonstrated that a sustainable solution to the pollution crisis of River Ganga requires environmental awareness and its proper understanding should be deeply rooted in the education of the youth of India today.

Through our primary and secondary research, we found that, though EE has a long history in India, it has not produced the impact among its people to protect their environment, especially their holy River Ganga. Even though the Supreme Court of India mandated the teaching of EE in schools, and there are numerous agencies and NGOs involved in promoting EE among school children, there is still room for significant improvement.

The absence of a Ministerial representative in the National River Ganga Basin Authority and the lack of a coordinating body in the implementation of EE across India has highlighted the need for more emphasis to be given to EE in schools.

The voices of the teachers in Varanasi call for a more locale-specific EE curriculum so that EE can have a greater impact in instilling good habits and interests in the school children. Similarly, there is a call for improved support and training for teachers who are teaching EE in schools.

As our primary research indicated, school children are also interested in EE activities but are hampered by the constraints of time and influenced by the perception that knowledge of EE is not beneficial to gaining entry to universities.

Thus, this study proposes key recommendations that can improve EE in Varanasi, elevating it to a higher status and importance, and enabling it to achieve the life-changing impact on the school children today for the sake of tomorrow.

Our findings and suggested recommendations are summarized in Table 7 below:

Table 7: Summary of Findings and Related Recommendations

Findings	Recommendations
1. Insufficient emphasis and priority of EE	1. Include an Education Ministerial Representative in the NGRBA 2. Establish a youth division within the NMCG
2. Overly-broad EE in school curriculum	3. Produce a formal Varanasi EE curriculum for infusion into subjects
3. Ineffective EE teaching	4. Establish EE teacher development alliances
4. Lack of impact of extra-curricular learning	5. Have student-led Eco-clubs to inform, inspire and mobilise the society.

To clean up River Ganga is indeed a mammoth task that requires significant political will and the coordinated efforts and resolve of the Indian people. To foster a sustainable transformation of River Ganga, the education of the youth of today must take centre stage.

> ***"The difference between a sustainable or a chaotic future is learning."***
>
> — Stephen Sterling (2001)

Acknowledgements

We are grateful to the following whose cooperation and kind help have made this study possible:

Teachers, Co-ordinators, Chairpersons from W.H. Smith Memorial School and the Sunbeam Group of Schools for their contributions during the focus group discussions in Varanasi in December 2014;
Mrs. Seema Rajgarihia, Director, and teachers and students from Delhi Public School, Varanasi, and two teachers from the Sunbeam Group of Schools, for their participation in the online survey;
Mr Navneet Raman, Kriti Gallery, for his kindest help in arranging for the focus group discussions, reviewing of our survey questionnaires, and introducing various schools for us to send the online surveys;
Prof. H.C.S Rathore, Dean of Faculty of Education, Banaras Hindu University (BHU), for his time in answering our queries over the phone;
Ms Preeti Kanaujia, Centre for Environment Education (CEE) North, for her kind assistance in answering our queries and providing invaluable information to help us understand CEE's work;
Mr Darren Tan, Senior Executive, Schools & Youth Department, 3P Network Division, National Environment Agency (NEA) Singapore, for helping us with our questions about NEA's involvement with the schools in Singapore;
Our professors, Prof. Ora-orn Poocharoen and Prof. Wu Xun, for their continued guidance and dedication;
Our Indian classmates, for their insights and help to our study.

Abbreviations

BHU	Banaras Hindu University
BVIEER	Bharati Vidyapeeth Institute for Environmental Education and Research

CEE	Centre for Environment Education
CPCB	Central Pollution Control Board
CPD	Continuing professional development
CTEs	Colleges of Teacher Education
CUREE	Centre for the Use of Research and Evidence in Education (UK)
DIET	District Institute of Education and Training
EE	Environmental Education
EEAs	Environmental Education Advisors (Singapore)
EVS	Environmental Studies
IASE	Institutes of Advanced Study in Education
MOE	Ministry of Education (Singapore)
MoEFCC	Ministry of Environment, Forests and Climate Change
MHRD	Ministry of Human Resource Development
NCERT	National Commission for Education, Research and Training
NCF	National Curriculum Framework
NCTE	National Council for Teacher Education
NEA	National Environment Agency (Singapore)
NGC	National Green Corps
NGRBA	National Ganga River Basin Authority
NMCG	National Mission for Clean Ganga
NUS	National University of Singapore (Singapore)
STIR	Schools and Teachers Innovating for Results
UGC	University Grants Commission
VIA	Values In Action (Singapore)

References

Alexandar, R and G Poyyamoli (2014). The Effectiveness of Environmental Education for Sustainable Development Based on Active Teaching and Learning at High School Level — A Case Study from Puducherry and Cuddalore Regions, India. *Journal of Sustainability Education*, 7. Accessed on May 29, 2015. Available at http://www.jsedimensions.org/wordpress/content/the-effectiveness-of-environmental-education-for-sustainable-development-based-on-active-teaching-and-learning-at-high-school-level-a-case-study-from-puducherry-and-cuddalore-regions-india_2014_12/.

Almeida, S and A Cutter-Mackenzie (2011). The Historical, Present And Future 'Ness' Of Environmental Education In India. *Australian Journal of Environmental Education*, 27 (1), 122–133.

Azam, M and Kingdon, Geeta Gandhi (2015). Assessing Teacher Quality in India. *Ideas for India: For More Evidence Based Policy*, February 16. Available at http://www.ideasforindia.in/article.aspx?article_id=412.

Barber, M and M Mourshed (2007). How the World's Best Performing School Systems Come Out On Top. *McKinsey & Company*, September 2007. Accessed on May 29, 2015. Available at http://mckinseyonsociety.com/downloads/reports/Education/Worlds_School_Systems_Final.pdf.

Barratt, R and BH Elisabeth (2011). Place-Based Education and Practice: Observations from the Field. *Children, Youth and Environments,* 21 (1), 1–13.

Caribbean Journal Staff (2012). Jamaican Prime Minister Sets Up Committee to Clean Up Environment. *Caribbean Journal*, April 23. Accessed on June 26, 2015. Available at http://caribjournal.com/2012/04/23/jamaican-prime-minister-sets-up-committee-to-clean-up-islands-environment/#.

Centre for Environment Education (CEE) (2007). Environmental Education in the Indian System: Status Report 2007. Centre for Environment Education, Ministry of Environment, Forests and Climate Change.

Centre for the Use of Research & Evidence in Education (CUREE) (2015). Building the Evidence Base for a Curriculum for the 21st Century. Accessed on June 28, 2015. Available at http://www.curee.co.uk/files/publication/[site-timestamp]/buildingevidencebasesummary.pdf.

Chawla, L (1999). "Life Paths into Effective Environmental Action." *The Journal of Environmental Education*, 31 (1), 15–26.

Cordingley, P, M Bell, D Evans and A Firth (2005). The Impact of Collaborative CPD on Classroom Teaching and Learning. Review: What do Teacher Impact Data Tell us About Collaborative CPD? In *Research Evidence in Education Library*. London: EPPI-Centre, Social Science Research Unit, Institute of Education, University of London. Accessed on June 6, 2015. Available at http://eppi.ioe.ac.uk/cms/Default.aspx?tabid=395&language=en-US.

Cutter-Mackenzie, A (2014). Environmental Education and Pedagogical Play in Early Childhood Education. In *Young Children's Play and Environmental Education in Early Childhood Education*, Cutter-Mackenzie, A, S Edwards, D Moore and W Boyd (eds.), 25–37. Cham: Springer.

Department for International Development Research and Evidence Division, UK. Pedagogy, Curriculum, Teaching Practices and Teacher Education in Developing Countries: Evidence Brief. Summary of *Pedagogy, Curriculum, Teaching Practices and Teacher Education in Developing Countries* by Westbrook *et al.*, DFID. Accessed on June 20, 2015. Available at http://r4d.dfid.gov.uk/Output/197015/.

EPPI Centre (2015). Collaborative Continuing Professional Development. Accessed on May 29, 2015. Available at http://eppi.ioe.ac.uk/cms/Default.aspx?tabid=274.

Gayford, C (2009). Learning for Sustainability: From the Pupils' Perspective, *WWF-UK*. Accessed on June 6, 2015. Available at http://assets.wwf.org.uk/downloads/wwf_report_final_web.pdf.

Gayford, C (2010). Learning for Sustainability in Schools: Effective Pedagogy, *WWF-UK*. Accessed on June 6, 2015. Available at http://assets.wwf.org.uk/downloads/wwf_pedagogy_report_final__no_back_tint__web.pdf.

Global Development Research Centre (GDRC) (2015). Tbilisi Declaration (1977). Accessed on June 3, 2015. Available at http://www.gdrc.org/uem/ee/tbilisi.html.

Gupta, S, SK Yadav and S Saini (2011). Innovative Approaches For Promoting Environmental Education In India. *Journal of Institute of Integrative Omics and Applied Biotechnology (IIOAB)*, 2 (3), 30–35.

Haigh, M (2008). The Sattvic Choice. In *Alternative Educational Futures*, Bussey M, S Inayatullah and I Milojevic (eds.), 235–252. Rotterdam: Sense Publishers.

Halder, S (2012). An Appraisal Of Environmental Education In Higher School Education System: A Case Study Of North Bengal, India. *International Journal of Environmental Sciences*, 2 (4), 2223–2233.

India Government. "Paryavaran Vahinis." Accessed on May 29, 2015. Available at http://envfor.nic.in/citizen/envedu/vahini.html.

India Education Review (2013). Govt Approved over 6,300 Cr to Improve Teachers' Education During 12th FYP. *IER*, August 21, 2013. Accessed on May 29, 2015. Available at http://www.indiaeducationreview.com/news/govt-approved-over-6300-cr-improve-teachers-education-during-12th-fyp/11108.

India Together (2013). Environmental Education — Revised Curricula. *India Together*, November 1. Accessed on April 17, 2015. Available at http://indiatogether.org/curricula-education.

Ministry of Education, Singapore (2015). Ministry Of Education, Singapore: Education System: Secondary Education: Community Involvement Programme (CIP) And Service Learning. Accessed on April 17, 2015. Available at http://www.moe.gov.sg/education/secondary/via/.

Ministry of Environment, Forests and Climate Change (MoEFCC). Paryavaran Vahinis. Accessed on May 29, 2015. Available at http://envfor.nic.in/citizen/envedu/vahini.html.

National Cadet Corps, India. Accessed on May 23, 2015. Available at http://www.psgtech.edu/ncc/02NccInfo.html#Activities.

National Council for Educational Research and Training. Accessed on May 29, 2015. Available at http://www.ncert.nic.in/departments/nie/dse/activities/advisory_board/print_material.html#.

National Council for Teacher Education, New Delhi (2009). *National Curriculum Framework for Teacher Education: Towards Preparing Professional and Humane Teachers*. Accessed on June 3, 2015. Available at http://www.azim-premjifoundation.org/pdf/NCFTE-2010.pdf.

National Environment Agency (NEA), Singapore. Accessed on June 30, 2015. Available at http://www.nea.gov.sg/, http://www.nea.gov.sg/events-programmes/programmes/learning-journeys, http://www.nea.gov.sg/events-programmes/programmes/youth/youth-for-the-environment-day, http://www.nea.gov.sg/events-programmes/programmes/schools-youth/clean-singapore-learning-trail-beaches, http://www.nea.gov.sg/events-programmes/programmes/schools-youth/environment-club-fund.

National Focus Group on Teacher Education for Curriculum Renewal (2012). *Position Paper*. National Council for Education Research and Training. Accessed on May 29, 2015. Available at http://www.teindia.nic.in/js/NATIONAL_FOCUS_GROUP_on_TE_for_curriculum_renewal.pdf.

National Library Board Singapore, Singapore Infopedia (2012). Keep Singapore Clean Campaign. Accessed on June 25, 2015. Available at http://eresources.nlb.gov.sg/infopedia/articles/SIP_1160_2008-12-05.html?utm_expid=85360850-6.qNOOYF40RhKK6gXsQEaAJA.0&utm_referrer=https%3A%2F%2Fwww.google.com.sg%2F.

National Service Scheme, India. Accessed on May 23, 2015. Available at http://nss.nic.in/.

National University of Singapore (2015). NUS — Office Of Admissions: Local Applicants. Accessed on June 30, 2015. Available at http://www.nus.edu.sg/oam/apply/local/poly/BYA-poly.html.

Percy-Smith, B and D Burns, (2009). Exploring the Role of Schools in the Development of Sustainable Communities. *Full Research Report ESRC End of Award Report*. Accessed on May 29, 2015. Available at http://www.esrc.ac.uk/my-esrc/grants/RES-182-25-0038/outputs/Read/4920a40d-c4a3-431f-b2ce-beffe6363e35.

Ravindranath, MJ (2007). Environmental Education in Teacher Education In India: Experiences and Challenges in the United Nation's Decade of Education for Sustainable Development. *Journal of Education for Teaching: International Research and Pedagogy*, 33 (2), 191–206.

Roberts, NS (2009). Impacts of the National Green Corps Programme (Eco-Clubs) on Students in India and their Participation in Environmental Education Activities. *Environmental Education Research*, 15(4), 443–464.

Seema, U, S Anju, S Bhagyashree and N Neelima (2014). Environmental Education and Practices in Schools. *Journal of Environmental Research and Development*, 9 (1), 271–283.

Sonowal, CJ (2009). Environmental Education in Schools: The Indian Scenario. *Journal of Human Ecology*, 28 (1), 15–36.

Srivastava, ABL (1999). *Study of the Support System and Process Which Underpins DPEP's Pedagogical Strategy in Six States: A Synthesis Report*, Ed. CIL: New Delhi.

The Times of India (2013). Education System Must Protect Natural Resources, Encourage Development. November 10. Accessed on June 25, 2015. Available at http://timesofindia.indiatimes.com/city/varanasi/Education-system-must-protect-natural-resources-encourage-development/articleshow/25525739.cms.

The Times of India (2015). Ganga Vahini to keep river free of pollution. May 1. Accessed on May 23, 2015. Available at http://timesofindia.indiatimes.com/city/kanpur/Ganga-Vahini-to-keep-river-free-of-pollution/articleshow/47118599.cms.

UNESCO-UNEP (1976). The Belgrade Charter: A Global Framework for Environmental Education. *Connect: UNESCO-UNEP Environmental Education Newsletter*, 1, 1–2.

UNESCO (2015). Teaching and Learning: Achieving quality for All. *Education for All Global Monitoring Report 2013/14 (2014).* Accessed on May 29, 2015. Available at http://en.unesco.org/gem-report/report/2014/teaching-and-learning-achieving-quality-all.

Worldometers (2015). Population of India (2015) — Worldometers. Accessed on May 23, 2015. Available at http://www.worldometers.info/world-population/india-population/.

Annex 1: Online Questionnaires

Environmental Survey for Clean Ganga Project - Teachers

Environmental Education Survey for Teachers

* Required

Name (Optional)

Email address (Optional)

Subject(s) related to Environmental Education taught *

School *

Curriculum followed (eg CBSC, ICSE etc) *

Survey Questions

1. To what extent do you feel that education can play a role in a sustainable clean-up of the River Ganga? *

1 2 3 4 5

To a small extent ○ ○ ○ ○ ○ To a large extent

2. How adequate is the current curriculum coverage on environment protection of the River Ganga? *

1 2 3 4 5

Extremely Inadequate ○ ○ ○ ○ ○ Very Adequate

3. How active are the students in using what they have learnt in school to protect the River Ganga? *

1 2 3 4 5

Not active at all ○ ○ ○ ○ ○ Very active

4. How well equipped and/or supported are you to teach the topics relating to keeping the River Ganga clean? *

1 2 3 4 5

Not able to ○ ○ ○ ○ ○ Very well

5. What are 3 main challenges that you face when conducting classes on environmental education? *

- ☐ Lack of state-specific curriculum, such as issues particular to the Ganga River
- ☐ Insufficient training for teachers
- ☐ Lack of funds
- ☐ Disinterested students
- ☐ Inadequate teaching resources and materials
- ☐ Parents and students prioritize other subjects over environmental education
- ☐ Other: ____

6. Choose 3 most interesting activities that you think students would like to take part in protecting and keeping River Ganga clean. *

- ☐ Participating in clean-up projects of the River Ganga e.g. picking litter from the river bank
- ☐ Working with older students (e.g. from universities) on projects related to the River Ganga
- ☐ Holding hands with classmates along the River Ganga on World Water Day to take an oath to protect the Ganga
- ☐ Create awareness among families and friends through clean up efforts and visitations
- ☐ Create awareness via social media, for example Facebook, twitter
- ☐ Joining an environmental club
- ☐ Arts and Drama to tell people about River Ganga
- ☐ Other: ____

7. Choose 3 most important reasons that you think may prevent students from doing what they have been taught in school when it comes to protecting the River Ganga. *

- ☐ Not interested in cleaning up the River Ganga
- ☐ Need to focus on other school subjects and examinations
- ☐ What is taught in school is not relevant to River Ganga
- ☐ River Ganga can protect and clean itself, there is no need for me to do anything more
- ☐ It is someone else's responsibility to clean and not school children
- ☐ As school students, their efforts make no difference and they feel powerless
- ☐ It is too troublesome to do what the school teaches
- ☐ Other: ____

8. Choose 3 activities that you think will be effective in cleaning the River Ganga. *

- ☐ Public education campaigns to increase awareness
- ☐ More money for environmental projects in school
- ☐ More emphasis on environmental education
- ☐ More environmental clubs in schools
- ☐ Greater coverage of Ganga River issues in school textbooks
- ☐ Other: ____

9. Who should be the one(s) cleaning up the River Ganga? Choose only 3 of the groups listed below: *

- ☐ The government
- ☐ The factories
- ☐ The schools and universities
- ☐ The priests
- ☐ The people who are living along the river
- ☐ The farmers
- ☐ Other:

10. Are there any other suggestions you can offer on how education can play a role in the clean up of River Ganga, such as good practices in your school, which others can learn from?

Submit

Never submit passwords through Google Forms.

Annex 1 – continued:

Environmental Survey for Clean Ganga Project - Students

In this survey, we are asking questions relating to the pollution of River Ganga. This is an anonymous survey. Please answer all of the following questions. Thank you for your participation!

* Required

Gender *

○ Male
○ Female

Class Level *

○ 7
○ 8
○ 9
○ 10
○ 11
○ 12

School *

1. How important is it to you in having a clean River Ganga? *

	1	2	3	4	5	
Not important at all	○	○	○	○	○	Very important

2. How much do you know about protecting the environment? *

	1	2	3	4	5	
Nothing	○	○	○	○	○	Know a lot

3. Do you think that learning about environmental protection will help you keep River Ganga clean? *

	1	2	3	4	5	
Not at all	○	○	○	○	○	To a large extent

4. Would you be interested in doing more to clean up the River Ganga? *

	1	2	3	4	5	
Not interested at all	○	○	○	○	○	Very interested

5. Choose 3 most important reasons why you think River Ganga is so polluted today. *

- ☐ I do not think that River Ganga is polluted as it is able to clean itself
- ☐ Dumping of sewage or litter into the river
- ☐ Factories dump harmful waste into the river
- ☐ Burning or disposal of ashes of dead bodies and dead animals into the river
- ☐ Bathing or washing of clothes with soap in the river
- ☐ Agricultural pesticides and fertilizers flowing into the river
- ☐ There is not enough water flow into the River Ganga
- ☐ Other: ______

6. What do you think will be the 3 most harmful effects of a polluted River Ganga? *

- ☐ I do not know of any harmful effects of a polluted Ganga
- ☐ The health of the people will be affected when they bathe in the River
- ☐ The dolphins and fishes in the river will be sick
- ☐ The farmers will have skin diseases when they use water that contain harmful chemicals
- ☐ A polluted River Ganga will prevent tourist from visiting Varanasi.
- ☐ There will not be enough clean water for the people living by the River Ganga
- ☐ Other: ______

7. Choose 3 most interesting activities that you would like to take part in protecting and keeping River Ganga clean. *

- ☐ Participating in clean-up projects of the River Ganga e.g. picking litter from the river bank
- ☐ Working with older students (e.g. from universities) on projects related to the River Ganga
- ☐ Holding hands with classmates along the River Ganga on World Water Day to take an oath to protect the Ganga
- ☐ Create awareness among families and friends through clean up efforts and visitations
- ☐ Create awareness via social media, for example Facebook, twitter
- ☐ Joining an environmental club
- ☐ Arts and Drama to tell people about River Ganga
- ☐ Other: ______

8. Choose 3 most important reasons that may prevent you from doing what you have been taught in school when it comes to protecting the River Ganga. *

- ☐ Not interested in cleaning up the River Ganga
- ☐ Need to focus on other school subjects and examinations
- ☐ What is taught in school is not relevant to River Ganga
- ☐ River Ganga can protect and clean itself, there is no need for me to do anything more
- ☐ The government should be the one cleaning and not school children
- ☐ As a school student, my effort makes no different and I feel powerless
- ☐ It is too troublesome to do what the school teaches
- ☐ Other: ______

9. Choose 3 activities that you think will be effective in cleaning the River Ganga. *

- ☐ Public education campaigns to increase awareness
- ☐ More money for environmental projects in school
- ☐ More emphasis on environmental education
- ☐ More environmental clubs in schools
- ☐ Greater coverage of Ganga River issues in school textbooks
- ☐ Other: ________

10. Who should be the one(s) cleaning up the River Ganga? Choose only 3 of the groups listed below: *

- ☐ The government
- ☐ The factories
- ☐ The schools and universities
- ☐ The priests
- ☐ The people who are living along the river
- ☐ The farmers
- ☐ Other: ________

11. How good are your teachers at teaching you about the environment protection of the River Ganga? *

	1	2	3	4	5	
Not good at all	○	○	○	○	○	Excellent teaching

Submit

Never submit passwords through Google Forms.

Annex 2: An extract of the NCERT Environment Science curriculum that could be used to teach about pollution in the River Ganga

Class Level	Topics that Could be Related to Ganga River Pollution	Activities Proposed	Teaching Learning Strategies and Evaluation
Class 7 Environment Education	**Environment and Natural Resources** Water(essential for life and life activities) **Man and Environment** (Consequences of human activities — stress on land use, water resources)	Collect information on industrial impact on environment. World environment day. (posters, charts, exhibitions, debates)	
Class 8 Environment Education	**Ecosystem** — agricultural impact on environment, Industrial waste management, — regional and national environment concerns	Collect water samples and test for impurities. List down local cottage industry's disposal of waste. Trips to garbage dumps.	Teaching learning strategies gives recommend multipronged evaluation by schools but no exam grade.

(*Continued*)

Annex 2: *(Continued)*

Class Level	Topics that Could be Related to Ganga River Pollution	Activities Proposed	Teaching Learning Strategies and Evaluation
	— environment pollution cause and effect (pollution of water) — role of individuals, community and government in prevention of pollution and improvement in environment (Harnessing resources and renewable sources)	Poster, discussions: water, waste water, water treatment system. Rule and regulation enacted by government. Agencies involve. (Street plays, campaigns, rallies) Provide opportunities to participate in campaigns organized by different NGOs.	
Class 9	**Understanding of Ecosystem** — Role of environmental impact assessment in maintaining quality of environment. — Conservation ecosystem – not so relevant **Depletion of resources** – not so relevant **Waste generation and management** (domestic, industrial, agricultural and commercial)	Plan and execute awareness campaigns through communities' participation on major environmental problems at local and national levels.	

(Continued)

Annex 2: (*Continued*)

Class Level	Topics that Could be Related to Ganga River Pollution	Activities Proposed	Teaching Learning Strategies and Evaluation
	— Methods of safe waste disposal — Reduction, reusing, recycling waste **Environmental Values and Ethics** — Human rights, fundamental duties and value education — Women and child welfare		
Class 10	**Restoring Balance in Ecosystem** (Need for conservation and management of water) **Pollution*** **Issues of the Environment*** **Striving for better environments***	Environmental pollution of air, water, soil, disposal of plastics and urbanization. Collect samples of water, quality of water. Visit to WTS, garbage dumping. (posters, charts, collages, cartoons, handouts, letters, street plays, rallies, campaigns)	Teaching learning strategies. Public Exam at class 10 allocates marks in proportion with other subjects.
Class 11	**Man and Environment*** (Population, urban and rural environment problem)	Greenhouse, environmental profile of a village, town.	

(*Continued*)

Annex 2: (*Continued*)

Class Level	Topics that Could be Related to Ganga River Pollution	Activities Proposed	Teaching Learning Strategies and Evaluation
	Environment and development* (Organization pattern in land use, deforestation) **Environmental Pollution and Global Issues** (a little on water pollution) **Energy***	Consumption of electricity. (not related). For 1 month, study of status of sanitary conditions and methods of waste disposal of a given locality and prepare action plan for making the condition more environment friendly. To investigate impact of an industry on the local environment-any waste is pollution. To study the impact of changes in agricultural practices and to suggest an action plan for modifying the prevailing practices so as to make them environment friendly and sustainable.	

(*Continued*)

Annex 2: *(Continued)*

Class Level	Topics that Could be Related to Ganga River Pollution	Activities Proposed	Teaching Learning Strategies and Evaluation
		Collect water sample and identify pollution and devise an action plan for mobilizing public opinion for checking the pollution.	
Class 12	**Biodiversity*** (Balance in nature) **Environmental Management** (Environment ethical, economic, technology, social) Legal provision etc **Sustainable development*** (definition, concepts, challenges, manpower, community)	-----Above-------	Teaching Learning Strategies. Public exam at end of class 12 allocates marks in proportion with other subjects. Evaluation will have both theory and practicum.

*Materials suggested by NCERT under these topics are mostly not related to Ganga. They could be modified to include more specific issues relating to Ganga.

(Grade 7 to 10 is Infusion model and grade 11 to 12 is project based study.)

Chapter 5

SOCIAL MOBILISATION: THE FIGHT FOR HEARTS AND MINDS IN GANGA REJUVENATION

David Geordie Wilks, Gilbert E. Lumantao, Giridharan Sengaiah, Gloria Wei Wanting, Karen Ann Leong Sun Lian and Khial Muhammad Afghanzai

1. Why Social Mobilisation

The Ganga suffers from high levels of pollutants caused by ever-increasing loads of dumped untreated industrial, agricultural and domestic waste. This problem is compounded by low levels of water flows due to dams, barrages and irrigation canals situated upriver. Accusatory fingers are often pointed at India's political and administrative logjams when technical fixes and river cleanups do not accomplish their promises.

The persistence of industrial pollution is a product of the powerful nexus between polluting industries and government agencies. The two entities deprioritise pollution because the incentives for disregarding the problem are strong, versus the costs to address it. This is why, despite strong legal provisions for punishing polluters, the implementation of the law is lackadaisical. Driven by profit motives, industries find it 'cheaper' to pay off corrupt regulators rather than bear the extra costs of complying with environmental regulations. The Supreme Court has thus pinned the blame on the deep-rooted corruption within the pollution control boards for failure to carry out verdicts on industrial polluters (Pandey, 2014).

Even without industrial pollution and the incentives provided by bribery, the government's behaviour shows that it does not prioritise cleaning up Ganga. The National Ganga River Basin Authority (NGBRA), entrusted with cleaning the river Ganga, met only three times in between 2009 and 2012 (Jolly, 2014). Prominent environmentalists resigned from the NGBRA citing government apathy and insincerity towards cleaning Ganga (eUttaranchal, 2012).

On deeper analysis, however, this state of affairs is allowed to persist as a result of *public apathy*. In particular, the public has not extracted accountability from political and administrative bodies for the lack of performance in cleaning up the Ganga. While official accountability mechanisms exist through elections, Ganga hardly features as an issue for which the politicians are castigated. As the Indian Institute of Technology (IIT) puts it, "As a result of this apathy in urban citizens, the political parties do not find the issue of pollution in Ganga worth investing their time and resources as there is no political dividend to gain." (Indian Institutes of Technology, 2011).

As to unofficial mechanisms, protest actions are also just concentrated in a handful of non-government organisations (NGO) and community-based organisations (CBOs), which have generally faced limited success in rallying the masses. Widespread social mobilisation in the likes of recent campaigns in India, such as the anti-corruption movement and anti-rape movement, is absent.

1.1. *Why does public apathy exist*

Apathy towards environmental resources is common, especially if the resource is widely available. A free rider effect is at play here, where Ganga's non-excludable and non-rivalrous nature, which means that it is common for people to feel not responsible for keeping it clean or pay for the pollution caused by their usage. Tirth purohits (religious 'specialists') who regulate the behaviour of people on the river, express a typical sentiment by believers, saying "…the water is everyone's, it has already become polluted and here it is dirty, so a little more dirt will not hurt" (Mohapatra and Mishra, 2009).

In Ganga's case, certain social misconceptions exacerbate the free rider effect.

Firstly, pollution effects are not always *salient* to the public. Effects of pollution on individuals include water poisoning, contamination of food source, illness. However, there is often a time lapse between contact with Ganga and the impact. While the quickest link may be drawn for waterborne illnesses, the causality of more severe effects like cancers and arsenic poisoning is more difficult to establish.

Furthermore, Ganga's pollution is not reported or quantified in terms of its effects. Scientific data such as levels of coliform bacteria are reported, but effects are not contextualised even though it is acknowledged that the pollution is 'toxic and harmful' (*Ibid.*). Public apathy and government apathy in this case strengthen each other in a self-reinforcing loop. The government has no incentive to disrupt the people's apathy and therefore does not make pollution indicators and policies more relatable.

The media also acts as a poor information source. Rather than providing accessible technical information for the public's edification, it plays up dramatic narratives of political conflict. This results in absolving the general public of interest or capacity to meaningfully deal with the situation (Nambiar, 2014).

Secondly, Ganga's pollution costs do not affect everyone evenly. The impact is disproportionate, depending on one's geographic proximity and usage of the river. Its severity falls most on those who live or work on it, or use it for everyday routine purposes. Consequently, for the public at large, there are few reasons to care about the river.

The third factor, which is observed by many academicians, is the entrenched religious views towards the river. To the Hindus, Ganga is a manifestation of the Goddess Ganga (reveredly referred to as Ma Ganga by the Hindus) who is a life giver and a sin-cleanser, whose blessings are received by bathing in the river. Ma Ganga is considered to be an indestructible being, a self-healing, self-regenerative mother who accepts Her children's troubles and makes them go away. It is, therefore, considered 'acceptable' to pollute the river as Ma Ganga will take care of it (McAnally,

2007). Ganga's holy nature enables the public to accept and reconcile both its divine purity and material impurity (Das, 2012) (Bhargava, 1992).

Ganga's self-purifying property also appears to be rooted in some truth. Scientists have found that its water has antimicrobial attributes (Nautiyal, 2008), which has provided the scientific basis for many to continue believing that Ganga is able to clean itself. However, this view neglects the increasing pollution load on the river and the insufficient e-flow, which limit the self-cleansing quality of the river. Studies show that bacteria in Ganga have today become more pervasive and virulent (Hamner *et al.*, 2007, pp. 2369–2372).

1.2. *A case for social mobilisation*

Overcoming these social misconceptions and reducing public apathy are, therefore, critical to any solution proposing to clean Ganga. Social mobilisation, as a process to engage and involve the wider population in order to raise awareness and motivate them to demand change, is well-suited to the task. It is a process with dual advantages. Firstly, it is apt for educating the population on the issue. Citizens may then shift habits away from littering, illegal dumping and use of hazardous fertilisers or pesticides. They will also be more receptive to proactive policies such as connecting their homes with sewage drains and increased utility costs. Secondly, it aims to empower people such that power in society is decentralised, and the public gains greater influence and will seek to extract accountability from powerholders. (Chenoweth, 2011, p. 80) Every successful campaign has the ability to create an incremental withdrawal of public cooperation and support from the powerholders' system and thereby shift the power balance in favour of the people. Even the least powerful communities (usually the most affected by the pollution) can then shake free of the sense of hopelessness and unite to achieve objectives.

Social mobilisation is particularly vital at this juncture, when India is in the midst of the third attempt to clean up the river (*Namami Gange*). At status quo, with an apathetic population, efforts could again be as half-hearted as the first two Ganga Action Plans (GAP) I and II were.

The new *Namami Gange* also purports to work with schools and communities in projects for the river, and has consulted civil society in the drafting of plans. An NGO, Sulabh International had also been appointed to clean Assi Ghat, one of Varanasi's most significant ghats (Sandeep, 2015). While this represents a positive step, gathering public participation in this way will only be skin-deep if there continues to be widespread apathy. Also, what we risk is further segregation between government institutions and society, where the public and civil society adapt to a diminished role scoped out by the government. They sit at the fringes around the core work run by the government which brooks no intervention or interference.

1.3. *About this chapter*

This chapter aims to develop social mobilisation as a solution to Ganga's woes. Its primary purpose is to assist NGOs that seek to galvanise public interest in Ganga, by providing them with strategies for design of a movement, that are grounded in theory, practices and India's context. Secondarily, it can also help public institutions understand and embrace the rapidly evolving psyche of environmental attitudes and behaviours in India's advancing society.

The chapter is built on the premise that only grassroots civil society activism can build massive social capital and create the political opportunity for transformative change in the way polluters and government agencies behave. It will draw on the Movement Action Plan framework postulated by Bill Moyer to explain the current state of activism and highlight the need for massive mobilisation. By analysing the environmental attitudes, socio-economic and political conditions in India, these insights are used to recommend strategies for activist groups.

Section 2 is an overview of historic and political landscape of activism around Ganga. Section 3 outlines a social movement framework and maps Ganga's activist history to the theoretical framework. The rest of the chapter focuses on the design features of the Ganga activist movement. Section 4 discusses the central message that the movement should embody. Section 5 describes the leading organisation of the movement in the form of a coalition. Section 6 talks about how the movement can successfully rally support by overcoming various barriers to participation. Finally, Section 7 presents an array of direct action tactics that can be used by activists to confront powerholders culminating in a massive trigger event.

2. What Do We Know: Learning from the Past

Activism has a long and distinguished history in India. Gandhi's 1930 salt march triggered the civil disobedience campaign that ended up expelling the British from India (Weber, 2002). The Chipko movement was a flagship environmental movement that resulted in the 1981 Forests Conservation Act (Gupta, 2008) and the failed Great Bombay Strike of 1982 represented a step-up in organisational leadership for activism (Malik, 2003). In more recent times, notable movements are the Mazdoor Kisan Shakti Sangathan's 53 day strike in Jaipur demanding access to government information (Boy, 2000), Anna Hazare's Janlokpal movement against corruption (Sengupta, 2014) and the anti-rape protests of December 2012 which culminated in new anti-rape laws being passed in March 2013 (Wilson, 2013).

2.1. *History of the movement*

Environmental activism against pollution in Ganga is also not new. Over many decades, Ganga activism has evolved from statements of concern to variegated forms including protests against the construction of dams, filing public interest litigations against polluters and ineffective government agencies, religious activism, and technical advocacy. In charting the future direction of the movement, it is useful to review how it has evolved in the past.

Before Mehta Case and Ganga Action Plan I

2.1.1. *Expressions of concern and social activism*

Nobel laureates Rabindranath Tagore (1861–1941) and Swami Vivekananda (1863–1902) have written and expressed their anguish about the deterioration of the river as early as in the 1900s (Basu, 2013).

The cause was kept alive by various personalities including Sunderlal Bahuguna, environmentalist and ardent follower of Mahatma Gandhi, who led the *Chipko* movement (for the preservation of forests in the Himalayas, 1970s) and the Anti-Tehri dam movement which started in the

1980s. He repeatedly went on hunger strikes on the banks of river Baghirathi to protest against government (Basu, 2013).

Another notable activist is Rajendra Singh who is also known as the "water man" of India (Basu, 2013). He did pioneering work in community-based water harvesting and management, building *johads* or water conservation structures that reversed desertification in over 1,000 villages in India. His work with Ganga Basin Authority led to his involvement in efforts to clean up river Ganga.

Also, Dr. Veer Bhadra Mishra, a hydraulic engineering professor at Banaras Hindu University, started the Swachh Ganga campaign in as early as 1975, after noticing the increasing pollution level of the Ganga, while performing daily rituals in the river. He published many articles in newspapers and magazines and generated public interest in the issue. He founded the Sankat Mochan Foundation in 1982 which continues to play a major role in the Ganga activism today.

Mehta Case and Ganga Action Plan I

2.1.2. *Rise of legal activism and policy response*

Legal activism started in the mid-1980s leading to a series of policy interventions. In 1985, M.C. Mehta, a public interest environmentalist lawyer, filed a petition in the Supreme Court against polluting factories, and later on broadened the petition to include all industries and municipalities in the river basin. In the following years, the courtroom was set aside every Friday just for Mehta's cases. In 1993, after a decade of court battles and threats from factory owners, the Supreme Court ordered 212 small factories around the Taj Mahal to be shut-down for not installing pollution control devices and served warning notices to another 300 industrial units.

In 1986, the Central Government of India launched the Ganga Action Plan (GAP), to counter and minimise pollution load on the river. A total of 462 crore rupees was budgeted towards improving water quality to acceptable standards (described as being suitable for bathing). The plan was to intercept and treat sewage in 25 towns located along the river in Uttar Pradesh, Bihar and West Bengal, all riparian states, before discharging into the river.

In the period after the launch of GAP I, there was a noticeable lack of public activism in relation to Ganga. Sankat Mochan Foundation, believing that the plan adopted was not appropriate, submitted an alternative plan but did not resort to public activism to insist on a change of direction.

The Central Government established the National Ganga River Basin Authority (NGRBA) in 2009 under the Environment Protection Act. In 2011, the Cabinet Committee on Economic Affairs approved a new project for the cleaning of river Ganga with an estimated cost of 7,000 crore rupees.

Responses to the failure of GAP I and II

2.1.3. *Public response to the failures of GAP, including the rise of religious activism*

In July 1998, Smt. Rama Rauta, founder of the Pune-based National Women's Organisation (NOW), launched the Save Ganga Movement, against pollution in river Ganga and its tributaries based on principles of non-violence (Basu, 2013). The movement tried to mobilise the public at large and attracted eminent Gandhians including Smt. Tara Gandhi Bhattacharjee, the grand-daughter of Mahatma Gandhi.

In 2008 and 2009, Prof. G. D. Agrawal went on a fast unto death to protest against building of dams in the upper stretches of Ganga. On both occasions, Dr. Agrawal succeeded in stopping construction in this sensitive and sacred portion of the river with the help of social activists like Anna Hazare. Prime Minister Manmohan Singh then publicly committed that there would be no more hydroelectric damming of the Ganga in Uttarakhand and instead called on the NGRBA to utilise the money budgeted to clean up the river (Ibid.).

In 2011, Swami Nigamanand Saraswati, a seer at the Matri Sadan Ashram, went on a fast unto death on the banks of Ganga to protest against illegal mining in the holy city of Haridwar. After 115 days, or almost four months of fasting, Nigamanand died at the Dehradun Himalayan Hospital. There were allegations that he died because of poisoning (Ibid.).

In the later part of 2011, Swami Nigamanand's guru, Swami Shivanand also went on an indefinite fast saying he would not break his fast until the state government issues a notification to stop quarrying in the Ganga. After 11 days, Swami Shivanand ended his fast when the state government of Uttarakhand banned stone and sand mining on the riverbed of Ganga in Haridwar (Ibid.).

In June 2012, the Ganga Mukti Sangram Samiti (an association of eminent saints and seers) launched a mega rally from Varanasi, to Allahabad, Kanpur, Mathura, Vrindavan and finally Delhi, gathering thousands of Ganga devotees along the way.

2.2. *Major players in the movement*

A review of Ganga activism landscape reveals the contribution of many organisations with differences in terms of organisational setup, focus, approaches and tactics.

Most of these organisations including the Sankat Mochan Foundation, Eco Friends, M.C. Mehta Environmental Foundation and the Panchganga Foundation are set up as non-governmental organisations (NGOs) and foundations. On the other hand, the Save the Ganga Movement is a coalition of many likeminded organisations while Ganga Action Parivar is set up with a strong network of supporting organisations (***Annex A*** provides further details).

These organisations have employed a range of approaches in promoting activism on the state of Ganga. All of the organisations employed non-confrontational approaches such as public awareness campaigns, doing voluntary work like clean-ups and monitoring, conducting scientific studies and researches, and engaging in coalition work and networking.

Some organisations adopted more confrontational tactics such as filing environmental public interest cases and seeking judicial remedies, and conducting rallies or public protests. On the other hand, some organisations were open to working with government agencies (by participating in government committees, meetings and other *fora,* and providing technical advocacy).

Figure 1 describes the tactics typically employed by the major activist players:

In particular, the Sankat Mochan Foundation and Eco Friends have constantly engaged government bodies; Eco Friends and MC Mehta

Major Players	*Public awareness campaigns*	*Scientific studies/researches*	*Coalition work/networking*	*Clean up drives/volunteerism*	*Active engagement of govt. bodies*	*Filing of cases/seeking judicial remedy*	*Rallies/peaceful demonstrations*	*Hunger strike/fast until death*
Sankat Mochan Foundation	■	■		■	■			
Eco Friends	■	■		■	■	■		
MC Mehta Environmental Foundation	■					■		
Panchganga Foundation	■			■				
Ganga Action Parivar	■	■	■	■	■			
Save Ganga Movement	■	■	■		■		■	
Religious Leaders and Environmentalists					■		■	■

Figure 1: Tactics Employed by the Major Players

Environmental Foundation are involved in public interest litigation; and the Save Ganga Movement has launched rallies and public protests to advance its objectives.

On the other hand, some of the movement's prominent individuals employed more direct approaches. Prof. G. D. Agrawal went on a fast unto death to protest building of dams in the upper stretches of Ganga, holy men like Swami Shivanand fasted to call for an end sand and gravel mining in Haridwar, and Rajendra Singh, together with two others, resigned as a non-government member of the NGRBA in 2012 saying it is not serious in cleaning up the Ganga.

2.3. *Lessons and observations*

Over the years, the approaches used by the major players have generated some positive responses from powerholders including the ban on sand and gravel mining in Haridwar and the stopping of construction of dams. It is notable that the more engaging and confrontational approaches like filing of cases, public protests and fasts, have resulted in more concrete responses from government.

Another observation is that, while there have been attempts to form a broad coalition (like the one attempted by the Save Ganga Movement and the inclusive 'family' of Ganga Action Parivar), there is, at present, no broad cross-sector coalition in the Ganga activism landscape.

Ganga activism has been led by a small group of environmentalists, academicians and religious leaders. There is hardly any mention of participation of grassroots communities such as the boatmen, dhobis, fishermen and farmers whose livelihoods depend on river Ganga. Nor are they represented in the leadership corps.

Also, while it is common in social movements in other countries for youths to be heavily involved and even playing major leadership roles, the youth are very noticeably absent in the movement to clean Ganga. The participation of the youth has so far been mostly limited to public awareness campaigns and voluntary clean-up drives. This is significant because the youth represent future generations who will ultimately suffer the greatest due to the deterioration of the river.

In addition, there is little coordination between the various organisations operating along the whole stretch of the river and in different riparian states. Except for a few instances, where there was an effort to launch coordinated protests along the banks of the river, organisations operating on various points in the river appear to be operating in isolation and not as a single coordinated movement.

Finally, despite all these organisations and their efforts, the movement has generated less interest from the public compared to other issues, and this interest stayed generally flat through the years, even during times of new policy announcements or protest actions such as fasts unto death.

A simple proxy measure for public interest in societal matters is the trend in search terms on the Internet of which the middle are the main users. Reviewing internet search trends, we found that issues such as 'corruption in India' and 'rape in India' trended far higher than 'pollution in Ganga'. (See Figure 2) This trend continued even in the last 12 months despite Modi's high-profile pronouncements regarding renewed efforts to clean the river.

More importantly, interest in "pollution in Ganga" remained relatively flat through the years, unaffected by events, which is in stark contrast with

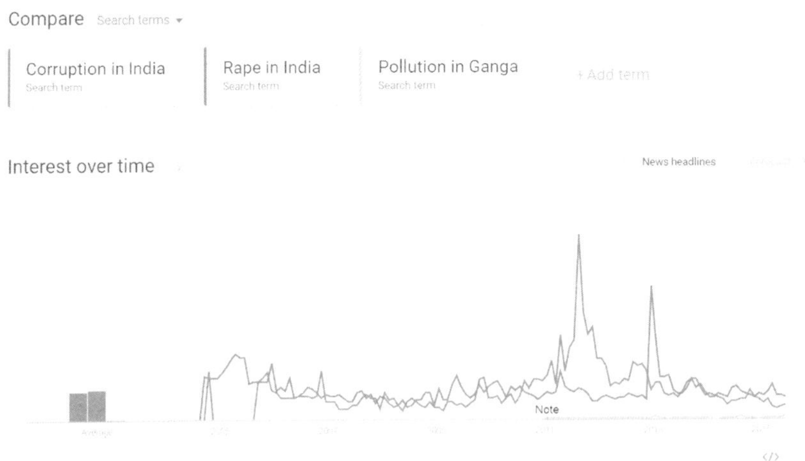

Figure 2: Google Trends on Corruption in India, Rape in India, and Pollution in Ganga, from January 2004 to June 2015

other search terms "corruption in India" and "rape in India" which had very noticeable peaks in recent years.

The issue of "corruption in India" peaked in August 2011 which was around the period when social activist Anna Hazare announced his hunger strike demanding an independent ombudsman to prosecute corrupt politicians and government officers. Interest stayed high in the succeeding months as Hazare ended his fast and decided to form a political party and enter politics (*The Economist*, 2011).

On the other hand, the issue of "rape in India" peaked in December 2012 and this coincides with much publicized rape of the Indian medical student on a moving bus. Interest on the issue also stayed high after the victim's death as widespread anger and demonstrations spread across India (Guardian, 2012).

We see none of these peaks in the trend line on the issue of pollution in Ganga, be it due to engineered events or unplanned crises.

Using these lessons from history, we propose to assess the current state of activism for Ganga using the framework explained in the next section.

3. What the Theory Says: Locating Ganga on MAP

The history of activism for Ganga shows a wealth of activity that started in the 1980s, with numerous players chiming in with similar causes along the way. Activism for Ganga is not starting from zero, but is currently situated at a point along a continuum, where certain gains have already been made, where awareness levels are at a certain point, and where powerholders and activists have had past altercations.

Movements are not sporadic outbursts of activity, but rather, evolve according to a certain sequence of events. Bill Moyer, a respected American activist, developed the Movement Action Plan (MAP), to describe such a lifecycle of social movements. The framework was informed by his experiences as a key player in activist movements in the United States. The purpose of MAP is to show the entire trajectory of a movement so that activists can understand their position in the lifecycle of their movement and manage their plans and expectations accordingly (Moyer, 1987). It is therefore useful for us to identify where Ganga's activism currently sits, according to MAP, in order to determine the next most optimal strategy for its activists.

3.1. *Elements of MAP*

The participants in the MAP framework include *powerholders*: people, organisations and institutions that have influence and control over an issue, including government bodies as well as business and political elites; *activists*: concerned citizens, local community organisations and NGOs that support meaningful change to upset powerholder interests but have little authority and direct power to influence powerholders; and the *public*: the population at large.

In MAP, power lies with the public; and powerholders derive power from the public's acceptance of authority and control. At the start of any movement, most of the public have no vested interests in the issue, but eventually, activists are able to mobilise them into a wave of activated citizens whose demands powerholders cannot ignore.

The activist's job is to educate the public about the issue so that they can develop a personal interest and desire to resolve such issue.

This serves two purposes. First, the public can pressure powerholders to make the changes that the activists are seeking and this will only happen when they are aware of the issues and are keen for change. Second, the education of the public benefits society as a whole because citizens can then take voluntary action to protect themselves from the effects of issues such as pollution on the River Ganga, or be a part of the solution.

3.2. *Eight stages of successful social movements*

MAP features 8 stages of successful social movements, depicted in Figure 3. Each stage of the lifecycle has different aims and calls for different activities to be completed.

The lifecycle begins with "normal times" where powerholders benefit from unfair practices that few are aware of. At this stage, the activist's job is to create awareness by openly demanding change and rebelling against the powerholder policies designed to maintain status quo. Activists also need to lay the groundwork for later stages by building organisational capacity and building connections with the general public.

After proving that powerholder resolution processes fail, rebellion takes place through "extraordinary" processes such as protests and strikes, leading to a trigger event (Stage 3) which becomes the catalyst for take-off of the movement (Stage 4).

Trigger events can be accidental or engineered. In the former, a crisis happens that stokes public emotions which are channelled into public outcry. In the absence of a crisis, a trigger event can be engineered. This can be done so by selecting the correct issue, publicising it through effective mobilisation, resistance tactics and messaging, and fronted by a strong organisation.

After the trigger event and take off, the public takes over the role of applying pressure to powerholders. Activists then take the lead in facilitating public's demands and strive to negotiate outcomes with powerholders.

3.3. *Ganga on MAP*

An analysis of the MAP framework allows us to understand where the Ganga activist movement is within the framework, so that appropriate

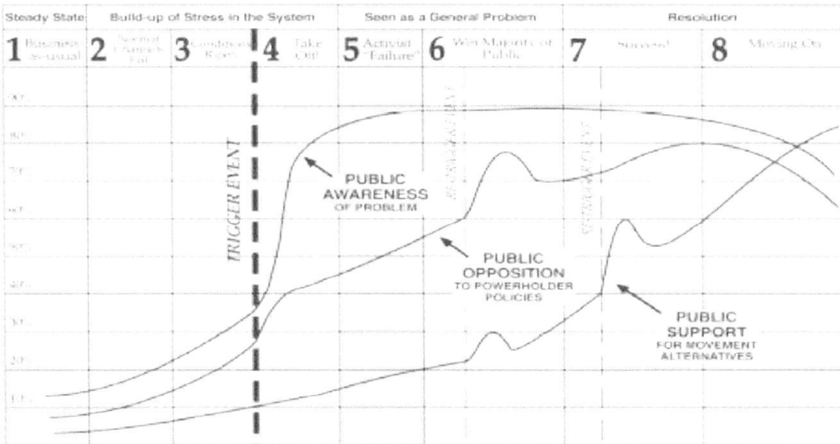

Stage 1	Stage 2	Stage 3	Stage 4	Stage 5	Stage 6	Stage 7	Stage 8
Activists must prove there is a problem	Activists must show that normal channels for resolving problems do not work	A window of opportunity opens for a trigger event to spark mass interest	The social movement takes off	Activists feel helpless after intense media activity which seems to have failed to produce results	The protest movement changes into a long term struggle to mainstream activist views into society	Powerholders now want to change policies to reflect the public's viewpoints, which reflect activists' stance	The movement switches its focus to other goals or subgoals

Figure 3: Bill Moyer's Movement Action Plan (MAP)

Source: Compiled by Authors.

strategies can be developed. We argue that the movement is in Stage 3 and that Ganga activists need to prepare the groundwork for the trigger event that will mobilise the public to demand meaningful change for Ganga.

The Actors

The key powerholders are the central, state and local government bodies that regulate Ganga and use it for political and bureaucratic purposes.

For example, Ganga-related funding provides resources for the Pollution Control Board, while the plight of the Ganga is often used as a political rallying call or a key component of a political platform.

Other powerholders include the businesses and organisations that use Ganga for economic gain. Both agricultural and industrial businesses use the river as a dumping ground for their waste products. Such businesses include tanneries and sugar cane producers. In addition, the agricultural sector also takes up water from the Ganga for irrigation purposes, reducing its flow. The organisations behind infrastructure projects in the upper reaches of Ganga also profit from large-scale water management practices.

The activists are the many organisations that were covered in the previous section of this chapter, which include Sankat Mochan Foundation, Eco-Friends and Ganga Action Parivar. As we have concluded earlier, the public is mostly apathetic about the Ganga. According to MAP, the activists' long term goal is to transform them into activated citizens keen to have the Ganga rejuvenated, so that they can play a key role bringing about the change.

The Case for Stage 3

Conditions have been ripening, setting the scene for Stage 3. Broad historical developments create window of opportunities for a new wave of grassroots activities to develop. Within India, four enabling trends have created such a window of opportunity for the Ganga movement.

First, the widespread high profile protests on social issues such as corruption and harassment of women have created a climate where protests are becoming common. This creates an expectation and gives permission to create more movements.

Secondly, these movements have been supported by cheap internet access and social media technologies, which are being used by many of the younger citizens in India.

Thirdly, the judiciary, while hampered by perceptions of being slow and being used as an excuse to delay resolution of pressing issues, has been playing a stronger activist role in the political landscape. This is evident through recent filings on the Right to Information movement and growing waves of Public Interest Litigation. In 2010, the National Green Tribunal was set up, providing a relatively fast legal track for environmental issues.

Finally the new prime minister, Narendra Modi, used Ganga as a key plank of his political messaging platform which drove subsequent initiatives such as the announcement of the new *Namami Gange* plan and the promise of a further 2,000 crore (approximately USD$3.17 Billion) investment into river plans. While the effectiveness of the plans may be doubted, it does put Ganga on the national stage.

The focus at this stage should be the development of strategies for a new wave of protest movements centred around Ganga. These are to be facilitated by several key visionaries that will guide the new movements. Conditions are ripe for a trigger event that will hit at the nerve of public consciousness and then the movement will take-off with a massive groundswell of public support.

4. Why the Need to Act Now: Identifying the Trigger Issue

Crises are often in the making for many years. They lay disguised, bubbling under the surface until some fail-safe breaks (like the safeguards around a nuclear plant), or a very public event which brings the issue to the fore (like a wrongful arrest or an extreme crime committed against the innocent). An analogy for this is Malcolm Gladwell's tipping point, where small things snowball and finally reaches a "moment of critical mass, the threshold, the boiling point" (Gladwell, 2006, p. 12). At the tipping point, the trigger event erupts.

When engineering a trigger event, therefore, it is possible to seek out issues that are bubbling beneath the surface and pick a suitable one to highlight to mass consciousness. *The purpose here is to leverage the severity of the issue as the content around which to develop a social movement.*

What matters are bubbling beneath the surface for Ganga? There are myriad crises waiting to happen — these might include the biological 'death' of the river, the collapse of the river economy, the extinction of riverine ecosystems, widespread health epidemic, drying up of the river, and contamination of water and food chain.

4.1. *Health risks are at tipping point*

Among the above, the issue most discussed in public consciousness is that of health risks. Most people have a basic understanding that Ganga's polluted waters are harmful to their health when they are exposed to it or ingest it. This threat to humans is often highlighted in conjunction with discussions about the river's pollution. In other words, there is already an unconscious correlation being made in people's minds.

However, what we do find is a gap in the depth of understanding about the severity of the issue, even though there have been studies about this aspect. One possibility that these have not come to the fore in entirety could be because the studies are academic and not disseminated to the mainstream. It is observed in India that "most research institutions have singularly failed to even publicise their findings of a serious nature, least aside the question of lobbying for policy changes." (Thakkar, 2009, 271). Or, the technicality of the studies did not interest mainstream media.

A more speculative reason could be official suppression — government officials are alleged to have halted some studies about the issue, when alarmed by 'superbug' discoveries in Ganga and the potential impact on India's medical tourism industry (*New York Times*, 2014).

The health risks posed by Ganga's pollution, as described in various studies, are alarming and there is an urgent responsibility to ensure public awareness about them. There are three kinds of diseases that can be caused by polluted water: water-borne diseases (spread by ingesting the pathogens in the water, e.g. gastroenteritis, hepatitis, cholera), vector-breeding illnesses (spread by mosquitoes which are increasingly insecticide-resistant, e.g. malaria, dengue), and serious direct infections (due to direct exposure to water pollutants, e.g. arsenic, causes of cancer and fluorites).[1] Studies reveal strong evidence of the physical harm arising from Ganga's pollution.

[1] Gupta, HK (2014). Public Health and the Ganges. Presentation to authors.

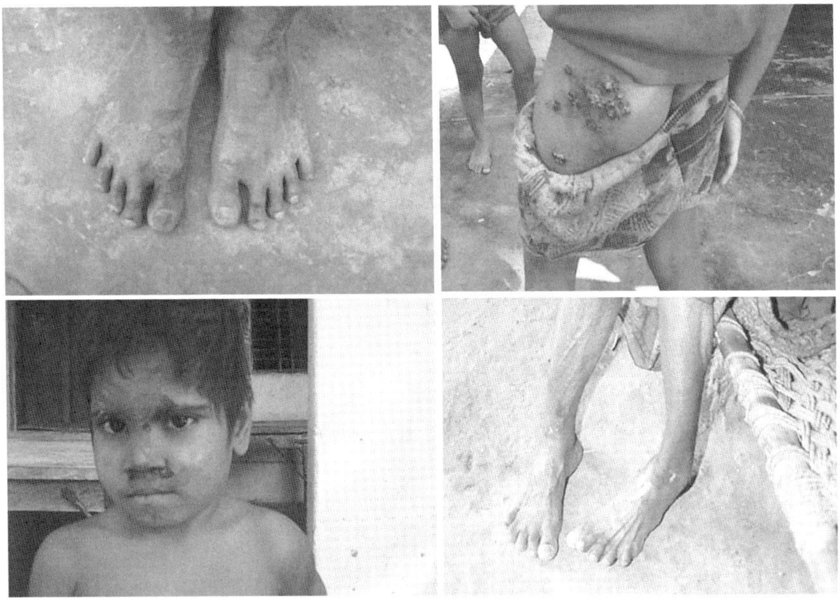

Sufferers of dermal conditions caused by exposure to Ganga's polluted water at Kanpur, India[2]

The current evidence throws up three key messages:

Firstly, that there is a very high correlation between Ganga's pollution and the illnesses suffered by river-side communities. In particular, the least advantaged suffer the worst odds.

Secondly, not only are the exposed communities affected, the river also acts as a conduit that transports health risks to the greater population. It does this by contaminating the ground water or the food supply (e.g. fish and farmlands). Thirdly, the problem is getting worse, as the bacteria breeding in the river get ever more virulent and resistant to antibiotics.

[2] *Source*: Jaiswal, R (2014). Eco Friends. "An Environmental Education, Protection, and Security". Presentation to authors.

Table 1 is a glimpse of the studies which have sought to uncover Ganga's health effects:

The findings highlighted present only the tip of the iceberg and are likely to increase. This is because new technologies and access to information will make it easier and easier to report health effects: for example, through new technologies that can detect pathogenic strains; new hospitals and clinics where cases can be logged (whereas previously ailments would be treated at home and go unreported); increased use of citizen journalism and public access to information; globalisation allowing for transboundary migration of virulent antibiotic resistant genes leading to international concern and reporting. A disaster like the Ganga-caused 1991 epidemic of Hepatitis E affecting 79,000 people in Kanpur (Naik, 1992),[3] would unlikely be as under-reported as it was back then. The increasing societal ability to collect, aggregate and access data about health effects contribute to an impression of a tsunami-like build-up in river-related diseases. In other words, a tipping point scenario is appearing. Such conditions present a strong opportunity for the activist organisation to trigger public awareness and concern.

4.2. *Appealing to India's evolution of environmentalism*

Health risks is an issue that has high likelihood to capture public attention when revealed. Its effectiveness stems from the saliency and potential of direct costs on one's own person. Studies reveal that perceived heightened *personal* risks related to the environment is a very strong predictor of environmental activist behaviour, more so than socioeconomic variables like being more educated or having a higher income (Wakefield, 2005; Seguin, 1998). If health risks posed by Ganga are thrown into the limelight, people are more likely to take on the activist mantle.

Another reason for the effectiveness of health problems as a rallying point, relates to the historical evolution of Indian mind-sets towards

[3]The epidemic was attributed to the presence of Ganga water used in drinking water. The incidence of hepatitis was higher in city wards that used drinking-water from Ganga and tube-well water than in those wards supplied only with tube-well water (5.6% versus 1.2%) (Naik, 1992).

Table 1: Health Effects of Ganga

	Key Message	Evidences	Source
A	Health issues of river-side communities are strongly correlated with Ganga's pollution	Incidence of waterborne and enteric diseases is as high as 66% among families that use Ganga water for routine purposes (e.g. washing of laundry and kitchen utensils, and personal hygiene like bathing and brushing teeth). 33 cases of cholera also occurred among families who bathe and wash laundry in the Ganga, compared to none in families who do not do so.	Yearlong study of 104 families in Varanasi (Hamner *et al.*, 2006)
		The impact of waterborne diseases falls heaviest on those who are the least advantaged as they rely heavily on Ganga for daily needs. Occurrence is as high as 93% among residents in a particularly poor community, compared to 37% in the relatively more affluent residents.	Yearlong study of 104 families in Varanasi (Hamner *et al.*, 2006)
		The average effect of a 1% increase in fecal coliforms in Ganga is an increase of 3–5 deaths per 100,000 births in a given month.	Study of infant mortality along Ganga (Do *et al.*, 2014)
		The extensive reach of industrial pollutants puts an estimated 50% of factory workers and 30% of Kanpur's population at risk of severe illnesses including typhoid, bacillary dysentery, botulism, poliomyelitis, hepatitis, fluorosis, and cancer. Morbidity level of tannery workers was significantly higher than those who were unexposed to tannery activity (40.1% versus 19.6%).	Study of industrial pollution in Kanpur (Singh, 2001) Study of 197 tannery workers in Kanpur (Rastogi *et al.*, 2008)

(*Continued*)

Table 1: *(Continued)*

	Key Message	Evidences	Source
B	Health risks extend beyond the localised communities exposed to the pollution	Industrial pollutants carried by Ganga seeps into the ground water, which affects the greater population. Areas of Kanpur receive Chromium-6 contaminated water; and they report "higher prevalence of health complaints, altered hematological parameters and spirometric defects" and face increased odds of ocular defects (18% versus 8%), gastrointestinal disease (39% versus 19%) and dermal conditions (25% versus 7%).	Study of 186 exposed subjects in Chromium-6 contaminated communities (Sharma *et al.*, 2012)
		Infant mortality risks persist even at significant distances downstream of the source of pollution, measuring at approximately 1–2 deaths per 100,000 births.	Study of infant mortality along Ganga (Do *et al.*, 2014)
		At a national level, infants in India are today born with bacterial infections that are resistant to most antibiotics. Ganga and other Indian rivers transmit drug resistance through its untreated fecal matter from humans and animals that use powerful antibiotics and improper disposals by hospitals. Pregnant women who ingest the water or fish then develop resistances, which are passed through the maternal genital tract to their newborns. An ongoing study shows that 70% of India's babies' infections are immune to numerous antibiotics. More than 58,000 babies died in 2013 as a result.	(*New York Times*, 2014)

C	Ganga's bacteria are becoming harder to cure as they gain in virulence and antibiotic resistance	Three bacteria groups were measured as indicators of Ganga's pathogenic (disease-causing) potential: Enterococcus, Escherichia coli ('E. coli') and Salmonella. The first two exhibit increased levels of intensity and virulence, and stronger resistance to multiple antibiotics. Salmonella increased in intensity, far exceeding the minimal dose to cause infection.	Study of enterococcus (Lata *et al.*, 2009) Study of E.coli (Hamner *et al.*, 2007; Ram *et al.* 2007) Study of salmonella (Jyoti *et al.*, 2009)
		Antibiotic resistance genes (ARG) escalate to 20-fold at pilgrim centers — Haridwar and Rishikesh during pilgrimage seasons.	Study of ARG (Ahammad *et al.*, 2014)

environmental issues. Early environmental movements like the *Chipko* movement (where simple villagers stopped loggers) and the 'Stop Tehri Dam' project in the 1970s and 1980s were the result of disadvantaged communities trying to protect their homes and livelihoods. They were what renowned historian Ramachandra Guha termed as 'environmentalism of the poor' (Guha, 2013), and is starkly different from the environmental ideals of protecting biodiversity and habitats that is prevalent in the West. Early environmentalists claim Gandhi as their main inspiration, whose motivation was borne out of the desire to preserve agrarian resources against impending industrialisation, rather than the romantic preservation of nature and all its wilderness (Kalland and Persoon, 2013, 80–81).

This, hence, characterises much of Indian environmentalism where concerns tend to be more practical rather than idealistic. The kinds of causes that have most exercised Indian environmentalists are resource-based like forest and land, and marine resources and fisheries; industrial pollution; development projects (dams, power projects, mining, industrial plants, military bases) and farming aid. (Shah, 2004, p. 258) It is, therefore, not surprising that the Ganga's pollution has not galvanised much

Village protest against Omkareshwar Dam on Narmada River involving 250 men and women standing in rising waters[4]

[4] *Source*: https://icrindia.wordpress.com/2012/08/28/250-men-and-women-continuously-standing-in-the-rising-waters-of-omkareshwar-dam/jal-satyagrah-2012-1/. Accessed on 15 may 2015.

activism, given that the framing of the issue thus far — either as a moral affront or a political debacle — is not as a practical concern in people's day-to-day lives.

One critical battle that the environmentalist fights today is ideological. Since the early days of *Chipko*, public support of environmental movements has peaked and then dropped into a trough. The prime victim of economic growth has been environmental sustainability (Guha, 2013) Environmental protection continues to be dampened by economic priorities today and, for the common man, basic needs.

The question of how best to engage the Indian public in environmentalism should recognise this spectrum of environmental attitudes that the country is still progressing along. "A persuasive environmental ethic... cannot be constructed *de novo*. It must be located in a more general cognitive context, and it must retain continuity with the moral ideas and ideals of the past." (Callicott, 1994, p. xiv).

While no longer in 'survival' mode for natural resources, the society is also not at the phase of Western ideals (a rallying call appealing to save the Gangetic dolphins might perhaps engage only the enlightened few). Ganga's health risks is, therefore, ideal as the type of problem which is somewhere in the middle of this spectrum, where practical, personal concerns and the altruistic need to alleviate societal harm can push the case for environmental change.

4.3. *A movement based on health issues — what does it mean?*

While health issues provide the foundational content, the trigger event itself is not the likes of a disease outbreak or a health emergency. Rather, the engineered trigger event is a high-visibility protest tactic, such as an industrial strike, a large-scale protest march, or a public sit-in, which uses mass mobilisation to highlight the issue (in our case, health effects) and ignite further awareness among the population. For example, in the anti-corruption movement, we saw in our Google trend analysis in Section 2 that when the Delhi Commonwealth Games organiser was arrested in a corruption investigation in April 2011, there was just a blip in public interest. The surge came about only in August 2011 after renowned activist

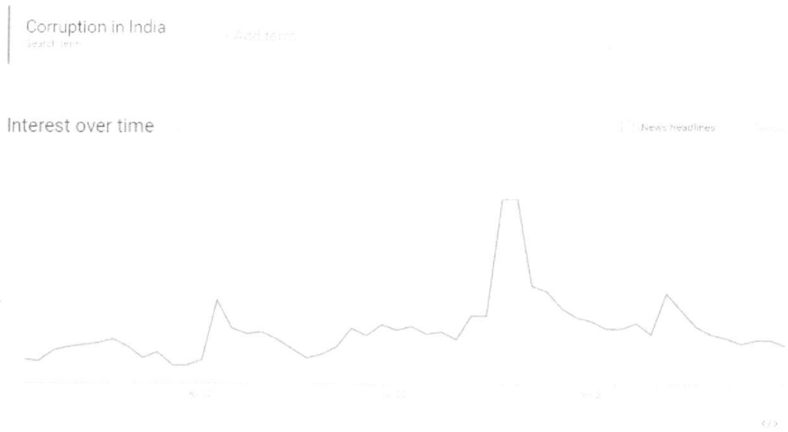

Figure 4: Google trends on Corruption in India in 2011

Anna Hazare announced the intention to fast-unto-death for demands upon the government to address corruption (see Figure 4). The corruption scandal was not itself sufficient to tip corruption issues into public consciousness. It was Hazare's announcement half a year later that did the tipping. His highly public and extreme act of self-sacrifice was the trigger event. The choice of tactics and publicity are, therefore, critical to staging the trigger event. These will be discussed in Sections 6 and 7.

In addition, the skillsets needed by an activist organisation in this approach will need to shift, from fighting based on moral grounds, to rational grounds. The organisation's language and modes of participation for people need to be more diverse, recognising that the "appropriation of rational modes of inquiry are becoming key organising strategies for environmental movements" (Couch and Kroll-Smith 2000, p. 389). We turn now to discuss the form of activist organisation needed to fight this fight.

5. Who to Involve: The Need for a Coalition

According to Moyer, movements derive their power from an upset, impassioned, and motivated populace. When conditions are ripening, subpopulations of victims gain a new level of understanding about the seriousness of the problem and form new local autonomous groups to become the new wave of grassroots opposition (Moyer, 1987).

In the context of Ganga, there is at present no broad cross-sector coalition, in the sense of one that unites all the organisations involved in the movement to clean up Ganga and accommodates new groups interested in joining the movement.

This absence of a broad coalition may be explained by a lack of appreciation for its need by some players in the movement. This may be because individual organisations believe that direct representations with powerholders will be sufficient to bring changes to the status quo and thus attain their objectives.

At the same time, the few attempts to create a broad coalition continue to be works-in-progress and have yet to obtain significant participation from relevant key sectors, including the aggrieved communities (boatmen, fishermen, sweepers, dhobis, even farmers relying on irrigation from the river), women, youth and others.

The failure to effectively engage these sectors may be attributed to the lack of saliency and urgency of the issue as currently framed, but may also be due to the present leadership mix of the movement which is heavily dominated by academics, scientists, engineers, and religious leaders, limiting the movement's ability to connect to lower-caste sectors of the Indian society.

5.1. *Lessons from Narmada*

One of the more successful environmental movements in India that involved the grassroots heavily is the Narmada Bachao Andolan (NBA) or the 'Save Narmada Movement' which protests against dams along the Narmada River. It was led by an influential activist, Medha Patkar, for over two decades. She was able to unite existing groups (specifically the Narmada Ghati Navnirman Samiti in Madhya Pradesh and Narmada Ghati Dharangrastha Samiti in Maharashtra) in the same cause (Narula, 2008).

Under Medha Patkar, the NBA flourished as a broad-based participatory movement. Her leadership allowed member organisations to flourish, including the Narmada Shakti Dal, a separate women's organisation within the NBA set up during the International Women's Day of 1988 and comprised female villagers from Maheshwar (Ibid.).

The NBA's campaign was successful because of innovative strategies of resistance at the grassroots, national and international level. One of the clear illustrations of its success is the formation of the Friends of River Narmada, an international coalition of individuals and organisations, primarily of Indian descent, in support and in solidarity with NBA.[5] As a result, NBA's success is said to extend beyond India's national borders. It is "regarded as one of the signature public contestations in the twentieth century that redefined the terms of development, democracy and accountability" (Ibid).

The experience of the Save Narmada Movement evinces the benefits of a strong coalition as the guiding force of the movement. Ganga activists should exert earnest effort to unite the existing organisations and open up to new groups.

5.2. *Defining coalition*

Coalitions are structures of collaborations undergirded by common vision, which facilitate shared decision-making power, influence, and material resources among individuals, groups and organisations. Through coalitions, members are able to achieve what they cannot achieve on their own, empowering them and making them less vulnerable to harassment and intimidation by vested interests (World Bank CommGap, 2010).

Groups in a coalition build strength and gain advantage by learning from each other, harnessing expertise from other member groups, combining voting power (such as in government bodies or in elections), mustering bigger numbers in public actions and getting more media attention, among others. These allow the groups to achieve more, especially on common objectives.

[5] The Friends of River Narmada maintains a web page at http://www.narmada.org/ which provides excellent documentation of the movement over the years.

5.3. *What the coalition should look like*

Mavens, Connectors and Salesmen

In the book *Tipping Point*, Malcolm Gladwell describes three types of individuals responsible in spreading a social epidemic, or achieving a tipping point (viral spread) (see Figure 5). The first are mavens, or information specialists who know everything about the issue and love to share their knowledge with others. They are important in attaining tipping point as they are cutting edge in acquiring new information, and thus know things before the rest, including potential system change (Gladwell, 2006).

The second are connectors, people who seem to know everyone, or have the uncanny habit and skill of collecting acquaintances or "weak ties." They are people who can share information through their networks

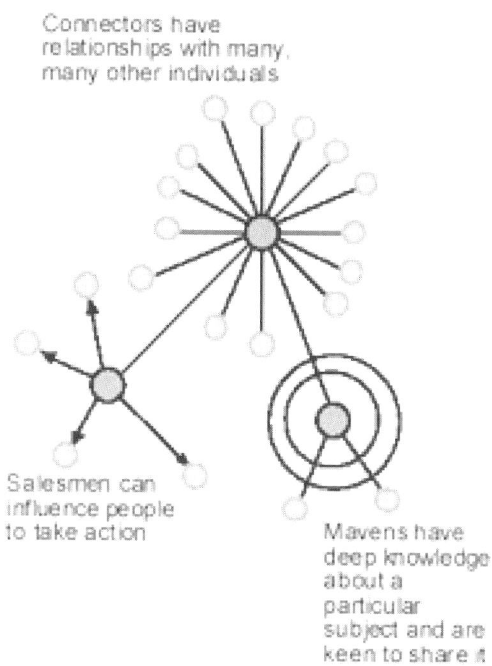

Figure 5: Gladwell's Mavens, Connectors and Salesmen

Source: Gladwell 2006.

at exponential speed and are thus important in achieving tipping point (*Ibid.*).

And the third are salesmen, or the quintessential persuaders who can get people to make decisions and take actions that they ordinarily would not take if left to themselves. They are people who can make others root for them in the same way that performers can make their audience to root for them and take their side. Their ability to produce action makes them the last important cog in producing a tipping point (*Ibid.*).

The group of academics, scientists, engineers, some religious leaders, who have been following and monitoring the pollution in the river, are easily the mavens of the movement. To a great degree, these people have kept the movement alive through the years. However, their expert knowledge will benefit from the local knowledge provided by the grassroots organisations. Examples of environmental movements in United States show that 'citizen-based' science are at times more effective at collecting data and more credible at presenting them (Couch and Kroll-Smith, 2000, 388–389).[6] Sectors like those whose source of income from the river, including boatmen, dhobis, fishermen, and even farmers who depend on irrigation from the river, are potential mavens because of their close relationship with the river. At the same time, other forms of technical expertise should be sought. Professional associations representing doctors and lawyers can bring strong value to the coalition, especially as we will discuss later the potential of using legal means as a protest tactic.

What seems to be missing in the movement is the ability to connect with the greater public. Accordingly, the coalition should be organised such that the present crop of leaders, with their maven-like skills, are able to link with groups that could be good connectors and salesmen (like the factions of media, artists, teachers, women associations and youth wings) in order to reach and convince the public to take a position and act.

---

[6] Notable environmental fights have been brought against a chemical dump by a community in Love Canal (1978), against the use of harmful chemicals by Multiple Chemical Sensitivity (MCS) victims whose claims were earlier debunked by medical experts (1980s–90s), and union workers proving the harmful effects of 2,4,5-T herbicide poisoning (1990s).

Common Vehicle

The key feature of a coalition is that it is a common vehicle for its members to pursue common objectives. To become such a vehicle, pre-existing groups and networks need to negotiate on what can be their common objectives.

While all the groups apparently support the cleanup of the river Ganga as the main goal, there seems to be no agreement yet on how best to achieve this goal. Some groups are focused on sewage, others on industrial pollution, still others are more into the flow of water in the river. There are also different interests, including religious and livelihood-related.

To form a broad coalition, the major players of the movement should strive to negotiate and agree on certain commonalities. It is found that even if ideologies (shared purpose) do not always cohere, organisations can come together based on the desire to share either identity (such as sense of community) or instrumentality (such as resources and political efficacy) (World Bank CommGap, 2010). In the early stage of coalition development, building trust is fundamental. This relates to fostering relationships and affinity. Concurrently, a clear administrative process can ease the effort of arriving at shared goals. Key steps in this process would include framing the problem at hand with key aspects and solutions broken down to identify overlapping interests; and stakeholder mapping to identify relational synergies (*Ibid.*) At some juncture, a small core group will likely coalesce, which will shape the agenda thereafter.

The broad coalition will then allow Ganga's activists to legitimately engage with powerholders, pool resources, and attract new members. It will also allow the group to recruit more mavens, connectors and salesmen, who can help spread their message and bring things to a tipping point.

Inclusiveness

One of the key features of successful movements is inclusiveness. For the broad coalition to be inclusive, it is important that all sectors of society who have a stake in the cleanup of Ganga are given equal opportunity to participate. By representing the interests of the majority of stakeholders, the coalition can also gain more legitimacy in dealing with powerholders.

Specifically, the movement must extend from its present core constituency of academics, scientists, engineers, religious leaders, and environmental activists, and should be open to and attract participation from stakeholder groups representing boatmen, dhobis, sweepers, farmers, village leaders, women, youth, lawyers, artists, teachers, media, among others.

In addition, the movement must extend to the whole stretch of the river Ganga and cover all the riparian rivers and states. This will allow the movement, through its broad coalition, to engage powerholders effectively at different levels, including the central, state and local level.

Tiers of Participation

While being inclusive, it is expected that members of the movement will have different levels of expertise, commitment and focus. To manage this, the broad coalition should be designed to allow tiers of participation.

A core of expert and committed individuals and organisations should take the leadership of the coalition, supported by mass-based and grassroots organisations that are committed to working towards the attainment of the coalition's common objectives. The main role played by the core group and supporting organisations is as mavens.

The broad coalition, with its core leadership and mass support organisations, should then partner with external, aligned individual and groups which can be effective connectors and salesmen for the movement. These groups may include the media, international organisations, and unions.

Designing the coalition into tiers of participation will allow the movement to avoid one of the pitfalls which is paralysis due to difference among the members of the coalition. With a core leadership comprising experts, committed activists and organisations, decisions can be made on operational matters on a day-to-day basis, minimising the chances of conflict between too many partners.

6. How to Get Supporters: Overcoming Barriers to Participation

In planning a movement, an activist doubtlessly wishes that it is as massive as possible and made up of people who as passionately cares for the cause as he does. The question, always, is how?

The preceding sections about health effects as the trigger issue and about building a coalition offer us half the answer. Health issues, as a deeply emotive and universal concern, provide the content around which to engage, recruit and rally people; while coalition as the organising body provides the knowledge, expertise and resources to carry out whatever the movement needs.

This section then attempts to provide the other half of the answer, by discussing the tactics for publicity and recruitment. How to promote awareness and stir interest? How should one engage and recruit supporters? For years, the lack of substantial public support to champion Ganga pollution limited the ability to challenge powerholders. The first order of business for Ganga activists should be to focus on building a public support base that is both massive and diverse.

The curiosity of India's situation is that it is a society *used to* activism. Cries for accountability can be heard from various civil society groups on countless issues. Do they all work in garnering interest? Unfortunately, not always so.

For an effective social movement, we believe that something more than the usual is needed. Innovative ideas are needed to jolt the general populace; and to be extraordinary is to look both inside and outside. Inside, to understand one's 'operating context', meaning India's socio-economic conditions, its political nuances and the general attitudes towards environmentalism. These aspects can pose as either barriers to, or opportunities for, participation and support. Outside, is to seek sources of inspiration for new ideas that can be applied or adapted.

These inside and outside lenses will be applied through the rest of this section, as we discuss the strategies for conversion.

6.1. *Environmental attitudes*

An understanding of the psychological factors that nudge people to care about environmental issues and what makes them take action enable us to determine the most persuasive approach to engaging the masses.

In an examination of Indian attitudes towards the environment, a noble portrait emerges.

Values driving environmental concern are distinguished as biospheric (concern for the well-being of all living things), altruistic (concern for other people e.g. family and friends, community, humanity) or egoistic (focused on oneself and own goals e.g. success, wealth) (Oskamp and Schultz, 2005).

Compared to several countries (see Figure 6), Indian citizens exhibit more biospheric concerns towards the environment rather than egoistic concerns (Oskamp and Schultz, 2005, p. 451). They also feel more responsible for environmental problems as well as acknowledging their severity than the citizens of several countries including Germany and New Zealand (Oskamp and Schultz, 2005, p. 466). However, what is interesting is that these beliefs do not translate to more action. In the same study, environmental behaviour (including civil society participation) registers lower, despite stronger beliefs in personal responsibility and the harm of environmental problems.

The infrastructural support and societal conditioning in these other countries are, of course, critical factors nudging environmental behaviour. It is *easier* to recycle, to purchase green products, or take part in environmental civil society within a system that favours such behaviour (e.g. where there is a corporate norm of permitting leave for volunteerism).

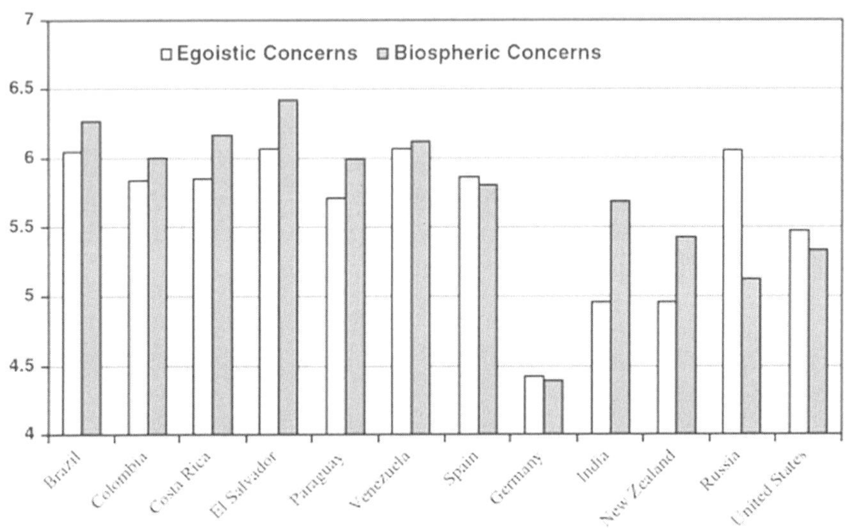

Figure 6: Environmental Attitudes of Countries Including India

Source: Oskamp and Schultz, (2005), p. 451.

In Section 6.4, we will discuss further the issue of social barriers to participation.

However, beyond external factors, internal psychic barriers to action also exist. While attitudes can lead to environmental behaviour, much more is needed to nudge someone to take on the more taxing behaviour required by civic action. An element observed in the development of environmental activists is *significant life experiences* such as childhood spent in natural outdoor areas, familial support for environmental concerns, and experience of losing a valued natural area to development (Oskamp and Schultz, 2005, p. 459). With the exception of the last, such experiences are growing scarcer in India's rapidly urbanising context. Amidst the exigencies of life in India, the average Indian child is also likelier to grow up with guidance on material success over environmental values. An instance of this at work is the oft-observed middle-class' materialism, which has led to civic indifference towards the environment (Mawdsley, 2004).

A national rhetoric prizing consumerism[7] further compounds the secondary role that environmentalism is playing. The safe-keep of environmental resources — in our case, a river — is secondary to the economic use it can be put to. For this reason, environmental fights against hydroelectric dams are harder to win in countries that are desperate to create new energy sources (as it was in the United States through the 1960s to 80s (Palmer, 2004).

Because of the dominance of this way of thinking in India — where life experiences do not foster civic action and economic progress is prioritised over environmentalism — a more 'economic' approach seems best suited to convince the Indian public, where costs and benefits of environmental actions can be quantified. The hitherto-discussed trigger issue of health risks will appeal to self-interest and it would be very easy to generate a publicity campaign solely about Ganga's health effects.

However, while the use of health issues is suited to trigger public attention, it should not be used to sustain awareness. Otherwise, it would be a negative campaign, with fear as its main instigator. Among the

[7]A description of the state of India reads: "In contrast with a previous ethic of austerity encouraged (in some ways) by the state, 'Consumerism [is now] sanctified because the middle class ability to consume [is] an index of progress'." (Mawdsley, 2004, p. 86).

responsibilities of an activist organisation engaged in environmental conservation, surely, is the duty to educate and change mindsets to foster greater care and concern for the environment. Within India, the strand of Gandhian morality flows strongly, and concepts of justice figure prominently in personal development. An appeal to morality is a strong advantage. The activist organisation's publicity strategy should therefore be multi-faceted, encompassing not only the hard facts of the pollution and the benefits of participation, but also employing emotive tactics allowing ideological development and affection for Ganga to grow.

Public participation in social movements is a voluntary process. It is a rational choice that is dictated by popular social values, beliefs and competing priorities. Mobilisation is a cognitive process that involves four distinct stages of overcoming barriers to participation as shown below in Figure 7.

The utility of the framework is in recognising that people choose not to participate for a variety of reasons — because they are opposed to the cause, lack awareness, belong to alienated social groups, have limited practical resources to participate or are simply not motivated to do so. Converting the public into supporters therefore requires a variety of techniques targeted at different parties, to address myriad barriers to participation, namely:

- Informational barriers
- Psychological barriers

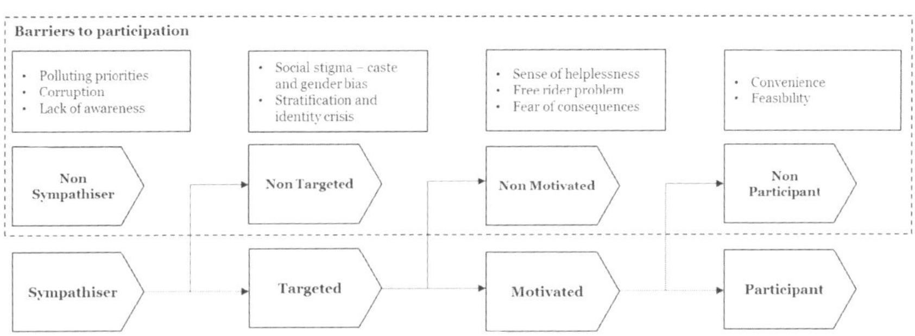

Figure 7: Klandermans' Barriers to Participation in Social Movements

Source: Klandermans (2010).

- Social barriers
- Practical barriers

This will each be examined in turn.

6.2. *Informational barriers — using publicity*

The lack of information is a common barrier to participation. It is simple to overcome but often overlooked by civil society groups. The crux lies not only in the delivery of adequate information, but also in framing the issue(s) in a salient and relatable manner. Successful movements owe much of their success to a strong publicity strategy.

Ganga NGOs have thus far employed limited tactics in generating publicity. The primary tactics are passive (website, information leaflets, reports), augmented with some active methods in recent years (Facebook, media engagement). Savviness in media engagement is observed to be low.

Good publicity can be a way of shifting the power balance in favour of the activist organisation. By creating emotive symbols and strong messages, the organisation can capture more of societal support. More importantly, a good strategy aims further up in the value-chain, to capture media agenda so that it can in turn influence public and policy agenda. Studies find a high correlation in the United States between newspaper articles and public opinion on those same issues by a factor as high as +0.9 (McCombs, 2014, p. 103). India's highly open political and media systems, in addition to the affordability of access (e.g. the low cost of newspapers) even for low-income groups (Freedom House, 2013), also provide a strong foundation for media's agenda-setting effects.

We will focus on two important aspects of publicity — the management of media and tonality of the messaging.

6.2.1. *Managing traditional and social media*

Media exposure impacts policy decision by firstly, shaping how the public view risk and secondly, shaping how policy makers perceive public opinion (Bakir, 2006). In other words, media plays a critical role in determining how the public should think and feel about a certain issue (a process termed as 'attribute agenda-setting'), and public opinion in turn is digested by policy makers as fodder for design of their policies.

At the same time, public agenda is highly limited — at any one time, the public is concerned about only two to six issues, and a level of concern at 10% is considered the 'threshold for significant public attention' (McCombs, 2014, p. 95). There is, hence, always intense competition among issues. The first battle that an activist organisation needs to tackle is getting the journalist — the 'gatekeeper' — to write about its stories.

How do journalists determine what news has 'news value'? On the one hand, one could view it as a rational exercise where the journalist weighs the relative importance of every story and decides which brings the most public value to his readers. However, shortcut indicators for decision-making are often used, such as the status of the people involved (e.g. celebrity, power elite), presence of magnitude (in scale or gravitas), degree of surprise or relevance to other issues of the day (Harcup and O'Neill, 2001), as well as catering to the taste, preferences and concerns of its target readers and, at times, advertisers. Such can veer towards the sensationalistic. A candid description of what is being taught as 'criteria' in journalism schools reads: "for a calamity to make it to the news pages the people who die have to do so in sufficiently large numbers, they should preferably be well-to-do, they have to die suddenly and all at once, in one place. There have to be good visuals, and the victims should speak English" (Dixit, 2012).

In India, 'over-commercialisation' of media has been observed, where media tends to pick or frame sensational content to increase readership (Campaign India, 2013). In turn, readers and viewers are unaware of the skewed social agenda being delivered to them, and form unfounded opinions about issues (Dixit, 2012). An interesting example is *The Times of India* coverage of the 2009 H1N1 outbreak in India. An analysis showed that the newspaper did not play the objective role of providing public service information, but tended to utilise the frames of fear and panic (through key words such as 'flu death march', 'govt presses panic button', and describing deaths in the public hospitals) and of attributing blame and responsibility (by covering the lapses of the hospitals and government health department extensively). Consequently, it may have had the effect of causing more panic than necessary and distrust in public facilities for infected individuals, which may have deterred them from seeking aid as soon as possible (Krishnatray and Rahul, 2014).

More heinous is the observation that some of India's media plays to the needs of its advertisers (Dixit, 2012), including both national and state-level governments (which use the placement or withdrawal of advertisements as rewards or punishments). Freedom House observes that the line between editorial and advertising departments at many news houses are blurred, leading easily to patronage and bribery (Freedom House, 2013).

Besides a skewed agenda-setting process, the news publications of India are tiered in an almost hierarchical fashion. Foreign media is observed to be a dominant influence on the content and style of Indian English newspapers, and in turn, Indian English newspapers dominate public agenda more so than vernacular newspapers. When the Delhi Bus Rapid Transit (BRT) corridor was launched, Delhi-based English newspapers in reflecting the interests of its car-driving readers, roundly criticised the system (as car drivers could not access the bus lanes). In contrast, the vernacular newspaper, the Hindi and Urdu, did not immediately respond with accounts from their own non-car-owning constituents (commuters, pedestrians and cyclists), even though their experiences were positive. They only did so when prompted by a survey by an English news channel (Sahay, 2008).

The above suggests to us a few insights: that media management in India is a game that needs to be played carefully for fear of sensationalism, that the role of foreign media in agenda-setting should not be underestimated, and that novel tactics of engagement are needed to secure interest from English language media due to the localised nature of health issues in Ganga's case.

'Subsidising' the Media

Studies have shown that 'subsidising' the media through public relations work helps significantly in securing coverage. Such measures include issue of press releases, and providing journalists with key contacts and informational and visual resources. In the United States, it was found that almost half of the stories in *New York Times* and *The Washington Post* over a 20-year period were highly based on information subsidies like press releases (McCombs, 2014, p. 131). These work because they make it easy for journalists to report stories, which is especially relevant in India's case where journalists face an immense flood of stories and pressing timelines daily.

Securing public relations expertise in the activist coalition mentioned earlier is therefore critical. Crafting press releases is also well worth the effort in ensuring that one's story is reported in a preferred manner. For every story, a journalist can choose from a multiple number of frames. As shown earlier, there is a tendency towards sensationalism, which would be good to avoid. Ganga's health risks may be framed as a failure of government (again) or, more constructively, as a collective problem caused by and to be addressed by society as a whole. The power to influence how the story is framed lies in the press release issued by the activist coalition itself.

Journalists also often choose frames based on their sources of information. Sometimes, the first version told to media becomes adopted as the primary frame ('primary definer') (Tuchman, 1978). Thus, the coalition should proactively prepare certain key messages that can be adapted to different issues that arise and which give consistent interpretations of events.

More often, however, official sources or authority figures determine the frames. One way to subvert the process of 'official' frame-setting, is to ensure that one has an easily-accessible representative who is able to speak with authority and clarity to media. The organisation should also garner a network of 'arbiters', which will be viewed as objective and authoritative figures that can be provided to journalists as key contacts (Deacon and Golding, 1994). They may include scientists not officially allied with the coalition, or non-environmental activists.

What may work against the activist's efforts at publicising health issues is the increasing degree of political confidence and public consensus that change for Ganga is coming, that the government is this time determined and able to clean up the river. Against the relatively smaller voices of civil society who say otherwise, journalists are likely to choose the more consensual framing. In such a case, the activist organisation will benefit from voices of authority that support its case (such as famous environmentalists from writers to celebrities) and using scientific proof and emotive case studies, in order to pitch the scales in favour of the Ganga movement. Foreign media can also be a strong ally. Their reportage can itself be an endorsement, nudging local media to frame the story similarly.

Besides editorial content, using advertisements is a possible option. Thus far, the government utilises advertising slots to highlight social issues (though they tend to be preachy and fail to make an impact) (Ninen, 1992). Negotiating with likeminded television channels for advertising slots is well worth the effort as a short high-quality one-minute video that emotively explains public apathy and government negligence can be effective at changing mindsets.

The Role of Online Media[8]

One good reason for using online media as a complementary tool relates to the sensationalistic tendency of traditional media described above. The latter has been found to rely on a 'protest paradigm' where the scale, tactics and spectacle of the protests are reported, rather than the underlying issues or significance (McLeod and Hertog, 1999). Online media is hence the 'counter public sphere' where activists can disseminate and discuss the key issues, enhanced by the virtual ability to communicate and mobilise across distances.

The reliance on online media for India activism however should be caveated. Its access is largely dependent on socioeconomic factors like income, class, education, causing a digital divide. If only those with access are able to voice their views on social change, then the online sphere is hardly representative of the larger population. "In such cases the Internet and social media do not democratise public conversations but perpetuate exclusion, substituting one ruling elite for another rather than overturning the system as a whole" (Chardha and Harlow, 2015, p. 675). Today, only 20% of India's 1.2 billion are Internet users (Internet World Stats, as at 2014) and 5% is on Facebook (*Ibid.* as at 2012).

---

[8]Online media refers to all digital content distributed over the Internet. In this section, 'social media' is also discussed, which refers to virtual communities where interaction is key, for example, Facebook and Twitter.

Social media as a tool is also limited by Indian habits and norms around its usage — few use it to share political views. Usage is mainly functional, to keep in touch with family and friends or share views about entertainment news, which is a pattern similar to other developing countries (Poushter, 2014). This could be why the primary challenge cited by Indian civil society groups in using social media is the *lack of response* (Chardha and Harlow, 2015, p. 680). The superficial state of engagement has led to fears of 'clicktivism', where the notional support given online (such as displaying 'protest avatars' as Facebook profile pictures) is mistaken as sufficient participation in the cause, thereby harming participation in off-line activism (*Ibid*., p. 684).

Nonetheless, online media is an important tool, and should be invested in for two specific reasons in Ganga's case: its potential to influence traditional media and its ability to reach youths.

Activity in online media is increasingly becoming a strong influence on traditional media. This form of inter-media agenda setting has been seen in several cases in India, where the sheer amount of online support for movements like Anna Hazare's anti-corruption campaign (Chadha and Harlow, 2015) and the 'Pink Chaddi' campaign[9] (Chattopadhyay, 2011) prompted strong coverage by traditional media. The blogosphere is also a potential source of influence (Meraz, 2011, p. 187).

Secondly, India's youths are highly engaged in the online sphere. Seventy-Six percent believe that "social media empowers them to bring change to the world we live in". They are empowered by their freedom of expression and of choice on the Internet, where they can engage in discourse and choose to accept or reject issues (*The Times of India*, 8 December 2011).

Online media engagement is therefore necessary, however, given the typical resource limitations faced by civil society groups, it bears both pros and cons — it is a cheap and quick tool for dissemination of news, but also a creature that needs 24/7 engagement. The Ganga activist

---

[9] In 2009, a female journalist started a Facebook group titled the 'Consortium of Pub-Going, Loose and Forward Women' in protest of ultra-conservative group, Sri Ram Sene. Its 'Pink Chaddi' campaign advocated sending pink underwear to the office of Sri Ram Sene.

coalition might do well to leverage different online media for different means. The more static online sources such as blogs can be used to discuss issues and solicit comments, whereas social media can be used as a more straightforward tool to mobilise participation for activities. When it is activity-based, social media activity need only be managed in spurts, before, during and after key events.

6.2.2. *Tonality and messaging*

Several environmental movements and issues in India often use Hindu myths and symbolism to mobilise the public. We see this in the campaign to stop Tehri Dam where Ganga and Hinduism are intentionally associated, so that people view the dam as an affront to collective history, culture, community and nationhood (Sharma, 2002, p. 13). The Chipko movement also demonised opponents through casting them as villains from the Hindu Epics, the Mahabharata and the Ramayana (Mawdsley, 2004, p. 95).

However, the religious rhetoric that has been most effective at engaging the populace's emotions about Ganga's pollution is, ironically, also the rhetoric that potentially undermines the pollution and over-simplifies the complexity of its causes.

(1) "Hindu understanding of sacred purity does not render purity and uncleanness mutually exclusive categories and conditions" (Alley, 2002, p. 232). Unclean waters are not understood as making Ganga impure. Hence, polluted waters is an affront to Mother Ganga, but only in the way that it is a moral offence, rather than being an impingement on its actual state of purity. "The essentialism of Mother undermines the efficacy of scientific and governmental statements about pollution and weakens activist discourses on environmentalism" (*Ibid*. 237).
(2) It is a red herring, which distracts people from the real issues. "(T)he reasons for ecological problems and disputes are being framed through a narrative of loss of Hindu purity. Instead of asking questions about political and corporate power, development agendas, class biases or urban consumption (none of which would be appealing

to the middle class supporters of Hindu-tva), the problem is blamed on the weakening of the Hindu nation…" (Mawdsley, 2003, p. 385) In Ganga, we have seen that this approach allows powerholders to appear to espouse similar values while sidestepping technical and political issues. It does not allow the people to think clearly about the real issues — failing to protect 'Mother Ganga' becomes a moral and religious failure, rather than a legislative and executive failure).[10] The failure to understand the key issues is illustrated in a recent survey by The Energy and Resources Institute (TERI) where 72% attributed the blame for Ganga's pollution to individual citizens, then followed by industries (15%) and local government/Municipal Corporation (13%) (*The Times of India*, 2015).

(3) Recommendations may be exclusionary and potentially regressive: this is a problem because of who leads the rallying call. An elitist approach appealing to Hindu nationalism tends to lead to exclusionary policies that pin the blame on practices of the poor or on population growth, and then proposing displacement forms of strategies, instead of redistribution or assistance (Mawdsley, 2003, pp. 92–93).

The above analysis does not mean to discredit the use of Hindu beliefs in advocacy for Ganga, but serves to highlight the implications of the approach.

Social Affinity Rather than Religious Association

'Place attachment' based on social affinity may serve as a possible alternative to a campaign with religious undertones. Numerous studies show that emotional connections with places can contribute toward pro-environmental attitudes. A study testing this in India shows that stronger place identity

[10]Alley argues that the Indian government is being forced to respond to the public and NGOs in religious language ("…their everyday engagements with the public necessitate some lip service to these views and a significant amount of bending to religious practices and movements."). (Alley, 2002, p. 234) However, we contend that their supposed 'powerlessness' cannot be the case as there are much more political benefits to espousing a religious rhetoric, especially for the Hindu nationalistic ruling Bharatiya Janata Party (BJP).

is correlated with greater agreement about the precarious balance between humans and nature, and weaker support for human dominance over nature (Budruk, 2009).

The nature of people's engagement with rivers tends to create strong place attachment. For the Ganga, generations of Hindus have visited it, bathed in it and worshipped it. Its association with rebirth and death also mean that the experience of being near or in the river is deeply spiritual and affective. Even for non-Hindus, the Ganga and its ghats are used for daily needs ranging from practical to recreational. The ability to leverage this sense of place attachment is the activist's greatest asset. Emoting place attachment through publicity materials may be done through a few ways:

- Imagery — Successful river conservation movements in the United States have found it useful to create an image for the river which represents the emotional bridge between the river and the public (specifically the target audience) (Bolling, 1994, p. 59)
- Icons — Reminders of the river can be developed through symbols that can be worn as everyday wear e.g. pins by sympathisers of the movement
- Neighbourhood ties — Creating activities which foster social capital enhances place attachment (Lewicka, 2005, p. 392). (For example, families in Varanasi living along the ghats clean their neighbourhoods together as a bonding experience (Faces of Varanasi, 2014)).

Visual and Emotive Cues

New approaches may well aid the effectiveness of the movement to spread the message.

'Image events' are increasingly used by environmental groups and other advocacy organisations. They are "staged acts of protest designed for media dissemination" (Delicath and Deluca, 2003, p. 315). An example: Greenpeace unintentionally composed such an image in 1975 when a harpoon ship fired at a whale over the heads of protesting Greenpeace activists on rubber dinghies. The image was captured and disseminated by multiple news networks, catapulting Greenpeace into mass consciousness. Image events typically capture the act of protest in order to habituate audiences to the act of challenging authority, and composed controversially to cause the viewer to question beliefs, thereby widening the possibilities for deliberation and debate. Furthermore, its visual form makes it easier to bring out emotive responses in today's 'distracted and disinterested' audience (Delicath and Deluca, 2003). Several environmental groups have also used using photo-essays and film. When captured in high quality with strong story lines, they will have the effect of tipping the viewers' emotions towards support.

Humour is another method of challenging paradigms. Establishment, institutions and policies are supposed to be serious. When humour is applied, it reformulates how people think about these entities and issues, and opens the door to alternative views to the world. Humour can also ignite the apathy that people may have about certain issues, by painting the incongruity between the official agenda and what really happens. For example, the group "Campaign against Conscription" in Norway infiltrated a false prosecutor into a trial, who despite hamming up the role, was not suspected to be a counterfeit. The trial was filmed and sent to the media, to push the point that the official court system was a sham (Sorensen, 2008, pp. 171–172).

In recruitment, humour helps increase outreach beyond the usual pool of sympathisers. Its power comes from a simple source — which is that when people laugh and feel good, they are more likely to want to be involved. A Serbian political activist organisation, Otpor, discovered this when it used intelligence and wit as a brand, which attracted many young people and students who felt that it was 'cool' to be part of Otpor. (Sorensen, 2008, pp. 175–176) If not humour, applying wit and irony can be the differentiator for Ganga's activist movement — which is all the more needed in India's over-cluttered scene of advocacy organisations which each compete for public participation. A good example is spotfixing volunteer organisation, 'The Ugly Indian', whose name is in itself an ironic self-reproach, and further translates well into a call for action in this form: "The Ugly Indian can take the world's best systems and find a way around it. Even outside India. Only we can save us. From ourselves." Since its inception in 2010, it has inspired countless volunteers to dedicate time and effort to cleaning filthy public spaces.

6.3. *Psychological barriers — mass recruitment strategies*

At the individual level, the decision to participate in social movement is a rational choice dictated by factors such as individual preferences, perception of costs and benefits of participation, and objective availability. Psychological barriers create a generally apathetic citizenry. The three most important among these barriers are discussed in this section (Passy, 2001).

- *Sense of powerlessness and fatigue*: Environmental activism against pollution in Ganga has a long history of more than 40 years. Despite many short wins, activism has not produced any tangible results. Even seasoned activists experience fatigue and their trust in activism wanes with time. Meanwhile, the powerholders have remained strong, corrupt and unmoved all along. As a result, scepticism sets in and people believe that no amount of activism can ever change the status quo. Their economic and social dependence on powerholders for public goods and services further reinforces this scepticism and instils a sense of powerlessness among the masses.
- *Perceived Effectiveness of participation*: People are more likely to participate when they believe that the movement has the capacity, resourcefulness and scale of impact to succeed against powerholders. They make a rudimentary assessment of the strengths of the activist movement based on the current size of its support base, quality of leadership, and endorsement from media, academicians and other credible social commentators. As a result, the modesty of Ganga-based activists in advertising their strengths may have worked against them in attracting supporters.

  People's participation will also depend on the perceived effectiveness of their individual contribution to the movement. If they believe that the marginal value of their participation is very low, and that the movement has the same chances of success with or without their contribution, then they are less likely to participate. This leads to a free rider effect wherein the people who want the movement to succeed choose not to participate.

- *Fear of powerholder response*: The goal of activism is to exert political pressure on powerholders and exact a favourable response from them. But powerholders' responses are sometimes violent and repressive, including police action to disperse or detain protesters. Other forms of non-violent and subversive tactics such as withdrawal of social support, court cases and confiscation of property have also been used to discourage people from participating.

  Powerholder response usually depends on the political climate, sensitivity of the demands, and legality of activist tactics. In India, the political climate is very conducive for activism as the constitution provides every individual the right to express his opinion. However, if protests turn violent, powerholders label it as a law and order issue and respond with violent subjugation. The fear of such a violent response is a psychological barrier for participation.

The current framework of activism needs a complete overhaul in order to address the psychological barriers to participation. First of all, Ganga activism has to refrain from being reactionary. The movement leadership team must be able to clearly define the short-term and long-term goals, and set immediate priorities for the movement. Secondly, the activists must have a good understanding of their constituency of support and their expected level of participation. Mechanisms for periodical assessment of the group morale and individual perceptions must be in place so that psychological barriers to participation can be identified and addressed on a continuous basis.

Short-term goals and priorities must be practical and achievable and should lead to a series of quick wins for the movement. The recent clean up of Assi Ghat by the NGO group, Sulabh International, is a good example of a quick win. The organisation gained greater visibility and credibility after the clean-up drive. Such quick wins, followed by 'show of strength' events such as rallies and public meetings, will instil a sense of hope among supporters and help them overcome activism fatigue.

Social networks have an important role in framing social values that people hold. The embeddedness of the activist movement in such formal and informal social networks has been shown in research studies as an important predictor of the intensity of participation (*Ibid.*). Formal social networks include the work environment, and social, political and religious affiliations. Informal networks are our friends, social circle, and extended family. Recruitment strategy must focus on leveraging these formal and informal networks by encouraging current participants to influence people in their social networks to join the movement. The level of involvement is also higher when people are recruited through activists who are part of their social network.

Advertising the strengths of the movement using media reports and other information dissemination platforms can help allay doubts about its efficacy. As discussed in the section on publicity, proper media management will ensure greater visibility for the movement and will have a positive influence on people's perception of the movement's effectiveness.

At one end of the spectrum of participation are the leaders of the movement who dedicate their lives to the cause and at the other end are people who only make financial contributions. Between these two extremes, there are many intermediary levels of participation. This differential participation can be explained by differences in motivational level, affordability, psychological readiness and objective availability. In order to be inclusive, it is unwise to foist the same handful of protest tactics on all kinds of participants. The Ganga movement must employ a wide array of innovative protest tactics that can accommodate the differentiated participation of supporters.

6.4. *Social barriers: Strategies for Target groups*

The Indian social structure, defined by age-old traditions, social values and norms, has stigmatised and distanced select social groups from actively participating in social movements. Our analysis of the Ganga activism landscape pointed to the social exclusion and non-participation of women, students and aggrieved communities that are dependent on the river ecosystem. This section is a snapshot of barriers faced by these social groups and strategies to overcome them.

6.4.1. *Women's participation*

Indian women have actively participated in political and environmental movements for many decades. The *salt satyagraha* in 1930 marked the beginning of this social trend. Gandhi declared, "To call women the weaker sex is a libel. If by strength is meant moral power, then woman is immeasurably man's superior" (Norvell, 1996). Of the 30,000 activists who courted arrest during *salt march,* 17,000 were women (Agarwal, 2008). India has, since then, produced eminent women activists such as Gauri Devi of the *Chipko movement*, Medha Patkar of the *Narmada Bachao Andolan*, Sunita Narain, Ela Bhatt, Rama Rauta and many others.

In India, women are expected to collect water, wash clothes, clean and cook. They are more frequently exposed to river pollution and consequently more vulnerable to water related dermal problems than men (Sharma, 2012). However, it is disheartening to note that women are not well represented in the Ganga activism. One plausible reason could be that the civil society leaders have failed to create an inclusive environment for women activists. It may also be coincidental that Ganga activism has not produced formidable women leaders like Medha Patkar of *Narmada Bachao Andolan*.

Whatever be the reasons, women's participation is crucial to trigger a public opinion avalanche against river pollution. Large-scale participation of women will add to the political potential and resilience of the movement. Given the social setting, the onus lies on the current leadership to proactively seek women's participation.

Normal social gatherings are usually dominated by men and it is difficult for women to contribute to the discussion. An account relates how women's views were sometimes regarded with disinterest by men within the Tehri Dam protest movement. (Drew, 2014, pp. 238–239) It is, therefore, important to create exclusive platforms for women activists to speak and be heard. The current leadership team should cede the proverbial floor to accommodate women leaders in the movement.

Eco-feminists believe that feminism and environmentalism are mutually reinforcing ideologies and hence should be jointly pursued. Participation of women in Ganga activism can be an opportunity for empowerment. Partnering with well-established feminist movements and organisations is a potentially effective strategy for bulk recruitment of women's participation against river pollution.

As women experience pollution differently than men, highlighting the pollution effects on pregnancy, childcare and women-centric issues will also contextualise the pollution issue as problems faced by women. Use of women-based stories and female leads in social-interest-advertisements, media reports and in other promotional material will be an effective communication strategy to encourage women's participation.

6.4.2. *Student participation*

Student groups in India have been politically active since early 1900s. When the non-cooperation movement launched against British rule, the central emphasis was on the participation of students and lawyers. After independence, student groups became more political and are today affiliated to mainstream political parties. Student groups such as the Student Federation of India (SFI) and All India Student Association (AISA) are left leaning socialists; and the Akhil Bharatiya Vidhyarti Parishad (ABVP) subscribes to right wing nationalist ideology. Another distinct characteristic of contemporary student movements in India is the deficiency of leadership. This was obvious during the anti-rape protests led by student groups that quickly turned violent and were subdued with tear gas and water cannons.

The influence of mainstream politics on student groups may have had a crowding out effect on other issues, causing student groups to be less interested in social and environmental issues. Political influence has also led to higher levels of criminal activities within campus. This prompted the Supreme Court of India in 2006 to commission a study on student groups which recommended various limitations on student activities, stripping them of their political connections (Ajmal, 2014).

All this has led academicians to believe that activism is an unnecessary and unaffordable distraction for students. In order to garner the support of academic institutions and to engage students more meaningfully, the form of participation expected from students must go beyond the shouting of slogans and hurling of stones. Student activism can be designed to be more non-confrontational, enriching and well within the scope of academic pursuit. In this regard, we propose two strategies for enlisting student groups.

Academic collaboration between activists and educational institutions can lead to meaningful student participation in creating a knowledge base for Ganga activism. Students from various disciplines such as sociology, environmental science, civil engineering, business administration and political science can be involved in projects such as to track the river's health effects on specific communities, to research legal issues in Ganga's history, to design a publicity campaign, or to create film and photography assets. In this way, students, too, can play the role of mavens and strengthen technical capacity of the movement.

Colleges can also provide the platform for awareness campaigns and host student-run Environmental chapters with a mandate to discuss and debate pollution issues. Student chapters can then organise non-confrontational protests, exhibitions and educational seminars that are designed to express solidarity for the cause and build awareness among the masses about river pollution.

6.4.3. *Aggrieved communities*

Aggrieved communities refer to the occupational groups such as the farmers, boatmen and dhobis that are economically dependent on the river, and the low-income communities living along the river which rely on it as a water source for everyday use. These groups are united by their high degree of exposure to the pollution and are, therefore, termed 'aggrieved' because the impact of pollution disproportionately falls upon them. In the social context, they are considered the most disadvantaged as they tend to face both class and caste based discrimination. Despite their dependence on the river for livelihood, these aggrieved communities are less involved in activist movements. We identify two major barriers to participation — bandwidth tax and social alienation.

For the low-income groups, their daily struggle to make ends meet is such a pressing concern that they do not have the bandwidth to worry about pollution in Ganga. Economists define this cognitive burden of poverty as the *bandwidth tax* (Low and Gill, 2014).[11] The future benefits of a pollution-free Ganga does not motivate them to forego today's wages to

[11] Harvard economist Mullainathan and psychologist Shafir use the term 'Bandwidth Tax' in their book "Scarcity: Why Having Too Little Means So Much" to explain why long term benefits do not induce behavioural changes among the poor.

participate in movement activities and, therefore, the cost of participation is relatively high for these groups.

A recurrent observation about public involvement in civil society within India is the apparent divide between the elites and the poor. 'Elites' have variously been described as the broad middle-class, or more narrowly, academia, environmentalists, religious leaders and so on, while the 'poor' are the working class, or the urban or rural poor. As we have observed, Ganga's NGOs have predominantly been led by elites. The demands of this better educated and resourced group are understandably different from that of the aggrieved communities; "roads, rather than public transport; garbage and pollution, rather than public housing: mosquitoes and public toilets rather than public health" (Ibid, p. 2720).

Thus social alienation of aggrieved communities by Ganga activists manifests in two ways. Firstly, over-representation of middle class elites in Ganga activism runs the risk of alienating the aggrieved communities who feel out of place or end up being 'voiceless' in the movement (Dreze, 2002, p. 30). Secondly, they find it harder to relate to the demands that are articulated, especially when highly technical terms are continually used, such as reducing Biochemical Oxygen Demand and sustaining environmental flow.

According to government estimates, low-income communities account for 28% of the population in the Ganga basin (GoI, 2012).[12] Without their participation, the mobilisation drive will be ineffective. To involve the very communities harmed by Ganga's pollution also goes beyond boosting numbers. It is a moral responsibility. Without their representation, the movement risks developing plans and objectives that further alienate or aggravate the situation faced by these communities.

Our recommendations to overcome these barriers are two-fold — to build an inclusive collective identity for the movement and to reduce cost of participation for the poor.

The aggrieved communities in our instance tend to be closely knit and have a hierarchical social structure. Ganga activists can leverage these

[12] Uttar Pradesh, Bihar and West Bengal account for more than 80% of the Ganga basin. The below poverty line population in these states is 30%, 34% and 20% respectively (Average 28%).

pre-existing social networks by inviting community leaders to join the movement's leadership team. The activist vocabulary must also be changed to reflect aspirations and demands of the aggrieved groups.

In order to reduce cost of participation for the aggrieved communities, Ganga activists should look beyond traditional forms of participation (protest marches and strikes). Participants must be able to pursue their occupational activities while at the same time protest against river pollution. For example, the boatmen can leverage their point of presence on riverbanks to become effective connectors and organise awareness campaigns for tourists. They can also be gainfully engaged by civil society to keep the riverfront litter-free.

7. How to Stage the Trigger Event: Protest Tactics

The engineered trigger event is *a sequence of high-pitch protest tactics* that serve to effuse urgency and propel the central issue of Ganga's health risks into mass consciousness. The scale of participation in these tactics and the choice of tactics are both critical factors for success.

The volume of people participating in protest tactics has been shown to improve the chances of success (Kuran, 1991, p. 17), with a single unit increase of active participants correlated with 10% increase in likelihood of success for a campaign to achieve its ultimate outcome (Chenoweth, 2011, p. 58). When public participation in activist movement reaches a threshold size, it raises fears of anti-incumbency among elected representatives and leads to change. We have therefore focused extensively in the preceding section on how a massive support base can be galvanised for the activist organisation.

In addition to sheer numbers, the choice of tactics needs to be strategic. A few principles should be kept in mind.

7.1. *Principles for protest tactics*

The primary principle is to choose tactics that *impose the most costs on powerholders*. Costs can be of different forms — for example, strikes impose economic costs, while street demonstrations impose political and administrative costs. Social movements have used techniques to amplify costs on powerholders like using the element of surprise, control of symbolic sites, seizing of physical resources, and timing the event to coincide with symbolic dates (Sharp, 2013, 66–70).

Diversity in tactics also matters. Too often, movements tend to rely on only a few methods which can lose impact with each repetition; and if they did not disrupt powerholders in the first place, then subsequent enactments are wasted efforts. A study of over 100 major nonviolent campaigns since 1900 show that successful movements employ high diversity in their tactics through a sequence that maximizes involvement by supporters and disruption to powerholders, while minimising costs to participation (Chenoweth, 2011, p. 75). Strong flexibility in tactic planning and innovation will allow the movement to adapt to different situations and to manoeuvre when powerholders choose to clamp down on specific tactics.

A type of tactical diversity is alternating between methods of concentration and methods of dispersion. Methods of concentration are based on the congregation of large numbers of people gathered together in public spaces, engaging in tactics like demonstrations. Methods of dispersion are diffused across society, where small pockets of supporters exhibit resistance in different areas, such as through staging protest assemblies within their area of work or study (Chenoweth, 2011, 74–75).

In order to keep up the morale of supporters, it is better to plan for a sequence of victories, meaning that each campaign should lobby for a limited issue instead of the entire movement objective. This was earlier discussed in section 6.3 on overcoming psychological barriers. Each victory will then weaken opponents while progressively strengthening the movement's supporters (Sharp, 2013, p. 59). Splitting the movement's goal into multiple achievable objectives also makes it easier to negotiate with powerholders, whereby at least some wins can be gained (Sharp, 2013, p. 74).

The key approach to protest tactics should be nonviolence. Nonviolent tactics should not be misunderstood as passive; rather, its power derives from "the belief that the exercise of power depends on the consent of the ruled who, by withdrawing that consent, can control and even destroy the power of their opponents." (Sharp, 1973, p. 1, 4).

Nonviolent tactics has a strong historical legacy in India, since the times of Gandhi's civil disobedience. They have been successful and are also protected by Article 19.b of the Constitution of India as a citizen's right to expression and is (Government of India, 1950). Indian judiciary and media have always defended this right. The Supreme Court of India has also issued *suo moto* notices to stop government from using violence against peaceful protesters. Furthermore, nonviolent protest tactics are more attractive to participants due to reduced potential of personal harm.

When peaceful protests turn violent, they are immediately branded as a law-and-order issue. This allows for violent police response against protesters. Therefore maintaining discipline is therefore critical, especially when participants lack experience and understanding. There are tools that can be used, such as pledges (to be signed by participants), strong communication channels, and even the presence of 'marshals' to instil discipline (Sharp, 2013, pp. 104–105). Publicizing the powerholders' attempts at repression (especially where violent) is also a strong counter-strategy that can alienate their support base.

7.2. *Repertoire of Tactics*

For many decades, only a handful of tactics have been used repeatedly by environmental activists in India to mobilize public dissent (Guha, 2013, p. 13). Gene Sharp, the foremost scholar on nonviolent struggle, identified 198 methods of nonviolent action[13] (Sharp, 1973, p. 2). These tactics are classified into three groups: (i) protest and persuasion, (ii) noncooperation and (iii) intervention. We will briefly describe them, as well as discuss the specific tactic(s) often practiced in India within each category. We advocate that the trigger event comprise a mix of tactics from each category.

7.2.1. *Protest and persuasion*

These tactics are mainly symbolic and serve to inject awareness of dissent. Examples include public speeches and marches.

[13]The 198 Methods can be found at http://www.aeinstein.org/nva/198-methods-of-nonviolent-action/.

Pradarshans (demonstrations) are a common manifestation of this form of tactic. Activists mobilize as many participants as they can and march through a prominent street displaying placards, shouting slogans and singing songs about the pollution issue. These are designed as a collective show of strength and to create awareness among the public. Marches usually end in a public meeting where prominent leaders address the public.

Such symbolic protests are important because of their simplicity, and ability to gather masses and build camaraderie. However, they can be exacting on participants if long drawn. To maximize the power of a *pradarshan* or any such symbolic protest, one should pick strategic timings to stage them where powerholders are vulnerable, such as at key events with foreign diplomatic presence and election rallies. The media should be told beforehand, and provided with press materials explaining the purpose of the demonstration.

Protest march organised during the Jan Lokpal movement and led by Anna Hazare.[14]

[14] *Source*: http://cryptome.org/info/india-protest/india-protest.htm.

As a river close to many hearts, Ganga also allows the activists to explore more affective tactics that can engage participants. The seriousness of conventional protests — public marches with shouting chants — can wear down participants. Creative protests can be more entertaining and easily weaved into everyday life of people. These tactics "can engage and intrigue ordinary citizens and, in this sense, have a low participation threshold" (Shepard *et al.*, 2008). "The joy experienced in solidarity with others often leads to sustained and ever-growing citizen commitment and concomitant actions for the cause" (Duncombe, 2004).

Different forms of creative protests: (clockwise from top left) campaign against airport expansion; flash mob demonstration against anti-abortion laws; demonstration in support of vegetarianism in Mexico; symbolic protests against air-pollution.[15]

---

[15] *Sources* (clockwise from top left): http://www.dailymail.co.uk/news/article-1120604/Heathrow-chaos-runway-activists-stage-airport-flash-mob-protest.html; http://www.vladpetri.ro/english/-photo/the_street.html; http://www.independent.co.uk/news/world/americas/activism-in-mexico-city-what-do-we-want-another-chance-to-shout-protest-slogans-in-the-street-9211073.html; http://www.airportwatch.org.uk/category/recentnews/publicity-politics/.

Creative protests are a product of innovation. Crowd sourcing is the best way to design such protests. Some indicative ideas for creative protests are:

- Organising flash mob performances, mock awards and elections, and guerrilla plays that use irony and art to communicate the seriousness of Ganga pollution. The preceding section on publicity discussed the benefits of using humour and wit to convey the movement's cause.
- Large artistic depiction can be made of the future condition of Ganga if pollution remains unchecked. The dead river can be artistically portrayed to nudge people to make this a personal priority. The 'Save the Irrawaddy Campaign' by artist, Myint Zaw, is an excellent example of how art, songs and poems were used to convey the cause.[16]
- Large number of protesters can "play dead" (in stretchers) for a short duration in one of the ghats decorated with messages about health risks of pollution
- Beautifying the river banks — volunteers can clean and reclaim dirty public spaces and beautify them with art work, statues and park benches. It creates a psychological barrier for people who pollute by trashing or spitting. These spots can also be used to launch awareness campaigns about the threat of pollution.

7.2.2. *Noncooperation*

These tactics aim at interrupting or stopping the normal operation or efficiency of a system. Examples include boycotts and strikes.

Bandhs and hartals (shut down) mushroomed during colonial India and emblematic of the anti-colonial struggle. It is a coercive tactic which

---

[16]The 'Save the Irrawaddy' campaign was a protest against China-owned Myitsone dam. Irrawaddy River serves as a unifying force for the many ethnicities in Myanmar and was a rallying point for many. It was also a cause taken up by Aung San Suu Kyi. In order to evade censorship, protesters used art exhibitions and artists wrote poems and songs. A very moving campaign documentary was created and disseminated underground, which was so affective that members of the public would illicitly make more copies to pass on. Myint Zaw won the Goldman Environmental Prize in 2015. http://asiapacific.anu.edu.au/newmandala/2015/04/23/art-activism-and-saving-sacred-rivers/.

calls for the complete or partial shutdown of the town or select services. The intent here is to capture the attention of general public by creating inconveniences to day-to-day life. The tactic works only when there is a large ideological support base that accepts the inconvenience as a form of showing solidarity with the movement.

One segment associated with Ganga's pollution immediately comes to mind as an arena for boycotts and strikes: the polluting industries releasing effluents into the river. A public boycott of their products will signal these industries to implement more environmentally sustainable practices. One cautionary note however is that boycotts and strikes can be taxing on the economic welfare of the people. What has at times happened is the loss of jobs as companies transition. In addition to boycotts and strikes, other forms of noncooperation which minimize the cost of participation for supporters can also be explored, such as the overloading of government administrative systems, or partnering likeminded judicial or administrative constituents for noncooperation.

7.2.3. *Intervention*

These tactics directly 'intervene' to change the pattern of normality. Examples include fasting (a psychological intervention) and establishing alternative systems.

Dharnas (nonviolent sit-ins) were a popular form of protest during the Indian independence movement and remain so today. They are designed to obstruct the function of targeted powerholders, such as by sitting en masse at the door of a public agency or court awaiting court decision. They can be used either against one individual or against an activity sanctioned by powerholders. Dharnas pose a political and moral dilemma to powerholders as the response is usually to arrest the leaders or disperse the participants using violence. This leads to negative publicity for the powerholders.

Bhook hartals (hunger strike) are yet another popular form of intervention. Instead of causing physical or political disturbance, *bhook hartals* cause moral dissonance among powerholders and their supporters. Led by prominent leaders, the tactic involves fasting until demands of the movement are met by powerholders. The courage and self-sacrifice of the activist draws public attention to the cause. As the health condition of the activists decline, powerholders are forced to engage with them.

Fasting has been a highly effective tactic in India because of the psychological impact of watching someone on hunger strike. In the anti-corruption movement, it was particularly effective because the face of the movement, Anna Hazare, was portrayed as an elderly next-door neighbour who was willing to risk his life for the cause. Says a supporter: "I came because corruption affects everyone, and I came because of Annaji. An old man is fighting for all of us. The whole country must join." (Singh, 2014, p. 22) The disadvantages are that it is highly fatiguing and, with each rendition, may attract less and less attention. It must therefore be strategized with other powerful tactics which can be used as the movement draws on.

Another strong intervention tactic is legal activism. There are a few legal activists who have used petitions and Public Interest Litigations (PILs) since 1985 to confront powerholders. Indian judiciary has always been sympathetic and responsive to environmental issues. In fact, the Supreme Court has donned the activist hat to hold government accountable for the pollution in Ganga. Responding to M.C. Mehta's Public Interest Litigation (PIL), the Supreme Court ordered closure of polluting industries and construction of Sewage Treatment Plants (STP) in more than 250 towns along the river. Its main constraints, however, have been frustrating delays and incapacity to follow through with the verdict.

Fortunately, conditions are ripening for legal activism to be a more effective tool. The government passed the Right to Information Act in 2005 to enable easy access to government information. The National Green Tribunal was created in 2010 to expediently address environmental cases. In a recent verdict, the Supreme Court directed National Green Tribunal (NGT) to take over the responsibility of preventing pollution of Ganga. There are reports that the current government is proposing to make river pollution a criminal offense. There is growing international support for use of legal tactics for environmental issues. Legal activist organisations such as Community Environment Legal Defence Fund,[17] Waterkeeper Alliance, and Environment Law Alliance Worldwide are working with

---

[17] The Community Environment Legal Defence Funds (CELDF) works towards empowering communities' rights to local self-government and to nature. It is currently collaborating

local partners on various environmental issues. These developments have reinstated hope in using legal tactics to fight pollution.

Legal activism requires a unique set of capabilities and resources. Firstly, the quality of the legal petition determines the success of the tactic. Legal experts have to be recruited to frame the petition well. The movement should partner with international groups such as Environment Law Alliance Worldwide (ELAW)[18] to seek technical guidance. The movement can also bring together eminent lawyers to form an environmental lawyers association to build a knowledge base.

Secondly, media coverage of legal petitions and hearings is very high and this ensures large-scale public attention for the issue. Activists should have a sound communication strategy to leverage this advantage without violating *sub judice* rules. It is to be noted that the PIL filed by M.C. Mehta and the judicial activism that followed has been the most successful in building public awareness for the pollution issue.

The third capacity aspect is about building a ground force for community policing of rivers and gathering evidence against polluters. There are international models such as the Waterkeeper Alliance[19] that provides guidance and support for creating a volunteer force to monitor river conditions. There are active Waterkeeper chapters for the river Yamuna in India. A similar community group should be created by the movement. These activists should be trained in investigative research and in collecting evidence. Such a community organised ground force will address limitations of the existing model.

Legal activism is relatively slow but a long-term strategy that produces effective results. Legal activists can also advocate legislative reforms which are long-term sustainable institutional changes that are

---

with NGOs to advance the 'National Ganga River Rights Act' wherein the Ganga River is recognised as a legal entity. http://celdf.org/section.php?id=395.

[18] Environment Law Alliance Worldwide (ELAW) is a global entity assisting communities with legal expertise for environmental justice. It has also been active in working with India NGOs. http://www.elaw.org/what-we-do.

[19] The Waterkeeper Alliance is a worldwide alliance comprising over 240 Waterkeeper Organizations on six continents. It operates on the principle of patrolling waterways to capture evidences of pollution which are then brought into legal action. http://waterkeeper.org/what-we-do/our-mission/.

more effective than ad hoc project-based promises. Examples of legislative reforms include increasing powers of NGT, creating an ombudsman for pollution control, and strengthening the Water (Prevention and Control of Pollution) Act.

In sum, the above tactics should be carefully planned in a sequence to create the trigger event. The decision factors for planning include the situation's available opportunities, the activist's available support base and anticipated response from powerholders. Generally, protest and persuasion tactics entail the least risk (as it imposes the least cost on powerholders), followed by noncooperation tactics, then intervention tactics. A straightforward method of escalating the movement's impact is, therefore, to employ a larger and larger proportion of noncooperation and intervention tactics as the movement progresses.

As India has seen protests and activism for more than 10 decades, powerholders have also started to build resistance to specific forms of protest tactics and have successfully deployed response tactics to derail and nullify activist movements. We shall now analyse some typical powerholder responses in order to build immunity against them.

7.3. *Managing powerholders' responses*

India enjoys a rich history of civil society and liberal democracy. The typical 'repression' tactics experienced by many nonviolent conflicts is almost unthinkable in India's context. Such would include violent repression, censorship of communication and information, seizure of properties and funds, economic sanctions, prohibitions of public activities including legislative restrictions (Sharp, 2013, pp. 89–92). Nonetheless, arrests and charges are common, which Indian activists have countered through *jail bharo andolan* tactics (where activists voluntarily get arrested so as to flood the jails and cause administrative problems).

However, a more formidable powerholder response in India's context is psychological pressures — such as co-optations and buy-outs, pretence of compromise or creating diversions, discrediting of the movement and/or leaders, and extreme ostracism. Setting up investigative committees as a response to protests is a commonly used pretence of compromise tactic. Anna Hazare called off his fast based on the powerholders' promise of setting up a joint committee to create the Lokpal. However, the government diluted the bill to such an extent that Anna Hazare had to return to protesting. Another appalling case of extreme ostracism is when the Government of Tamil Nadu filed more than 300 court cases against 50,000 participants of the Kudankulam Anti-Nuclear Plant movement. Even after two years since the filing, charges were brought in only two cases. The intent of powerholder response was to threaten people with legal consequences and dissuade them from participating in the movement (Human Rights Watch, 2014).

This type of tactics subvert the activist movement without resorting to violence, and are hence the most effective since they do not provoke negative public response while being able to dismantle the activists' strategies through covert actions. Knowing this, activists can anticipate and plan appropriate responses. The control of specific feelings — fear, fatigue and alienation — is critical in countering these powerholder tactics.

To address death threats, spreading the responsibility of leadership or important tasks across several individuals can help diffuse the fear that threats can have on a single individual. Enhancing community support at this time also helps. For example, activists in feminist movements often face public ridicule and threats, and strong movements have developed a system of mutual support among the women such as through travelling in numbers, frequent communication and sharing of resources for familial responsibilities.

Co-optation in this context is a powerholder tactic used to assimilate the activist movement by including the activist leaders in the decision making process. Activists are given positions in government committees as an institutional opportunity to influence policy and decision-making. However, this method has not usually led to favourable circumstances for activists in India. In the absence of positive outcomes, the credibility of activist leadership takes a beating as they are now seen as part of the powerholder's faction. This is the intent of co-optation in most cases.

In cases where co-optation, defection or arrest of a key leader has taken place, movements have sometimes lapsed into inaction. It is therefore critical that, as part of organisation development, leadership succession as a process is clear and automatic. Some decentralisation of leadership can also help address this issue, while helping to disseminate expertise and authority throughout the organisation.

Most importantly, the likelihood of powerholders causing defections or drop-outs increases as the following types of factors decreases: members' sense of affection and loyalty to the organisation, sense of camaraderie to other members, and state of emotional and physical health. To address the first two issues, ensuring transparency in communication through all ranks of the organisation and addressing concerns is essential. Demonstrations of group unity such as meetings, parades, wearing symbols of the organisation are useful in fostering group solidarity. The third has become a critical issue of late, as activists find themselves battling for over 10 years without significant progress while their own bodies and mental health deteriorate. Given the severity of Ganga's issues and anticipated long-term fight, practising 'self-care' (caring for physical health, creating strong support networks) among activists is absolutely vital for preventing fatigue and eventual desertion.[20]

Forms of capitulation

There are four possible ways in which the activist movement can result in victory (Sharp, 2013):

(i) Conversion: is the opponent coming around to the movement's point of view.
(ii) Accommodation: is the agreement of powerholders to meet demands (because of the difficulty in managing the conflict, or to cut losses quickly) but without actually agreeing on fundamental principles.
(iii) Nonviolent coercion: the actions of the activist or circumstances are such that the powerholders have no choice but to give in.
(iv) Disintegration: the movement or situation further causes the relevant powerholder institutions to fall apart.

Of the above, conversion is the most sustainable and preferred form of capitulation. The coalition can focus special efforts on winning over sympathizers among the powerholders or people who work within their institutions, such as the administrative staff (Chenoweth, 2011, p. 65).

[20] New Tactics for Human Rights provides a list of resources to help activists maintain strong physical and emotional welfare at https://www.newtactics.org/conversation/self-care-activists-sustaining-your-most-valuable-resource.

These sympathizers can slowly tip the balance within the powerholders' position from the inside. The ability to reach these powerholders increase when diversity of the participants is wider, as there may exist kinship ties and other social networks linking the elite and one's support base (Chenoweth, 2011, p. 69).

Activists can also enhance their chances of fostering conversion by reducing the 'social distance' between them and the powerholders. This will mean a willingness to be open and truthful with the powerholders through personal contact, being constructive in resolving issues and willing to make sacrifices, and also refraining from actions that intentionally shame or embarrass the powerholders (Sharp, 2013, p. 123).

8. Conclusion

Social mobilisation for Ganga has been in the process for several years. We have attempted to propose alternative approaches to social mobilisation for the river, to disrupt the monotony of tactics used thus far.

We believe that the innovation in our research lies in the high degree of customisation to India's socio-political-economic conditions and Ganga's unique context. While principles and strategies for organising movements abound, there are few normative studies, which purposefully design a movement for a specific environmental asset. Research and writings in this area can go a long way in helping activists, by customising broad principles to their needs at hand.

In our treatment of activism for Ganga, we have chosen to focus on and discuss a few aspects.

- At the core of our recommendations is the necessity of a trigger event to propel Ganga's pollution issues into public awareness.
- The trigger event will be based on a certain impending urgent threat.
- We identified health risks as this threat, due to its power to invoke greater empathy and moral responsibility to act.
- To deliver the movement well, the activist organisation must represent a broader swathe of the population to adequately reflect concerns from the ground and to draw on new sources of skills.

- A more inclusive leadership structure in the form of a broad coalition is the most ideal form, given the existence of already-strong activist groups for Ganga.
- For the trigger event to be effective, a massive diverse support base is needed. Special efforts must be made engage meaningfully and reduce the cost of participation for the population at large and to reach out to target groups.
- Beyond fasts and dharnas, a wider repertoire of tactics including legal activism and creative protests should be drawn on, to employ the element of surprise on powerholders and to reduce protest fatigue.
- Remembering the needs of one's supporters is critical. Periodic wins and accommodating differentiated participation are essential to sustain participant morale in the long run.

For years, activist groups have worked tirelessly to highlight pollution issues and to take the government to task. They have been instrumental in bringing awareness of Ganga pollution to the inflexion point it is at today. It is our hope that the tactics informed in this chapter will serve as a useful resource to escalate the movement beyond that point and take it on an exponential path.

While our writing focuses only on Ganga, the recommendations may be applicable to other environmental issues in India. A ripple effect may also be caused by successful social mobilisation for Ganga to change mindsets and behavioural traits in India towards ecological protection. The plight of environmental degradation may well turn on its head with the success of the Ganga movement.

Annex 1: Major Players in the Clean Ganga Movement

Organisation/Background	Declared Mission/ Objectives	Demand/s	Approaches and Tactics	Accomplishments
Sankat Mochan Foundation — Established in 1982 and included both citizens and engineers in Varanasi http://www.sankatmochanfoundationonline.org/	• No sewage water in Ganga • River's physical purity • Holding government/ contractors accountable for their actions • Informing the public of environmental awareness	• Stop the flow of sewage water into Ganga • Improve the condition of the Ganga • Raise the environmental awareness	• Engaging government bodies (i.e. submission of proposal for waste water management) • Conducting water quality research • Public outreach campaigns (i.e. World Water Day) • Education and health care programs for the less privileged members of society	• Proposed an innovative solution to manage wastewater • Submited 37 mld Ramana — Detailed Project Report to government in 2010 • Point sources of pollution survey about pouring in River Ganga, River Asi and River Varuna

- Jointly established an initiative "Swatcha Ganga Environmental Education Centre" with Oz-GREEN/ FOG, Australia in 1998
- Jointly prepared a report on "Feasibility Study of Interceptor Sewers and AIWPS Technology for Prevention of Pollution of Ganga at Vatranasi" with Oswald-Green LLC recommended and submitted to the Government for GAP Phase- II

(*Continued*)

Annex 1: (*Continued*)

Organisation/Background	Declared Mission/ Objectives	Demand/s	Approaches and Tactics	Accomplishments
Eco Friends — A Kanpur-based NGA founded in 1995 by Rakesh K. Jaiswal http://www.ecofriends.org/	• Protecting River Ganga from various pollution • Raising mass awareness and deepening public debate about pollution-related issues at both national and international level	• Act on polluting industries	• Legal activism through public interest litigation • Public awareness campaigns • Mobilisation of local community in sustained high-pitched campaigns	• Exposed the failures of GAP implementation and put civic agencies on their toes • Formed Ganga Ambassadors in various educational institutions of Kanpu • Established Ganga Vahini (Ganga Task Force) in 20 villages • Sought judicial intervention successfully in many cases involving environmental abuse and violation

				• Created a broad network of concerned policy makers, religious leaders, opinion builders/eminent people, media and NGOs • Obtained landmark judicial pronouncements to check river pollution and initiation of remedial stages by industry and government alike
MC Mehta Environmental Foundation — A non-profit NGO founded by MC Mehta, a public interest lawyer and environmentalist http://www.mcmef.org/	• Create an interactive movement for environmental and social justice at national level	• Act on public interest cases	• Helping victims of pollution environmental degradation through public interest litigation	• Courtroom aside every Friday only for Mehta's cases

(*Continued*)

Annex 1: *(Continued)*

Organisation/Background	Declared Mission/ Objectives	Demand/s	Approaches and Tactics	Accomplishments
	• Provide a forum for concerned citizens, NGOs and activists working for the survival of living beings • Strengthen environmental law and policy through training and sharping the skills of young lawyers		• Creating public awareness and implementing community-oriented research and training programs	• 212 small factories nearby the Taj Mahal were ordered to close by Supreme Court because there were no pollution control devices installed • Factories along the river directed to install pollution control devices and 300 factories were closed. Around 250 towns and cities in the Ganges Basin have been ordered to set up sewage treatment plants.

Panchganga Foundation — An organization working for protection and preservation of environment http://www.panchganga.org.in/	• Generating awareness of the ills that are stealing away the sheen of our rivers • Promoting the significance of the rivers, its tributaries through evaluating the environmental status based on the ecological impact on the rivers • Involving the public in activities of fostering the protection, enhancement and establishment of clean rivers and sustainable development along the river banks	• Providing a cleaner environment through clean rivers	• Identifying, evaluating and addressing threats to biological, cultural, and economic components of conservation of the river and riparian areas

(*Continued*)

Annex 1: *(Continued)*

Organisation/Background	Declared Mission/ Objectives	Demand/s	Approaches and Tactics	Accomplishments
Ganga Action Parivar — Founded in 2010 and a broad combination including professionals, engineers, scientists, activists, spiritual leaders, environmental specialists http://www.gangaaction.org/	• Restore, protect and maintain River Ganga and her tributaries in their free-flowing and unpolluted state and to explore feasible solutions to both pollution and obstruction problems	• Waste management • Restoration of flow • Agricultural management • Renewable energy • Ecological management • Awareness and education	• Connecting differences to work as one family • Providing responsible and sustainable solutions and action-based plans • Rights-based approach	
Save Ganga Movement — A non-violent movement and founded in 1998 by Smt Rama Rauta, president of the National Womens Organization (NOW) and member of the NGRBA http://www.savegangamovement.org/	• Generate mass awareness for an eco-friendly non-violent culture of development for protecting life-sustaining natural systems in general	• Ten Demands http://www.savegangamovement.org/index.php?option=com_content&task=view&id=30&Itemid=46	• Holding rallies, yatra and marches • Conducting of seminars, workshops, panel discussions	• Presented the charter of Ten Demands in the form of an appeal to Save The Ganga & Himalayas • Government accepted in totality the Charter of Ten

	especially the sacred Ganga and the Himalayas • Put moral pressure on government to take significant steps to fully and permanently save the Ganga		Demands and took up a National Mission for Saving the Himalayan Ecosystem as part of the National Action Plan for Climate Change (Activity in 2009)

Source: Compiled by Authors Base on Information from the Organisations' Websites.

Bibliography

4 ministries to work together for cleaning up the Ganges (Deccan Herald). Available at http://www.deccanherald.com/content/411972/4-ministries-work-together-cleaning.html. Accessed on 8 June, 2015.

A Magsaysay Laureate Wins The Siwi 2015 Stockholm Water Prize (Rmaf). Available at http://www.rmaf.org.ph/newrmaf/main/community/announcement/page/1/view/55. Accessed on 8 June, 2015.

Abraham, W-R (2011). Megacities as Sources for Pathogenic Bacteria in Rivers and Their Fate Downstream. *International Journal of Microbiology*, 1–13.

Agarwal, MG (2008). *Freedom fighters of India, Volume 4*. Gyan Publishing House.

Agrawal, A *et al.* (2010). Water Pollution with Special Reference to Pesticide Contamination in India. *Journal of Water Resource and Protection*, 2 (5), 432–448.

Ahammad, *et al.* (2014). Increased Waterborne $bla_{\text{NDM-1}}$ Resistance Gene Abundances Associated with Seasonal Human Pilgrimages to the Upper Ganges River. *Environmental Science & Technology*, 48 (5), 3014–3020.

Ajmal (2014). Regaining the power of Student Activism. *The Companion*. http://thecompanion.in/regaining-the-power-of-student-activism/.

Alley, KD (2002). *On the Banks of the Ganga: When Wastewater Meets a Sacred River*. University of Michigan Press.

Amudha (2008). Civil Society Activism in India: Political Opportunities and Participation of Women in Empowerment Processes: A Study of CASA in Tamilnadu, India. Paper presented at International Political Science Association Conference, Montreal. Available at paperroom.ipsa.org/papers/paper_36881.pdf. Accessed on 15 May 2015.

Bakir, V (2006). Policy agenda-setting and risk communication: Greenpeace, Shell and issues of trust. *Harvard International Journal of Press/Politics*, 11 (3), 67–88.

Battacharya, S (2014). Effects of Idol Immersion on the Water Quality Parameters of Indian Water Bodies: Environmental Health Perspectives. *International Letters of Chemistry, Physics and Astronomy*, 39, 234–263.

Bhargava, DS (1992). Why the Ganga (Ganges) could not be cleaned. Environmental Conservation, 19 (2), 170–172. doi:10.1017/S0376892900030666.

Bhattacharyya, D (2004). *Interrogating Social Capital: The Indian experience*. Thousand Oaks, California, New Delhi: Sage.

Bilgrami, KS and S Kumar (1998). Bacterial Contamination in Water of the River Ganga and Its Risk to Human Health. *International Journal of Environmental Health Research*, 8 (1), 5–13.

Bolling, DM (1994). *How to Save a River: A Handbook for Citizen Action*. Island Press.

Boy, B (2000). The Right to Information: India's Struggle Against Grassroots Corruption. *United Nations Chronicle*, 37 (1), 86.

Budruk, M *et al.* (2009). Urban Green Spaces: A Study of Place Attachment and Environmental Attitudes in India. *Society & Natural Resources: An International Journal*, 22 (9), 824–83.

Callicott, JB (1994). Earth's Insights: A Multicultural Survey of Ecological Ethics from the Mediterranean Basin to the Australian Outback. University of California Press.

Campaign India (12 October 2011). Double Standards: Why is 'Agenda Setting in Media' Significant? http://www.campaignindia.in/Article/276406,double-standards-why-is-8216agenda-setting-in-media8217-significant.aspx. Accessed on 29 June 2015.

Chardha, M and H Summer (2015). The Writing Is on the Wall, or Is It? Exploring Indian Activists' Beliefs About Online Social Media's Potential for Social Change. *International Journal of Communication*, 9, 672–693.

Chattopadhyay, S (2011). Online Activism for a Heterogenous Time: The Pink Chaddi Campaign and the Social Media in India. *Proteus: A Journal of Ideas,* 27 (1), 633–68.

Chaudhary, J (2014). India's plan to clean up the Ganga River is flawed, say experts." *Chinadialogue*. Available at https://www.chinadialogue.net/article/show/single/en/7395-India-s-plan-to-clean-up-the-Ganga-River-is-flawed-say-experts. Accessed on 9 June, 2015.

Chenoweth, E and J Maria (2011). *Why Civil Resistance Works — The Strategic Logic of Nonviolent Conflict*. New York: Columbia University Press.

Couch, SR and K-S Steve. Environmental Movements and Expert Knowledge: Evidence for a New Populism. *Illness and the Environment: A Reader in Contested Medicine*, Steve, K-R and P Brown (eds). NYU Press.

Cut water, power supply to industries polluting Ganga: SC. *The Hindu*. Available at http://www.thehindu.com/sci-tech/energy-and-environment/cut-water-power-supply-to-industries-polluting-ganga-sc/article6545930.ece. Accessed on 3 June 2015.

Das, P and K Tamminga (2012). The Ganges and the GAP: An Assessment of Efforts to Clean a Sacred River. *Sustainability*, 4 (8), 1647–1668, doi:10.3390/su408164.

Das, S (2011). Cleaning of the Ganga. *Journal of the Geological Society of India*, 78 (2), 124–130, doi:10.1007/s12594-011-0073-9.

Das, S (2014). Ganga — Our Endangered Heritage. In *Our National River Ganga: Lifeline of Millions*, Sanghi, R (ed.), pp. 45–71. Cham:Springer.

Deacon, D and G Peter (1994). *Taxation and Representation: The Media, Political Communication and the Poll Tax*. London: John Libbey.

Delicath, J and K Deluca (2003). *Image Events, the Public Sphere, and Argumentative Practice: The Case of Radical Environmental Groups. Argumentation*, 17, 315–333.

DeLuca, KM (2005). *Image Politics: The New Rhetoric of Environmental Activism*. Psychology Press.

Department Associates, Lucknow (1996). NGOs/Civic Societies and Urban Environmental. In Advocacy. *Urban Environmental Management, The Indian Experience*, Singh BN, S Maitra and R Sharma (eds.), p. 70. Human Settlement Management Institute (HUDCO).

Dixit, K (2012). Journalism's Important Public Service Remit. *Media Helping Media*. Available at http://www.mediahelpingmedia.org/training-resources/advanced-journalism/641-the-importance-of-journalisms-public-service-remit. Accessed on 29 June 2015.

Do, Quay-Toan *et al.* (2014). *Pollution Externalities and Health*: A Study of Indian Rivers. *Working Paper*, 3 July 2014.

Drew, G (2014). Mountain Women, Dams, and the Gendered Dimensions of Environmental Protest in the Garhwal Himalaya. *Mountain Research and Development*, p. 239.

Dreze, J and A Sen (2002). *India: Development and Participation*. Oxford University Press.

Duncombe, S (2004). Carnivals against Capitalism: Culture and Politics in Contemporary Activism. *Pugwash Mexico Magazine*, 1, 1–14.

East Boston Times-Free Press. Available at http://www.eastietimes.com/2015/05/06/dr-gene-sharp-nominated-by-nobel-peace-prize-comm/. Accessed on 17 May 2015.

eUttaranchal (2012). Three Ganga authority members resign against government inaction. *www.euttaranchal.com*. Accessed on July 7, 2015. Available at http://www.euttaranchal.com/news/government-and-politics/threeo-ganga-authority-members-resign-against-government-inaction.html.

Faces of Varanasi, Arts/Humanities. Available at https://www.facebook.com/facesofvaranasi?fref=ts. Accessed on 10 June 2015.

Freedom House (2013). Freedom of the Press: India 2013. Available at https://freedomhouse.org/report/freedom-press/2013/india#.VZEMyWCZbww. Accessed on 28 June 2015.

Friends of Narmada. "Who are we?". Available at http://www.narmada.org/about-us.html. Accessed on 30 June 2015.

Gandhi, M (2008). *The Essential Writings*. Oxford University Press.

Ganga, Its Activism and Media, Issue Brief, Asia Foundation, September 2013.

Gerbaudo, P (2015). Protest Avatars as Memetic Signifiers: Political Profile Pictures and the Construction of Collective Identity on Social Media in the 2011 Protest Wave. *Information, Communication & Society*, 1–14.

Gladwell, M (2006). *The Tipping Point: How Little Things Can Make a Big Difference*. Little, Brown and Company.

GoI (2012). *Press Note on Poverty Estimates, 2011–12*. Press Note, Government of India.

Guha, R (1998). Ideological Trends in Indian Environmentalism. *Economic and Political Weekly*, 23 (49), 2578–2581.

Guha, R (2013). The Past and Present of Indian Environmentalism. *The Hindu*, 27 March 2013.

Guha, R and JM Alier (1997). Varieties of Environmentalism: Essays North and South. London: Earthscan Publications.

Guin Seguin, *et al.* (1998). Toward a Model of Environmental Activism. *Environment and Behaviour*, 30 (5), September 1998, 628–652.

Gupta, PH (2008). From Chipko to Climate Change. *Mountain Research and Development*, 28 (1), 4–7, doi:10.1659/mrd.0968.

Hamner, S *et al.* (2006). The Role of Water Use Patterns and Sewage Pollution in Incidence of Water-Borne/Enteric Diseases along the Ganges River in Varanasi, India. *International Journal of Environmental Health Research*, 16 (2), 113–132.

Hamner, S *et al.* (2007). Isolation of Potentially Pathogenic Escherichia coli O157:H7 from the Ganges River. *Applied and Environmental Microbiology*, 2369–2372.

Hamner, S, SC Broadaway, VB Mishra, A Tripathi, RK Mishra, E Pulcini, BH Pyle and TE Ford (2007). Isolation of Potentially Pathogenic Escherichia coli O157:H7 from the Ganges River. *Applied and Environmental Microbiology*, 2369–2372.

Harcup, T and D O'Neill (2001). What is News? Galtung and Ruge Revisited. *Journalism Studies*, 2 (2), 261–280.

Harriss, J (2007). Antinomies of Empowerment: Observations on Civil Society, Politics and Urban Governance in India. *Economic and Political Weekly*, 42 (26), 2716–2724.

Human Rights Watch (2014). *India: Stop Harassing Anti-Protesters.* News Report, New York: Human Rights Watch. Available at http://www.hrw.org/news/2014/09/23/india-stop-harassing-anti-nuclear-protesters. Accessed on June 8, 2015.

Indian Institutes of Technology (2011). Prevention of River Pollution by Urban Sewage. Recommendations from Policy and Governance Perspective Based on a Model Case Study. "Prevention of River Pollution by Urban Sewage. Recommendations from Policy and Governance Perspective Based on a Model Case Study." 29.

Internet World Stats. Asia Internet Use, Population Data and Facebook Statistics. Available at http://www.internetworldstats.com/stats3.htm. Accessed on 30 June 2015.

Jolly, A (2014), "indiatoday.in." *Can India's sacred river Ganga be saved?* July 3. Available at http://indiatoday.intoday.in/story/ganga-river-varanasi-contamination-narendra-modi-cremation-uma-bharti/1/369741.html.

Jurgen, G and D Rucht (1992). *Mesomobilization: Organizing and Framing in Two Protest Campaigns in West Germany.* American Journal of Sociology, University of Chiago.

Jyoti, A *et al.* (2009). Contamination of Surface and Potable Water in South Asia by Salmonellae: Culture-Independent Quantification with Molecular Beacon Real-Time PCR. *Science of the Total Environment*, 1–8.

Kalland, A and G Persoon (2013). *Mahatma Gandhi and the Environmental Movement in India. Environmental Movements in Asia.* Routledge.

Klandermans, B (2008). The Demand and Supply of Participation: Social-Psychological Correlates of Participation in Social Movements. *The Blackwell Companion to Social Movements.* Snow, David A. *et al.* (eds) John Wiley & Sons.

Klandermans, Jacquelien van Stekelenburg and Bert (2010), "The social psychology of protest." *sociopedia.isa.*

Krishnatray, P and R Gadekar (2014). Construction of Death in H1N1 News in *The Times of India. Journalism*, 15 (6), 731–753.

Kuran, T. (1991). Now out of never: The element of surprise in the east European revolution of 1989. *World Politics*, 44 (1), 7–48.

Lata, P *et al.* (2009). Enterococci in River Ganga Surface Waters: Propensity of Species Distribution, Dissemination of Antimicrobial-Resistance and Virulence-Markers Among Species Along Landscape. *BMC Microbiology*, 9 (140), 1–10.

Lee D, (2015). The Role of Social Media in Occupy Protests, on the Ground and Around the World. South China Morning Post, International Edition. Available at http://www.scmp.com/news/hong-kong/article/1628305/role-social-media-occupy-protests-ground-and-around-world. Accessed on 10 June 2015.

Lewicka, M (2005). Ways to Make People Active: The Role of Place Attachment, Cultural Capital, and Neighbourhood Ties. *Journal of Environmental Psychology*, 381–395.

Low, D and A Gill (2014). Singapore must ease 'bandwidth tax' on the poor. *todayonline.com.* http://www.todayonline.com/singapore/spore-must-ease-bandwidth-tax-poor. Accessed on 6 June 2015.

Lynds, J (2015). Dr. Gene Sharp Nominated by Nobel Peace Prize Comm. East Boston Times — Free Press. Available online www.eastietimes.com/2015/05/06/dr_gene-sharp-nominated-by-nobel-peace-prize-comm/. Accessed on 1 May 2017.

Malik, A (2003). Ghanshyam Shah. Social Movements in India: A Review of Literature. *Pakistan Development Review*, 42 (2), p. 175.

Mallet, V (2015). The Ganges: Holy, Deadly River. FT.Com.

Manjari, M (2014). Rs 20,000 crore spent in 28 years, Ganga still a flowing mess. *The Times of India*, May 14. Available at http://timesofindia.indiatimes.com/india/Rs-20000-crore-spent-in-28-years-Ganga-still-a-flowing-mess/articleshow/35090392.cms. Accessed on 1 June 2015.

Mawdsley, E (2003). Hindu Nationalism, Neo-Traditionalism and Environmental Discourses in India. *Geoforum*, 37, 380–390.

Mawdsley, E (2004). India's Middle Classes and the Environment. *Development and Change*, 35 (1), 79–103.

Mazumdar, S and S Mazumdar (1993). Sacred space and place attachment. *Journal of Environmental Psychology*, 13 (3), 231–242, doi:10.1016/S0272-4944(05)80175-6.

McAnally, EA (2007). Toward a Philosophy of Water: Politics of the Pollution and Damming Along the Ganges River. Ph.D. Dissertation, ProQuest, UMI Dissertations Publishing 28.

McCombs, M (2014). *Setting the Agenda: Mass Media and Public Opinion*, 2nd Edition. John Wiley & Sons.

McLeod, DM and JK Hertog (1999). Social Control, Social Change and the Mass Media's Role in the Regulation of Protest Groups. In *Mass Media, Social Control and Social Change: A Macrosocial Perspective*, Demers D and Viswanath K (eds.). Iowa State University Press.

Meena, M. Uma warns Ganga polluters. *The Hindu*. Available at http://www.thehindu.com/news/national/uma-warns-ganga-polluters/article6482423.ece. Accessed on 30 May, 2015.

Mishra, NK and SC Mohapatra (2009). Gangetic Pollution on Water Borne Diseases in Varanasi: A Case Study. *Indian Journal of Preventive & Social Medicine*, 40 (1 & 2), pp. 39–42.

Moyer, B (1987). The Movement Action Plan: A Strategic Framework Describing The Eight Stages of Successful Social Movements. *History is a Weapon*. http://www.historyisaweapon.com/defcon1/moyermap.html.

Naik, SR *et al*. (1992) A Large Waterborne Viral Hepatitis E Epidemic in Kanpur, India. *Bulletin of the World Health Organization*, 70 (5), 597–604.

Nambiar, P (2014). Media Construction of Environment and Sustainability in India. SAGE Publications India.

Narula, S (2008). The Story of Narmada Bachao Andolan: Human Rights in the Global Economy and the Struggle Against World Bank. *New York Public Law and Legal Theory Working Papers*, New York University School of Law, 17 December 2008.

Nautiyal, CS (2008). Self-Purificatory Ganga Water Facilitates Death of Pathogenic Escherichia coli O157:H7. *Curr Microbiol* (Curr Microbiol), pp. 25–29.

Nautiyal, CS (2009). Self-Purificatory Ganga Water Facilitates Death of Pathogenic Escherichia coli O157:H7. *Current Microbiology*, 58 (1), 25–29.

New Tactics in Human Rights. Self-Care for Activists: Sustaining Your Most Valuable Resource. Available at https://www.newtactics.org/conversation/self-care-activists-sustaining-your-most-valuable-resource. Accessed on 1 June 2015.

Ninen, S (1992). DD Social Service Ads Preach, But Rarely Convert. News article, Down to Earth.

Norvell, L (1996). Gandhi and the Indian women's movement. *British Library Journal* 14.

Oskamp, S and WP Schultz (2005). Attitudes and Opinions. Psychology Press.

Palmer, T (2004). *Endangered Rivers and the Conservation Movement*, Second Edition. First Rowman & Littlefield Publishers.

Pandey, S (2014). Ganga Clean-up: Pollution Boards a Story of Complete Failure, Frustration, Says Supreme Court. *ndtv.com*. Accessed on 2 July 2015. Available at http://www.ndtv.com/india-news/ganga-clean-up-pollution-boards-a-story-of-complete-failure-frustration-says-supreme-court-686032.

Parker-Pope, T (2008). Psychiatry Handbook Linked to Drug Industry. *The New York Times*. Available at http://well.blogs.nytimes.com/2008/05/06/psychiatry-handbook-linked-to-drug-industry/?_r=0. Accessed on 26 May 2015.

Passy, F and M Giugni (2001). Social Networks and Individual Preferences: Explaining Differential Participation in Social Movements. *Sociological Forum* (University of Geneva), pp. 123–153.

Paul, A (2011). Anti-corruption drive gathers storm in Bangalore. News, Bangalore: DNA. PTI. Delhi Metro stations near PM's residence closed. News, New Delhi: Times of India.

Poushter, J (2014). In India, Few Use Social Media to Share Political Views. *Pew Research Centre*. Available at http://www.pewresearch.org/fact-tank/2014/

04/08/in-india-few-use-social-media-to-share-political-views/. Accessed on 9 June 2015.

Ram, FA Singh and U Ram (2013). Human rights approach to maternal & child health: Has India fared well? Vol. 137, Medknow Publications and Media Pvt. Ltd. Available at http://lawmin.nic.in/olwing/coi/coienglish/Const.Pock%202Pg.Rom8Fsss(6).pdf. Accessed on 9 June 2015.

Ram, S *et al.* (2007). Prevalence of Multi-Antimicrobial-Agent Resistant, Shiga Toxin and Enterotoxin Producing Escherichia coli in Surface Waters of River Ganga. *Environmental Science & Technology*, 41 (21), pp. 7383–7388.

Ramos, H. (2008). Opportunity for Whom?: Political Opportunity and Critical Events in Canadian Aboriginal Mobilization, 1951–2000, Vol. 87. Oxford University Press.

Rastogi, SK *et al.* (2008). Occupational Health Risks Among the Workers Employed in Leather Tanneries at Kanpur. *Indian Journal of Occupational and Environmental Medicine*, 12 (3), 132–135.

Rodrigues, U (2014). Social media's impact on journalism: A study of media's coverage of anti-corruption protests in India. *Global Media Journal*, 8 (1), 1–10.

Rodrigues, UM (2013). Intermedia Agenda Setting. *The Hoot.* Available at http://thehoot.org/web/Intermediaagendasetting/6546-1-1-9-true.html. (Accessed on 29 June 2015).

Save the Mekong Coalition. Available at www.savethemekong.org. Accessed on 10 June 2015.

Sengupta, M (2014). Anna hazare's anti-corruption movement and the limits of mass mobilization in India. Social Movement Studies, 13 (3), 406–413, doi:10.1080/14742837.2013.860025.

Shah, G (2004). *Social Movements in India.* Sage Publications India.

Sharma, P *et al.* (2012). Groundwater Contaminated with Hexavalent Chromium [Cr (VI)]: A Health Survey and Clinical Examination of Community Inhabitants (Kanpur, India) PLOS ONE. 7 (10), October 2012, e47877, 1–7.

Sharma, P, V Bihari, SK Agarwal, V Verma and CN Kesavachandran (2012). *Groundwater Contaminated with Hexavalent Chromium [Cr (VI)]: A Health Survey and Clinical Examination of Community Inhabitants (Kanpur, India).* Article, PLoS ONE.

Sharp, G (1973 - 2). *Politics of Nonviolent Action, Part Two: The Methods of Nonviolent Action.* Boston, Porter Sargent.

Sharp, G (2013). *How Nonviolent Struggle Works.* The Albert Einstein Institution.

Shepard, B, LM Bogad and S Duncombe (2008). Performing vs. the Insurmountable: Theatrics, Activism, and Social Movements. *Liminalities: A Journal of Performance Studies*, 4 (3), 1–30.

Sidney, MS (2001). Images of Race, Class, and Markets: Rethinking the Origin of U.S. Fair Housing Policy. *Journal of Policy History*, 13(2), 181–214. doi:10.1353/jph.2001.0006.

Singh, B (2015). Water quality of river Ganga has an impact on human health: TERI, *The Times of India*, June 3. Available at http://timesofindia.indiatimes.com/city/varanasi/Water-quality-of-river-Ganga-has-an-impact-on-human-health-TERI/articleshow/47529071.cms. Accessed on 10 June, 2015.

Singh, R (2014). New Citizen's Activism in India — Moments, Movements, and Mobilisation. Centre for Democracy and Social Action (CDSA).

Singh, RP (2001). Effect of Wastewater Disposal and Extent of Industrial Pollution In and Around Kanpur, Uttar Pradesh, India. *Bulletin of Engineering Geology and the Environment,* 60 (1), 31–35.

Sorensen, MJ (2008). Humor as a Serious Strategy of Nonviolent Resistance. *Peace & Change*, 33 (2), 167–190.

Taylor, A (2012). *Violent Protests in India Over Rape Case.* News Report, Delhi: The Atlantic.

Terra, G (2014). Persistent Organic Pollutants: The Silent Killers. 7 (6), 26–33.

Thakkar, H (2009). Water Pollution Control: Role of Community and Public Action. In *Environmental Issues in India: A Reader*, Rangarajan, M (ed.). Pearson Education India.

The Economist (27 August 2011). I, the people. Available at http://www.economist.com/node/21526904. Accessed on 27 June 2015.

The Guardian (28 December 2012). Indian victim of gang rape dies in hospital in Singapore. Available at http://www.theguardian.com/world/2012/dec/28/indian-gang-rape-dies-singapore. Accessed on 27 June 2015.

The New York Times (2014). 'Superbugs' Kill India's Babies and Pose an Overseas Threat, 3 December 2014.

The Times of India (8 December 2011). Social Media is the Space for Change, Says the Youth of India. Available at http://timesofindia.indiatimes.com/tech/social/Social-media-is-the-space-for-change-says-the-youth-of-India/articleshow/11031467.cms. Accessed on 20 June 2015.

The Ugly Indian. "Is There Any Hope?". Available at http://www.theuglyindian.com/intro1.html. Accessed on 5 June 2015.

Tuchman, G. (1978). *Making News: A Study in the Construction of Reality.* New York, NY: Free Press.

Uday, S (25 June 2008). Agenda-Setting in Indian Newspapers?. Available at *IBN Live*. http://www.ibnlive.com/blogs/india/uday-sahay/agenda-setting-in-indian-newspapers-10957-742571.html. Accessed on 29 June 2015.

Wakefield, SEL (2005). Taking Environmental Action: The Role of Local Composition, Context, and Collective. *Environmental Management*, 37 (1), 40–53.

Weber, T (2002). Gandhian Nonviolence and the Salt March. *Social Alternatives*, 21 (2), 46–51.

Wilson, A (2013). India's anti-rape movement: redefining solidarity outside the colonial frame. Available at https://www.opendemocracy.net/5050/amrit-wilson/indias-anti-rape-movement-redefining-solidarity-outside-colonial-frame. Accessed 20 May, 2015.

World Bank (2010). *Communications for Governance and Accountability Program*, Coalition Building. Available at http://siteresources.worldbank.org/EXTGOVACC/Resources/CoalitionBuildingweb1.pdf. Accessed 10 June 2015.

Zeynep (2014). *Online Social Change: Easy to Organize, Hard to Win:* TEDGlobal 2014. Available at http://www.ted.com/talks/zeynep_tufekci_how_the_internet_has_made_social_change_easy_to_organize_hard_to_win?language=en. Accessed on 10 June 2015.

Chapter 6

WHITHER PPPs?

SUSTAINING SEWAGE INFRASTRUCTURE IN GANGA BASIN

Vishal Kapoor, Xuan Xiao, Zhang Jiapeng, Zhao Huaning

This chapter discusses the problems confronting India in executing the public-private partnerships that are identified as a major instrument in tackling untreated human wastewater pollution in Ganga. Based on the analysis and identification of core issues through a framework of cause-effect diagram and vicious loops, the authors suggest broad-based solutions from the macro policy perspective to ensure sustainability of the 3Ps.

Untreated sewage disposal into the river system has been identified as one of the main sources of point source pollution to the Ganga River. The Central Pollution Control Board (PCB), which regulates environmental quality in India, estimates that there is a 55% gap between municipal sewage generation and current installed capacity. In addition, evidence is available that even the currently installed sewage treatment plants (STPs) are not operating efficiently due to resource constraints of the local governments which are responsible for operating them.

Given the resource constraints of public agencies, there is general consensus that public-private partnerships (PPPs) may bring benefits, offering private investment finance and greater efficiency in the construction, operation and maintenance of STPs over conventional public procurement. However, the water and sanitation sector in India has historically seen very low take up rates of PPPs in comparison to other sectors. Indeed, many previous attempts to engage private parties in this sector have been beset with problems of high risk perceptions on the part of investors, intense political opposition and on-going public debates. The financial stability of wastewater treatment PPP projects and suboptimal risk apportionment between the public and private partners undermine the stability of PPP contracts. At the same time, the conventional approaches of treating sewage and water as conjoined twins and the resulting poor cost recovery has only exacerbated the problems of low take-up rates of PPPs for wastewater.

As the government of India unveils its plans to inject INR 200 billion (US$3 billion) into the "Namami Gange" (Clean Ganga) mission, this chapter identifies the vicious cycles that impede the development of wastewater PPPs and suggests possible solutions for large scale rollout and sustenance of PPPs that would help restore the glory of the national river of India — the Ganga.

1. Introduction and Context

The Ganga is held to be of great spiritual value to the majority of India's people. The rivers of the Ganga basin provides for human water consumption needs of 46% of the national population.[1] The Ganga basin also forms 47% of the overall irrigable area of India.[2] However, the water quality data of the river indicates that in respect to certain parameters, the river water is not fit for human contact, let alone for potable use. According to a 2013 study by the Central Pollution Control Board (CPCB) on the pollution levels of river, the parameters Total Coliforms (TC), Fecal Coliform (FC), biological oxygen demand (BOD) and conductivity over the period 2002–11 have been consistently poor, that too by a several orders of magnitude.[3]

[1] Upadhyay, P (2014). National Mission for Clean Ganga, Address to LKYSPP/ NUS MPA students, New Delhi.

[2] Richard, H and I Hespanhol (1997). Water Pollution Control — A Guide to the Use of Water Quality Management Principles, WHO/UNEP, 1997. Available at http://www.who.int/water_sanitation_health/resourcesquality/wpccasestudy1.pdf.

[3] CPCB, (2013). Pollution Assessment: River Ganga, New Delhi: Central Pollution Control Board. Available at http://cpcb.nic.in/upload/NewItems/NewItem_203_Ganga_report.pdf.

Out of the four parameters listed, the first three are direct indicators of presence of untreated sewage and human waste in the river.

A few higher order effects in the ecosystem of the river Ganga can also be traced to sewage pollution. Low dissolved oxygen (DO) levels are a direct indicator of presence of sewage in the water. According to 2013 river water pollution data, dangerously low values of DO levels (less than 4 mg/dl) have been noted in many tributaries of the Ganga such as Gomti and Hindon rivers. [4] Scientific studies have indicated a significant decline in the species of fish present in the lower reaches of Ganga[5], where CPCB confirms that DO levels have reached dangerously low levels.[6] The anecdotal evidence of dead fish found in the middle reaches of Ganga in Uttar Pradesh is worrisome.[7]

It is plausible to infer that the Ganga is being polluted by untreated sewage discharge flowing into the river. Sewage treatment and disposal is a public utility service which has traditionally been managed by the government through public ownership of sewage networks and treatment plants through traditional public procurements (TPPs). Generally, operations and maintenance (O&M) of networks and STPs are directly carried out by local government. Ostensibly, the point source wastewater of Ganga is due to the failure of the conventional models of delivery of wastewater collection and treatment services.

Public-private Partnerships (PPPs) in public utilities are often touted as an alternative model to conventional procurement and delivery of public services. However, despite long-standing efforts of the Government of India (GoI) since the Ganga Action Plan initiatives of the 1980s, the take-up rate of the wastewater PPPs in India has been extremely limited. This chapter delves into the underlying reasons for this.

---

[4] Uttar Pradesh Pollution Control Board, River Water Quality in Uttar Pradesh; Averages Jan–Dec 2013, *UP Pollution Control Board*, n.d. Accessed on February 13, 2015.

[5] Vinod T, G Roy and P Bose, Ganga River Basin Management Plan: Interim Report Kanpur, Indian Institute of Technology, September 2013.

[6] CPCB, Pollution Assessment: River Ganga.

[7] Bhuvan, B (2014). Can Modi Clean the Ganges, India's Biggest Sewage Line?, *Yahoo News*, July 17, 2014. Available at http://news.yahoo.com/modi-clean-ganges-indias-biggest-sewage-line-032922309.html.

1.1. *Why untreated waste water discharge? — Limitations of the TPPs*

The problem of sewage pollution of water bodies in India is not specific to the river Ganga. National estimates suggest that only 5900 MLD of 22900 MLD (equivalent to 26%) generated by the country every day is treated before discharge into river.[8] In the Ganga basin, the CPCB evaluations confirm that almost 2723 MLD of sewage generated by 50 cities located along the river accounts for about 85% of Ganga's pollution load.[9]

Sewage Treatment Plants built by the government under traditional public procurements (TPP) fail to address the problem on two counts. Firstly, there is a huge gap between installed capacity and wastewater generation. In two separate assessments by the CPCB in 2009[10] and 2012[11], the gap between STP capacity and the actual requirement was around 55%. In addition, there is a large utilisation gap. In a 2013 assessment, 51 out of 64 STPs studied were found to be working at less than 60% of their capacity.[12] The reasons range from lack of reliable power supply to the plant, sewer line connectivity issues, and the lack of financial resources at the municipal level.[13]

Furthermore, the quality of wastewater discharged from STPs does not meet the requisite standards in many cases. In one evaluation by CPCB, 26 out of 92 STPs did not meet the requisite BOD standard.[14]

---

[8] MOEF GOI (2015). National Status of Waste Water Generation & Treatment, *National Status of Waste Water Generation @ Treatment*, June 8, 2015. Available at http://www.sulabhenvis.nic.in/Database/Nationalstatusofwastewatergenerationandtreatment_2090.aspx?format=Print.

[9] Narain, S (2014). Ganga: The River, Its Pollution and What We Can Do To Clean It. New Delhi; Centre for Science and Environment. Available at http://www.cseindia.org/userfiles/ganga-the-river-pollution.pdf. Accessed May 1, 2017.

[10] MOEF (2009). State of Environment Report, India 2009. New Delhi: Ministry of Environment and Forests, Govt of India.

[11] Central Pollution Control Board (July 2013). Pollution Assessment: River Ganga. New Delhi: Ministry of Environment and Forests, Governemnt of India. Available at http://cpcb.nic.in/upload/NewItems/NewItem_203_Ganga_report.pdf.

[12] Sunita, N (2014). Ganga: The River, Its Pollution and What We Can Do to Clean It. New Delhi: Centre for Science and Environment.

[13] Ibid.

[14] Kaur *et al.*, Wastewater Production, Treatment and Use in India, n.d.

Secondly, on the capacity-generation front through the TPP models, the governments again face severe funding constraints, in particular for the extension of sewerage networks. The competition for public funds in a developing country like India is very high and it would be unrealistic to assume a big shift in the funds allocated to wastewater in the future.

The Ganga River Basin Management Plan 2015 (GRBMP) estimates that the overall capital expenditure (capex) needed to provide sanitation services for the Ganga Basin, including the sewage network infrastructure and treatment facilities, is INR 2962 billion (US$44 billion) upfront that translates to an annualized figure of INR 709 billion (US$10.6 billion).[15] The GRBMP similarly estimates annual operating expenditure (opex) of INR 1151 billion (US$17 billion). In contrast to this, through the *Namami Gange* mission, the union (federal) government of India has pledged INR 200 billion (US$3 billion) to upgrade infrastructure.[16] The gap between the requirement and availability of funds remains very wide. Therefore, GoI has either to limit the scope of work and select only the most effective package of measures or find the means to mobilise other sources of funds. Nevertheless, the inherent issues of inefficiency of public provision of sewage infrastructure and delivery of services will still remain.

This brief overview points to the need to draw on private investment finance and to improve efficiency. It is here that PPPs enter the policy arena.

1.2. *The context of public-private partnerships*

Public-Private Partnerships are an alternative model for procurement of services, distinct from the traditional public procurement in that they often transfer design, financing and operating responsibility to the private party in addition to construction and equipment provision. According to the World Bank, PPPs in the water and sanitation sector have led to

---

[15] Consortium of 7 IITs, Ganga River Basin Management Plan-2015 | Mission 2: Nirmal Dhara Consortium of 7 IITs, January 2015. Available at http://gangapedia.iitk.ac.in/sites/default/files/Mission%202_ND.pdf.

[16] '*Namami Gange*,' Down To Earth. Accessed on June 8, 2015. Available at http://www.downtoearth.org.in/content/namami-gange.

improvements in affordability, equity, better project performance and improved coverage in developing countries.[17]

In India, the oft-repeated justifications for PPPs for water and waste-water services are:[18] (i) reduce upfront costs of government of public infrastructure (ii) lower whole-life costs of projects when construction and operation are bundled together (iii) greater efficiency and innovation by private companies (iv) reduce time and budget over-runs.

Despite continued high-level backing for PPP,[19] India's steps to introduce PPP have been slow. The Database of Infrastructure projects in India[20] indicates that out of a total of 3104 PPP projects undertaken in India up to mid-2015, only 20 have been in for wastewater, which translates to a mere 0.6% of the total number of projects. Wastewater projects even make up a small share of activity in the water and sanitation sector, accounting for 18% of PPP in the sector and just 6% in value terms.

This limited take-up contrasts with the very rapid development of PPP in China, Brazil and some other emerging markets. The World Bank–PPIAF Private Participation in Infrastructure database reports that India's share of water and sanitation PPPs in 2001–2012 constituted a mere 2% of the total for all developing countries in project numbers and 0.2% of investment value.[21]

Focusing specifically on the Ganga, Ganga Action Plans I and II envisaged the use of PPPs for development of STPs as early as the 1980s, yet like elsewhere, this was not taken up and STPs continued to be

[17] Philippe, M (2009). *Public-Private Partnerships for Urban Water Utilities: A Review of Experiences in Developing Countries*, Vol. 8, World Bank Publications.

[18] Dwivedi, G (2010). *Public-Private Partnerships in Water Sector: Partnerships or Privatisation?* Badhwan (MP) Manthan Adhyayan Kendra.

[19] DEA Govt of India (2015). Overview: Public Private Partnerships in India, *Public Private Partenrships in India*. Accessed on June 9, 2015. Available at http://www.pppinindia.com/overview.php.

[20] Department of Economic Affairs GOI (2015). Database of Infrastructure Projects in India, *Database of Infrastructure Projects in India*. Available at https://infrastructureindia.gov.in/search.

[21] Wu, X, RS House and R Peri (2016). Public-private partnerships (PPPs) in water and sanitation in India: Lessons from China. *Water Policy*, 18(S1), 153–176.

procured by conventional means. This commitment to the PPP model has been reaffirmed in the current initiative.[22] The National Mission for Clean Ganga (NMCG) initiated a market outreach effort with various private operators in April 2015 to explore PPP models.[23]

How can the low take-up rate of wastewater PPPs in India, and in the Ganga basin in particular, be explained? Even more surprisingly, why is PPP not taking off despite specific support for this under the Ganga Action programmes that have been in place for the last 25 years? In the following section, we suggest some of the factors that account for this, and how they interlink to create "vicious loops."

2. The Vicious Loops of 3Ps of Water and Sanitation Sector

2.1. *PPPs are not problem-free*

Certain factors are specific to the national context. The Indian Ministry of Finance identifies constraints on PPP development as follows:[24] "(i) Policy and regulatory gaps, specially relating to specific sector policies and regulations; (ii) Inadequate availability of long term finance (10-year plus tenor) — both equity and debt; (iii) Inadequate capacity in public institutions and public officials to manage PPP processes; (iv) Inadequate capacity in the private sector — both in the form of developer/investor and technical manpower; (v) Inadequate shelf of bankable infrastructure projects that can be bid out to the private sector; and, (vi) Inadequate advocacy to create greater acceptance of PPPs by the public."

---

[22] This has been reaffirmed by the Additional Mission Director/ NMCG on an email vide: Pushkal Upadhyay, "Re: Request for Information in Relation to PPP Projects on Ganga Cleanup Program 'Namami Gange,'" July 6, 2015.

[23] For details on the market outreach initiative brochure, please see: NMCG, "Clean Ganga | Market Conference," *National Mission for Clean Ganga*. Accessed on June 29, 2015. Available at https://nmcg.nic.in/writereaddata/fileupload/8_CleanGanga.pdf.

[24] DEA GOI, Scheme and Guidelines for Infrastructure Project Development Fund, Department of Economic Affairs, Ministry of Finance, Government of India. Accessed on June 9, 2015. Available at http://www.pppinindia.com/pdf/guideline_scheme_IIPDF.pdf.

These cross-sector national issue combine with sector specific factors. In a global context, the OECD identifies some of the common obstacles in water and sanitation PPPs as follows:

> *"(i) Conflicting aims (e.g., financial, macroeconomic, social and environmental); (ii) Award procedures (e.g., lack of transparency); (iii) Regulatory frameworks (e.g., the existing legislation in developing countries is usually absent in regulating private participation in public services); (iv) Public governance (e.g., conflicts of interests between central and local governments and poor coordination mechanism among regulators); (v) Incumbent existing service providers; and (vi) Political commitment in carrying out meaningful reforms (e.g. tariffs reforms aimed at cost recovery)".*[25]

In addition, first best solutions may often be unattainable due to lack of political acceptability, public opposition and administrative capacity.[26]

The next section maps out cause and effect of the various problems, highlighting the presence of vicious cycles — those that are self-reinforcing in nature, where effects feed into the causes of problems.

2.2. *Cause-effect diagrams and vicious cycles in Ganga PPP*

Figure 1 presents a cause-effect diagram of wastewater PPPs in the context of efforts to restore the Ganga. Two self-reinforcing vicious cycles are evident in the green and red loops. It can be seen that the cycles are interconnected to each other, creating further amplification effects. The subsequent sections describe the loops and identify possible interventions to break them.

[25] OECD (2005). Encouraging Public-Private Partnerships in the Utilities Sector: The Role of Development Assistance. In *International Investment Perspectives 2005*, OECD Publishing, Paris. DOI: http://dx.doi.org/10.1787/iip-2005-4-en.

[26] For details on the practicability of second best approaches over the first best approaches, please see: Dani Rodrik, A Practical Approach to Formulating Growth Strategies. *From the Washington Consensus Towards a New Global Governance*, 2008.

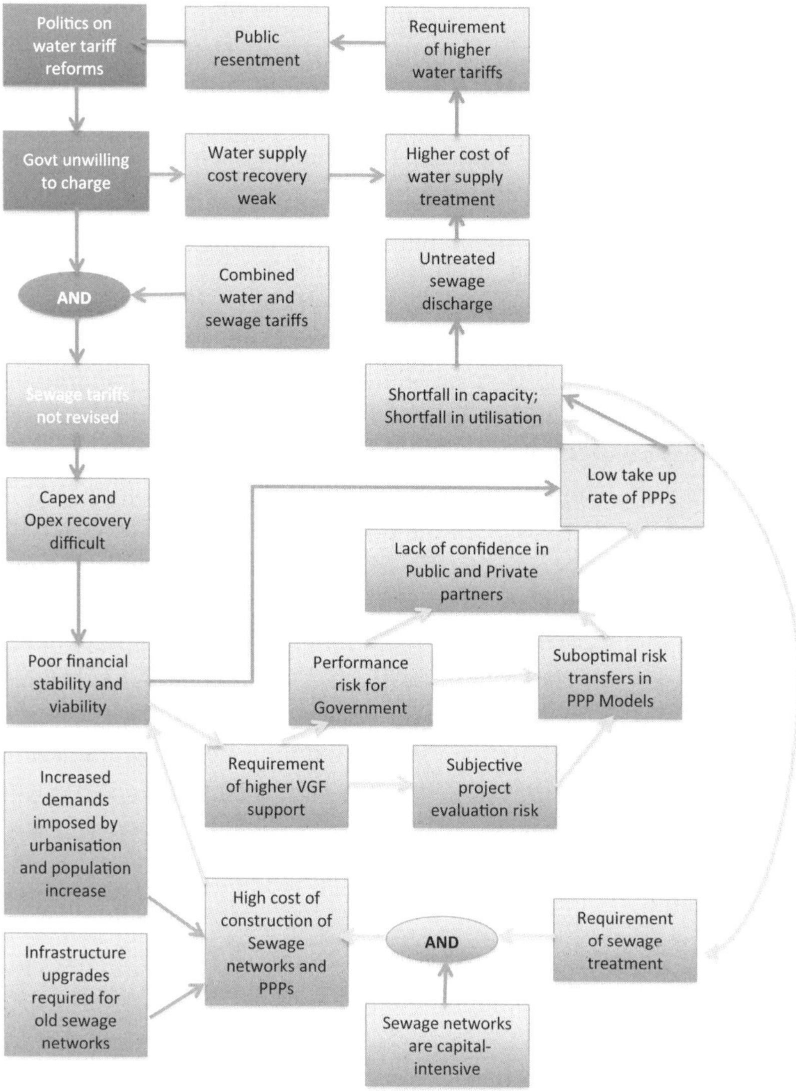

Figure 1: Cause-effect Diagram of Wastewater PPPs

2.2.1. *The red loop: Adverse effect of combined sewage and water tariffs*

The red loop relates to inadequate cost recovery for wastewater services. A low take-up rate of PPPs and the limitations of conventional procurement leads to insufficient treatment capacity, which leads to the untreated discharge and further deterioration of river quality. This, in turn, raises the costs of water treatment in localities which rely on the Ganga as a raw water source.[27] A higher cost of water treatment increases pressure to raise water tariffs to allow utilities to recover operating costs. Charging for water is resisted by some sections of the public and interest groups which characterise water as a human right and argue that the government should not charge for water.[28]

This, in turn, engenders political resistance or "unwillingness to charge," blocking the revision of the combined water and wastewater tariff. The lack of cost recovery tariffs undermines the financial stability of the utility and potential viability of PPPs, reinforcing the low take-up rate. This closes vicious cycle.

Combined water and sewage tariffs and unwillingness to charge by government makes the loop self-reinforcing. A possible intervention to break the cycle is to remove the external reinforcement of combined tariffs. This disruption will break the connection between government's unwillingness to charge for water with the level of the wastewater tariff. Section 3.2 develops this proposal.

2.2.2. *The green loop: financial viability & risk of PPP model*

The *green loop* identifies a self-reinforcing loop tying the high cost of construction of sewerage networks to the low take-up rate of PPPs.

---

[27] Sunita, N and P Pandey (2012). *Excreta Matters*, Vol. 7, New Delhi: Centre for Science and Environment.

[28] Dwivedi, G (2010). *Public-Private Partnerships in Water Sector: Partnerships or Privatization?*. Badwani (MP). Manthan Adhyayan Kendra. Available at http://www.manthan-india.org/wp-content/uploads/2015/04/PPPs-In-Water-Sector-Final-Book.pdf. Accessed May 1, 2017.

The double treatment capacity and capacity utilisation gaps were noted in Section 1.1. This signals a need for investment in both treatment capacity and in sewerage networks, potentially leading to competition for investment funds. At the same time, where sewerage networks do exist in India, they are often dilapidated and require major rehabilitation, repair and replacement.[29] Increasing urbanisation and population growth are impose even greater demands for extension of sewage facilities, leading to higher cost implications.[30]

Low tariff revenues and high investment costs, therefore, restrict the financial viability for PPPs. This implies a need for other sources of funding to fill this 'viability gap,' for example through transfers or subsidies from the federal government. The GoI has indicated that the viability gap funding (VGF) requirement for these projects could exceed 40%,[31] the current limit for VGF funding that applies to PPPs. It is not difficult to deduce that a higher VGF exposure could lead to a performance risk for the government subsequent to the completion of construction phase. This would also open a parallel subjective project evaluation risk to the private sector. An ensuing lack of confidence on both the private and public partners in the PPP would lead to a low take-up rate of PPP, and a shortfall in capacity generation essential to ensure treated wastewater discharge in the river Ganga. The resulting requirement for investment would feed into the high cost implications, thereby completing the vicious loop.

To break the loop, two possible interventions are feasible, which can be thought of in terms of supply and demand. Firstly, on the supply side, the costs of wastewater infrastructure construction need to be brought down to ensure financial viability. Keeping in view the desired metrics of second best outcome and quick interventions outlined in Section 2.1, the scope of projects could be limited to developing STPs and interception & diversion of sewage before discharge into rivers. The sewage transportation networks, being time consuming and resource intensive should be tackled through separate government initiatives. This is discussed in greater detail in Section 3.1.

---

[29] Narain, SP and P Pandey (2012). *Excreta matters: State of India's environment: A Citizens' Report 7*. New Delhi: Centre for Science and Environment.

[30] The World Bank, "National Ganga River Basin Project FAQ."

[31] Email communication with NMCG, Govt of India : Upadhyay, "Re: Request for Information in Relation to PPP Projects on Ganga Cleanup Program '*Namami Gange*.'" (July 6, 2015).

Secondly, on the demand side, it is imperative to address the allocation of risks between private and public partners to the extent that the feedback effects, even if not completely removed from the vicious loop, would be reduced in magnitude. This would call for a PPP model designed to deliver proper apportionment between the government and the private partners. This is discussed in greater detail in Section 3.3.

3. The Way Forward for PPPs for Clean-Ganga

The previous section elucidates the vicious loops holding back the use of PPPs for Ganga clean-up. While these are surely not the only problems confronting PPPs, they are issues that will need to be addressed to achieve long-term sustained success of PPPs. This section looks in more detail at possible interventions aimed at breaking the loops.

3.1. *Focus PPPs on interception & diversion of sewage and STPs*

Section 2.2.2 identifies investment cost reduction as a possible intervention. While choice of treatment technology plays an important role in limiting costs, this is relevant only to the STP, not to the network, and low-cost technologies may run the risk of reduced reliability and effectiveness. The scope of the intervention, and specifically whether or not the collection network is included, is, therefore, critical to cost control.

Past policy reflects the tension between STP and network investment. GAP-I and GAP-II both emphasised STP development and neglected networks but later on, in wake of criticism, the government changed strategy to cover both collection and treatment infrastructure.[32]

Critics pointed out that the poor state of sewerage networks had led to a situation in which approximately 80% of STPs built are under-utilised.[33]

---

[32] As per email communication with Asst. GIS Specialist NMCG: Bhawna Sharma, "Annexure: Comparative Analysis of Previous Programs with '*Namami Gange*,'" June 9, 2015.

[33] Consortium of IITs, "SWOT Analysis of Ganga Action Plan," GRB EMP: Ganga River Basin Environment Management Plan, Kanpur: Consortium of IITs, December 2011.

This is corroborated in a number of studies, which pointed out that in the absence of collection networks, wastewater would continue to flow to rivers through open drains and pollution levels would not be abated.[34] As a result, since the inception of NGRBA, the focus of the government has shifted towards an integrated approach for simultaneous construction of STPs and sewerage networks.[35]

Is another approach to the problem possible? Given that the objective is to prevent untreated wastewater entering the river, an expedient alternative to urban sewerage networks would be to cite interception and treatment infrastructure at drain discharge points.

Secondly, resource availability, including time and money, are imperative. An upper bound estimate for the costs of wastewater infrastructure comprising all collection networks, pumping and treatment is approximately INR 60 million (US$0.9 million) per million litres per day (MLD); the cost of setting of STPs is estimated to be about 17% of this total.[36] In addition to direct monetary costs, laying sewerage networks in densely developed cities imposes indirect economic costs from traffic disruption and congestion, inconvenience for local residents and reduced custom for businesses. Technicalities, procedural clearances and public opposition also hold back implementation.[37]

In addition, given the shortfall of treatment capacity compared to the overall requirement, the limited available resources could be more efficiently utilised to cover a larger number of STPs instead of the

---

Available at https://nmcg.nic.in/writereaddata/fileupload/50_006GEN.pdf.

[34]For example, in a 2013 assessment, 51 out of 64 sewage treatment plants were found to be working at less than 60% of their capacity; Please see: Narain, "Ganga: The River, Its Pollution and What We Can Do to Clean It." Narain S, P Pandey, New Delhi, *Excreta Matters*, Vol. 7, New Delhi: Centre for Science and Environment, 2012.

[35]As per document attachments from Assistant GIS specialist NMCG: Sharma, "Annexure: Comparative Analysis of Previous Programs with 'Namami Gange.'"

[36]Narain, SP and P Pandey (2012). *Excreta matters: State of India's environment: A Citizens' Report 7*. New Delhi: Centre for Science and Environment.

[37]For example, please see TNN, "Lucknow Nagar Nigam Discusses Development Plan with Various Departments," *The Times of India*, July 1, 2014. Available at http://timesofindia.indiatimes.com/city/lucknow/Lucknow-Nagar-Nigam-discusses-development-plan-with-various-departments/articleshow/28488802.cms.

comprehensive coverage option. Keeping in mind the capex requirement estimated by GRBMP[38] and assuming the 1:6 cost ratio found in Narain *et al.*'s 2012 study, the financial implication of the capex for new STPs would be around INR 420 billion (US$6.3 billion). The INR 200 billion (US$3 billion) allocated to the *Namami Gange* project by the central government[39] clearly falls far short of just the capex requirement for the Ganga basin. Adding to this annual opex of INR 1151 billion (US$17.2 billion),[40] it is evident that a large funding gap would remain.

With the relative immaturity of PPP market in water services and the lack of VGF funds, at present, the most effective use of available funds would be to follow the approach of GAP-I and II to concentrate on the interception and diversion of sewage at the point of discharge into the river and construction of STPs, in contrast to the consolidated infrastructure approach adopted under NGRBA in 2009.[41]

Thirdly, the focus on interception and treatment offers advantages from the perspective of administrative capacity. Previous experience in constructing sewage networks, as pointed out earlier, has shown the long lead time for this type of project.

Construction of wastewater infrastructure under the Namami Gange mission intersects with the goals of the Jawaharlal Nehru National Urban Renewal Mission (JNNURM) launched in the year 2005 and its successor, the Atal Mission for Rejuvenation and Urban Transformation (AMRUT) of the Modi government, which provide for central government funds for urban infrastructure investment. JNNURM provided for investment of INR 1200 billion (US$18 billion) of which 21% was to be allotted for

---

[38] Consortium of 7 IITs, "Ganga River Basin Management Plan -2015 | Mission 2: Nirmal Dhara."

[39] "Namami Gange."

[40] Consortium of 7 IITs, "Ganga River Basin Management Plan -2015 | Mission 2: Nirmal Dhara."

[41] The policy shift in the NGRBA from the earlier GAPs to include comprehensive Sewage infrastructure plans has been confirmed on personal communication through email from Asst GIS Specialist NMCG vide: Sharma, "Annexure: Comparative Analysis of Previous Programs with 'Namami Gange.'"

sewage infrastructure.[42] Thirteen major cities of the Ganga basin were covered under JNNURM.[43] Additional funds are also expected to be available under the AMRUT programme, to be channelled through state governments.

The discussion suggests that unbundling STP construction from sewage network roll-out would be a more cost-effective approach to improve environmental quality in the Ganga.

3.2. *Remove external reinforcement of a combined water and sewage tariff*

Section 2.2.2 outlines a possible policy intervention of removing the external reinforcement of combined water and sewage tariffs to break the red vicious loop.

Water and sewage services are often treated as a single policy field and public debate on these services tends to assume that they have similar characteristics.

Intuitively, water supply and sewage services do seem to have an inseparable connection as part of the water cycle from source back to river. Indeed, to estimate wastewater generation, GoI assumes that 80–85% of water supplied will be returned as sewage[44].

Be that as it may, this physical connect between water and sewage does not imply that the water and sewage should be subject to identical policy interventions. In the Indian context, it may be argued that water and sewage differ in their underlying characteristics and therefore should be subject to different policy interventions, especially from PPP perspective of cost recovery.

---

[42] "JNNURM and Environmental Sustainability | IGC," Accessed June 7, 2015. Available at http://www.theigc.org/blog/jnnurm-and-environmental-sustainability/.

[43] The cities in the Ganga Basin have been worked out from: MOUD GOI, "Jawaharlal Nehru National Urban Renewal Mission: Overview", Ministry of Urban Development, Ministry of Urban Employment and Poverty Alleviation, Govt of India. Accessed on June 7, 2015. Available at http://jnnurm.nic.in/wp-content/uploads/2011/01/PMSpeechOverviewE.pdf.

[44] Bharati Lal Seth, "Status Paper for River Ganga | Past Failures and Current Challenges", CSE India, June 2013. Available at http://www.cseindia.org/userfiles/Status%20Paper%20Ganga%202013.pdf.

3.2.1. *Rationales for separation of the conjoined twins — water and sewage*

This section examines the rationale for removing combined water and sewage tariffs from the vicious loop, the obstacles, and the ways to achieve this objective.

3.2.1.1. Rationale 1: water and sewage are different and distinct economic goods

From the economic stand-point, water has multiple characteristics, as a common pool resource, essential public service and depository for waste. Each of these constitutes a different type of market failure and thus a different associated policy prescription.

Water and wastewater services can be classified within the typology of economic goods: Public goods, private goods, merit goods (toll goods) and common resource goods.[45] Water services are private goods with large economies of scale; they are partly excludable, most clearly in the case of bottled water or water supplied through tankers but not for standpipes, and rivalrous under conditions of resource scarcity. The benefits of water supply fall mainly on the user but there may be additional benefits in terms of public health. Wastewater services are typically public goods with large externalities: the environmental and public health benefits of safe disposal are non-excludable and non-rival. Sanitation may be considered a merit good — one that individuals are likely to under-consumer compared to the social optimum. Raw water is a common pool resource, rival and excludable; allocation of ownership and use rights over water sources must be defined guided by some principle and enforced by the government in order to ensure that the use of the resource is efficient and socially optimal.

It is important to examine this classification especially in the Ganga basin context. Northern India, where the Ganga river basin is located, is considered rich in water resources, but average annual figures of water

[45] World Bank, *Initiating and Sustaining Water Sector Reforms (South Asia Development Series)*, South Asia Rural Development Series (World Bank Publications, 2000).

availability disguise high seasonal variability. Thus many locations experience severe resource shortages in the pre-monsoon period. [46] In such a situation, water as an economic good is highly rivalrous. In terms of excludability, river water is treated as a free natural resource for agricultural, religious and human needs. At the same time, the historic commitment to provide universal free access to water in many cities,[47] reinforces public perceptions of water being non-excludable.

At the same time, water tankers and bottled water are sold in abundance, demonstrating the excludability of some forms of potable water. These multiple facets of water are often conflated in political debate, making it more difficult to identify appropriate policy solutions.

On the other hand, wastewater services have low rivalrous nature, as usage of the sewage facilities by one person does not diminish the usage of the service by others. With a relatively low take up rate of organised sewage networks and household access in the Ganga basin, sewage also classifies high as a merit good.

As a result of their economic characteristics, we would expect the government to play a larger role in the provision of wastewater services than water services.

These different characteristics underscore the need to look at water and wastewater services as distinct and to consider different roles of the market and the government in each as well as different pricing structures.

3.2.1.2. Rationale 2: water supply costs outweigh sewage treatment costs

In India, the delivery of water and wastewater services is the responsibility of municipal administrations known as "Urban Local Bodies" (ULBs) under the oversight of state governments. For example, in the state of

[46]This was cited in an email reply by the Additional Mission Director, NMCG Upadhyay, "Re: Request for Information in Relation to PPP Projects on Ganga Cleanup Program 'Namami Gange.'" July 6, 2015.

[47]Narain, SP and P Pandey (2012). *Excreta matters: State of India's environment: A Citizens' Report 7*. New Delhi: Centre for Science and Environment.

Uttar Pradesh, which is a major source of pollutants to the Ganga, the "Uttar Pradesh Jal Nigam" deals with the planning and execution of capital works for both water supply and wastewater; at the same time, city-level "Jal sansthans" conduct day-to-day operations and maintenance of water supply and sewage services. [48]

Given the federal nature of India, there is a wide diversity in the way charges for these services are levied across the spectrum of cities but generally there is no clear separation between the agencies responsible for water and wastewater services.

Almost all cities in India price water services below cost recovery levels; in fact, there are some cities that do not charge at all.[49] A study by the World Bank Water and Sanitation Program covering 23 cities of India found that (i) average revenue against the water services across the cities studied ranged from INR 1 to 11 against the costs of production from INR 2 to 22; (ii) none of the 23 cities recovered the full CAPEX and OPEX for provision of Water services; and, (iii) 14 out of the 23 cities could not recover the OPEX costs of operation.[50] A further challenge on the cost recovery front in that non-revenue water (NRW, the volume of water supplied to the system for which no tariffs are collected, measured as a proportion of total water supplied) can be 50% or more. This compares to below 5% in the world's most efficient utilities like Singapore.[51]

A variety of pricing structures are in employed across the country. Some cities use volumetric pricing based on household metering, with either a fixed tariff per cubic metre or an increasing block or "slab" structure in which higher volume blocks are charged at unit higher rates. Cities that do

---

[48] JNNURM (2006). "Final Report | Kanpur City Development Plan", Jawaharlal Nehru National Urban Renewal Mission. Available at http://jnnurm.nic.in/wp-content/uploads/2010/12/CDP_Kanpur.pdf.

[49] Narain, SP and P Pandey (2012). *Excreta matters: State of India's environment: A Citizens' Report 7*. New Delhi: Centre for Science and Environment.

[50] Anjali Sen Gupta, "Cost Recovery in Urban Water Services: Select Experiences in Indian Cities", The World Bank, March 1, 2011. Available at http://documents.worldbank.org/curated/en/2011/03/14893347/cost-recovery-urban-water-services-select-experiences-indian-cities.

[51] Ibid.

not have metered connections charge for water on the basis of the house area or the number of taps installed in the house or on a flat fee basis.[52]

Most cities do not charge for wastewater services but do charge a one-time charge for connection to the sewer. Cities that charge for wastewater may do so through a levy based on a percentage of volume of water consumed. Delhi has some of the highest charges at 60% of the water bill. Alternatively, wastewater fees may be based on a fixed payment per month per sanitary fixture, or as a percentage of the property tax as in Allahabad and Lucknow, which charge 4% and 3% of the property tax respectively.[53]

Government infrastructure cost estimates published in 2011 found a capital investment requirement of INR 5099 (US$77) and INR 4704 (US$71) per capita for water and wastewater respectively (2009–2010 prices). Costs were estimated on the basis that all cities would meet government standards for continuity of supply and coverage. The annual per capita operating costs were estimated to be INR 501 (US$7.5) and INR 286 (US$2.3) respectively.[54] To fund the investment requirement for both water and wastewater infrastructure through user fees seems politically unrealistic at this time. Focusing exclusively on opex, a combined cost recovery tariff would be equivalent to a monthly tariff of INR 262 for a household with four members, a very significant increase on current rates for most households and one which would be likely to encounter significant public resistance.

We now combine the above findings with the policy recommendation in Section 3.1. If the Sewage PPP initiatives focus only on interception and diversion of sewage and setting up of STPs, capex costs could be reduced to approximately a one-sixth fraction of the water supply costs.[55]

Therefore, on the basis of substantial cost differences between the water supply and the sewage service and the sensitivity of charging for water, it

---

[52] Ibid.

[53] Narain, SP and P Pandey (2012). *Excreta matters: State of India's environment: A Citizens' Report 7*. New Delhi: Centre for Science and Environment.

[54] GOI HPEC, Report on Indian Infrastructure and Services, Govt of India, March 2011. Available at http://icrier.org/pdf/FinalReport-hpec.pdf.

[55] Please see the previous heading of "Unbundling of Sewage Infrastructures from STPs — Sustainability?"

could be beneficial to de-link water and sewage user charges. There are several different charging options that could be used. One possibility is to bill jointly with water but to distinguish clearly between the two services on the bill or to impose an "environmental levy" with property tax. It would be important to ring-fence revenues from the new charges in any case for the exclusive purpose of providing wastewater services.

At the same time, this will need the policy makers to bring these facts in public domain and engage the public in a direct public information campaign before notifying the increase in tariffs. Since the public is already anchored to proposals for water tariff increases, a relatively smaller increase in sewage tariffs linked to a clear programme of investment linked to a strong environmental improvement message may be feasible.

The government should increase sewage charges in gradual steps spreading over a period of time. This could even mean a complete subsidy of the OPEX without increase in sewage levies for several years, within which period the government could hope to show perceptible improvements in river water quality. Increased public confidence through action and branding would help the government gain legitimacy from the public in increasing sewage levies.

3.2.2. *The political connection between water, wastewater and PPP*

Water is a sensitive political issue in India, bound up in discourses of water as a commons and as a human right. Private sector participation in the sector has, therefore, given rise to virulent and sometimes violent opposition, mobilised behind slogans such as "Our mother Ganga is not for sale." The idea of a private company earning profits on the provision of water services is anathema to some interest groups and social movements, as seen in attempts to introduce private sector participation in Delhi.[56] Opponents argued further that privatisation would not deliver greater efficiency and that the process lacked democratic accountability.[57]

---

[56] A Bhaduri and A Kejriwal (2005). Reforming the Reformers, *Economic and Political Weekly*, 5543–45.

[57] Ibid.

One of the chief opponents of water "privatisation" in Delhi took office as the capital region's Chief Minister in 2015.

The anti-PPP discourse often equates PPP to privatisation and assumes that private benefits from the "partnership" would far outweigh those to the public. For example, a Madhya Pradesh based civil society organization, *Manthan,* opines that privatisation and PPPs have no apparent differences; it argues that the real nature of PPP is shrouded under cover of higher levels of community-orientation, accountability, transparency and governmental controls, whereas they essentially remain the same as privatisation at the operational, legal and structural levels.[58] Others use emotive phrases such as "water-for-profit" to describe the international consequences of private participation in water PPPs. [59]

The privatisation opponents draw on examples of "failed" PPPs both internationally and in India to bolster their case, such as the concession contracts for Jakarta (Indonesia),[60] and numerous cases of disputes and contract cancellations from Latin America and Africa.[61]

In India, only a small proportion of planned water PPPs have reached the operational phase. For example, one of the first attempted water PPPs was an ambitious INR 7.35 billion project for Pune municipal corporation in the year 1998, which was abandoned before award. One of the major reasons for this, according to a World Bank report, was an increase in the water tariffs by 25% by the Pune Municipal Corporation just before tenders were called for the PPP contract.[62]

---

[58] Dwivedi, G (2010). *Public-Private Partnerships in Water Sector: Partnerships or Privatization?*. Badwani (MP). Manthan Adhyayan Kendra. Available at http://www.manthan-india.org/wp-content/uploads/2015/04/PPPs-In-Water-Sector-Final-Book.pdf. Accessed May 1, 2017.

[59] For details on the market outreach initiative brochure, please see: NMCG, "Clean Ganga | Market Conference," *National Mission for Clean Ganga*. Accessed on June 29, 2015. Available at https://nmcg.nic.in/writereaddata/fileupload/8_CleanGanga.pdf.

[60] Bill, G (2007). Indonesia Losing Its Thrist for Privatization, *Asia Times Online*. Available at http://www.atimes.com/atimes/Southeast_Asia/II22Ae02.html.

[61] For details on the market outreach initiative brochure, please see: NMCG, "Clean Ganga | Market Conference," *National Mission for Clean Ganga*. Accessed on June 29, 2015. Available at https://nmcg.nic.in/writereaddata/fileupload/8_CleanGanga.pdf.

[62] Zerah, M and K Graham-Harrison (2000). *The Cancellation of the Pune Water Supply and Sewerage Project — Challenges in Private Sector Participation*. Water and Sanitation Program case study. South Asia. Washington, D.C. : World Bank Group. Available at http://

Despite clear political support at the outset, the discourse around the PPP veered to concerns about the private sector profiting from the delivery of a basic human need. As a result, many elected officials withdrew their support. The project covered both water and sewage services; however, the emotive issue of water supply became an impediment to the sewage related issues as well. This is similar to the case in Delhi, where the Delhi Jal Board failed in its initial attempts to procure sewage treatment plants via PPP due to opposition from public coupled with lack of financial viability.[63]

In order to secure public support for PPP models, it will be necessary for proponents to distinguish clearly between "privatisation" and PPP.

Private sector involvement can be seen as a continuum. On one end of the spectrum is complete divestiture of assets, in which the private sector takes full ownership and control of assets, although pricing and service standards would usually be set by a regulator; at the other end of the spectrum stands direct service provision by a government department. However, even at this end of the spectrum, the private sector plays a major role as contractor and equipment supplier. Of course, water and sewage treatment plants and networks completely owned and operated by the government are built, refurbished and expanded by private parties under conventional procurement contracts. In this procurement method, competitive bidding processes are expected to limit the profits earned by contractors.

Another way to look at the spectrum is in terms of risk transfer to the private sector. Under complete divestiture, the private sector takes on full operating and financing risks. Under government control with conventional procurement, the private party takes construction risk but operating and financing risks remain with the government. PPPs, in the middle of the spectrum, embody a wide variety of risk transfer arrangements but all configurations imply some sharing of risks by the parties. Specifically for wastewater PPPs, Chiu and Bosher (2005) identify the following risks: design and construction (D&C) risks, operation and maintenance (O&M)

documents.worldbank.org/curated/en/812071468269415567/The-cancellation-of-the-Pune-Water-Supply-and-Sewerage-Project-challenges-in-private-sector-participation.

[63] Asian Development Bank (2013). *Water in Asian Cities: Utilities Performance and Civil Society Views*, Asian Development Bank. Available at http://www.adb.org/publications/water-asian-cities-utilities-performance-and-civil-society-views.

risks, compliance risks, market risks, tariff risks, financial risks, transaction risks, legal and regulatory risks, and political risks.[64]

Communicating to the public the potential benefits of these often complex arrangements and the safeguards in place to protect public interests is undoubtedly a challenging task, but one that the government would have to take on to secure long-term public acceptance of PPPs as part of the *Namami Gange* programme.

3.3. *An optimal risk apportionment model for sewage PPPs*

The green vicious loop discussed in Section 2.2.2 establishes that an appropriate risk sharing arrangement between the public and the private partners can reduce the leverage of the vicious loop. This section examines these risks in more detail and suggests a suitable risk-sharing model for STPs in the Ganga basin.

Financial stability and risks are closely related in any PPP; these form the cornerstones of any investment decision in an infrastructure project. Investors have a hurdle rate of return based on perceived riskiness of the contract. In China, an indicative range of 8–12% internal rate of return for water and wastewater PPPs is set out in national guidelines.[65]

Compared with traditional procurement, PPP projects involve the transfer of more risks and therefore imply a higher hurdle rate for the private party.[66] The rate will depend on the attribution of risks under the contract. Getting the risk allocation framework right will be critical to the success of PPPs for Ganga clean-up.

---

[64] Chiu, TYC and CB Bosher (2005). Risk Sharing in Various Public Private Partnership (PPP) Arrangements for the Provision of Water and Wastewater Services. In *Conference on Public Private Partnerships–Opportunities and Challenges* (Vol. 22). Hong Kong. Available at http://www.civil.hku.hk/cicid/3_events/32/papers/11.pdf.

[65] Tao, F, M Chang and L Zhong (2008). *Reform of China's Urban Water Sector.* IWA publishing.

[66] Tarek MZ and L-M Chang (2002). Prototype Model for Build-Operate-Transfer Risk Assessment. *Journal of Management in Engineering*, 18(1), 7–16.

Financing risk will need to be shared between the parties. This has been recognised by GoI which has committed to providing investment funding for the construction of STPs and sewage interceptors under the Clean Ganga action plan.[67] The government has also signalled its intention to consider allowing an increase in the VGF funding higher than the levels of 40%.[68] However, upfront government subsidy comes at the cost of incurring further risks — subjective project evaluation risk and subsequent completion risk. In a world in which cash flows and financing costs could be predicted perfectly, there would be no difference between an upfront capital grant subsidy and an annual payment. However, this is generally not possible. Future cash flows are subject to changes in tariff, varying operation and maintenance cost, technological and managerial innovation and so on, while the discount rate is subject to inflation, the project company's financial performance and credit market conditions. In the context of a long-term PPP contract of 15–25 years, reliable prediction of costs and revenues is impossible. Facing these uncertainties, any amount of upfront capital grant proposed whether by the private sector or the public sector is likely to include a large element of subjective project evaluation risk. Private parties may perceive risks to be so high that even a grant of 40% of capital costs cannot compensate for the risk. This could be one reason of the low take up rate of PPP projects of sewage industry.

On the other hand, the problem also arises with a 100% upfront capital investment by the public sector. This approach can crowd out private investment and transfer unnecessary risks to the private parties, depending on the PPP model adopted. For example, an operations and maintenance (O&M) contract in which the government designs, finances and implements the capex programme while transferring O&M responsibility to the private party, the private sector may face additional risks such as mismatch in technical calibre, obsolete or inappropriate technology, delay in delivery and low asset quality which will in turn have an impact on O&M

[67] "Namami Gange."

[68] Puskal Upadhyay, "Re: Request for Information in Relation to PPP Projects on Ganga Cleanup Program '*Namami Gange*.'"

costs.[69] This will not be the case when the private sector constructs and operates the asset, as under a design-build-operate (DBO) contract and its variants. In order to maintain efficiency incentives for the private party, capital grants should not be paid upfront.

Therefore, a suitable risk allocation structure for wastewater projects under the *Namami Gange* programme could be a DBO-type contract with deferred governmental subsidies for capital costs in the form of an annuity payment, access to subsidised finance or a credit guarantee to the private sector.

It was shown above in Section 3.1 that the allocation of INR 200 billion (US$3 billion) would not be sufficient to cover capex costs for all planned STPs in the Ganga basin. The deferred subsidy approach would enable the Government to spread financial outlays over a period of time to reduce the impact on the annual budget. This would help in breaking the green vicious loop by replacing the upfront viability gap funding with annual payments to PPP partners.

Variants of this model can be adopted by the government depending upon local technical and social conditions. For instance, annuity payments could be calculated on the basis of costs plus an agreed rate of return in the early years of the contract and then shift to a capped payment in later years, with a gradual shift from transfers to user fees as the source of funds. This would reduce risks to the private sector in the initial years, encouraging private sector engagement, while retaining operating efficiency incentives in later years. Phased structures would also allow time for the benefits of projects to be observed, helping to assuage public scepticism of PPPs and increase willingness to pay user fees.

Subsidised finance or a loan guarantee from the government can reduce financial risk imposed on the private sector to promote the incentive of participation in PPPs. International experience demonstrates many successful applications of this approach. For example, the Australian government provided long-term loans and operating costs subsidies to the

[69] Chiu, TYC and CB Bosher (2005). Risk Sharing in Various Public Private Partnership (PPP) Arrangements for the Provision of Water and Wastewater Services. In *Conference on Public Private Partnerships–Opportunities and Challenges* (Vol. 22). Hong Kong. Available at http://www.civil.hku.hk/cicid/3_events/32/papers/11.pdf.

concessionaire in the Harbour Tunnel Project.[70] The government of India has set up many financial institutions such as IIFCL and funds India infrastructure Project Development fund (IIPDF) that can engage actively in PPPs in the sector, bringing down financing costs while building up financiers' understanding of the sector.

4. Conclusions

The Indian government has identified the development of public-private partnerships as one of the central policy tools in its mission to clean up the Ganga. The government is backing the mission with an unprecedented level of political commitment and funding, well beyond what was offered under the earlier Ganga Action Plans. However, India's past experience with PPPs for water and wastewater has been very mixed and public opposition to the involvement of the private sector has sometimes been virulent, undermining otherwise promising projects. The reasons for these failures are many and they are interlinked, posing considerable challenges for policy design.

By mapping out the self-reinforcing vicious loops that characterise the sector, this chapter has sought to identify policy recommendations specific to the Indian national context that could help the PPP approach to succeed and be sustained.

Firstly, it will be critical to use scarce resources to maximum effect. With the goal of improving environmental quality in the Ganges river within a short period, the most effective investments are likely to be effluent discharge point interceptors linked to STPs. Of course, investing in urban sewerage networks would provide many additional benefits to urban residents and is a worthy policy objective in itself, but the direct and indirect costs of network refurbishment and extension are such that network coverage cannot realistically be achieved within the timeframe envisaged for Ganga clean-up.

Secondly, given the great political sensitivity associated with water tariffs, government should consider de-linking water and wastewater

---

[70] Kong, DQ and BS He (1999). Analysis of Key Factors of BOT Project Financing. *China Harbour Engineering*, 5, pp. 57–61 (in Chinese).

tariffs and assess a range of options to increase revenues from users or taxpayers. Volumetric tariffs based on estimates of effluent discharged may be appropriate in some place but in others a tax or levy may be more suitable and easier to enforce. De-linking policy-making for water and wastewater would also help to assuage public concerns about the "privatisation" of water.

Thirdly, contracts need to be structured in a way that recognises the high degree of uncertainty about the future development of the sector. In order to ensure that public funds are leveraged to maximum effect, the government should consider subsidised finance and loan guarantees as well as transfers. Payment structures can be phased in such a way as to preserve incentives for the private party to invest and operate efficiently without driving up the private party's risk premium.

The scale of the Ganga Basin creates an immense challenge for policymakers but it also opens up opportunities to experiment, pilot and learn. Supportive policy, coordination and financial support from the centre are clearly essential for PPPs to achieve their objectives, but state and local governments should at the same time be given the flexibility to develop project structures which are appropriate to local conditions and to try out different approaches to charging. The centre can then play a valuable role in evaluating these different approaches and supporting the wider adoption of successful models.

Bibliography

Asian Development Bank (2013). *Water in Asian Cities: Utilities Performance and Civil Society Views*. Asian Development Bank. Available at http://www.adb.org/publications/water-asian-cities-utilities-performance-and-civil-society-views.

Bagga, B (2014). Can Modi Clean the Ganges, India's Biggest Sewage Line? *Yahoo News*, July 17, 2014. Available at http://news.yahoo.com/modi-clean-ganges-indias-biggest-sewage-line-032922309.html.

Bhaduri, A and A Kejriwal (2005). Reforming the Reformers. *Economic and Political Weekly*, 5543–45.

Bing, L, A Akintoye, PJ Edwards and C Hardcastle (2005). The Allocation of Risk in PPP/PFI Construction Projects in the UK. *International Journal of Project Management*, 23(1), 25–35.

Chiu, TYC and CB Bosher (2005). Risk Sharing in Various Public Private Partnership(PPP) Arrangements for the Provision of Water and Wastewater Services. In *Conference on Public Private Partnerships–Opportunities and Challenges* (Vol. 22). Hong Kong. Available at http://www.civil.hku.hk/cicid/3_events/32/papers/11.pdf.

Consortium of IITs (2011). SWOT Analysis of Ganga Action Plan. GRB EMP: Ganga River Basin Environment Management Plan. Kanpur: Consortium of IITs. Available at https://nmcg.nic.in/writereaddata/fileupload/50_006GEN.pdf.

Consortium of 7 IITs (2015). Ganga River Basin Management Plan -2015 | Mission 2: Nirmal Dhara. Consortium of 7 IITs. Available at http://gangapedia.iitk.ac.in/sites/default/files/Mission%202_ND.pdf.

Central Pollution Control Board (July 2013). Pollution Assessment: River Ganga. New Delhi: Ministry of Environment and Forests, Governemnt of India. Available at http://cpcb.nic.in/upload/NewItems/NewItem_203_Ganga_report.pdf.

Dias Jr, A and PG Ioannou (1995). A Desirability Model for the Development of Privately-Promoted Infrastructure Projects. *UMCEE Rep*, no. 95–09.

Dwivedi, G (2010). *Public-Private Partnerships in Water Sector: Partnerships or Privatization?*. Badwani (MP). Manthan Adhyayan Kendra. Available at http://www.manthan-india.org/wp-content/uploads/2015/04/PPPs-In-Water-Sector-Final-Book.pdf. Accessed May 1, 2017.

Fu, T, M Chang and L Zhong (2008). *Reform of China's Urban Water Sector*. IWA Publishing.

GOI, DEA. FAQs: Public Private Partnerships in India. *Public Private Partnerships in India*. Accessed on June 10, 2015. Available at http://www.pppinindia.com/newFAQs.php.

———. Scheme and Guidelines for Infrastructure Project Development Fund. Department of Economic Affiars, Ministry of Finance, Government of India. Accessed on June 9, 2015. Available at http://www.pppinindia.com/pdf/guideline_scheme_IIPDF.pdf.

GOI, Department of Economic Affairs (2015). Database of Infrastructure Projects in India. *Database of Infrastructure Projects in India*. Available at https://infrastructureindia.gov.in/search.

GOI, MOEF (2015). National Status of Waste Water Generation & Treatment. *National Status of Waste Water Generation @ Treatment*. Available at http://www.sulabhenvis.nic.in/Database/Nationalstatusofwastewatergenerationandtreatment_2090.aspx?format=Print.

GOI, MOUD (2015). Jawaharlal Nehru National Urban Renewal Mission: Overview. Ministry of Urban Development, Ministry of Urban Employement and Poverty Alleviation, Govt of India. Accessed on June 7, 2015. Available at http://jnnurm.nic.in/wp-content/uploads/2011/01/PMSpeechOverviewE.pdf.

Govt of India, DEA (2015). Overview: Public Private Partnerships in India. *Public Private Partenrships in India*. Accessed on June 9, 2015. Available at http://www.pppinindia.com/overview.php.

Guerin, B (2007). Indonesia Losing Its Thrist for Privatization. *Asia Times Online*. Available at http://www.atimes.com/atimes/Southeast_Asia/II22Ae 02.html.

Gupta, AS (2011). Cost Recovery in Urban Water Services: Select Experiences in Indian Cities. The World Bank. Available at http://documents.worldbank.org/curated/en/2011/03/14893347/cost-recovery-urban-water-services-select-experiences-indian-cities.

Helmer, R and I Hespanhol (1997). Water Pollution Control — A Guide to the Use of Water Quality Management Principles. WHO/UNEP. Available at http://www.who.int/water_sanitation_health/resourcesquality/wpccase study1.pdf.

HPEC, GOI (2011). Report on Indian Infrastructure and Services. Govt of India. Available at http://icrier.org/pdf/FinalReport-hpec.pdf.

JNNURM (2006). Final Report | Kanpur City Development Plan. Jawaharlal Nehru National Urban Renewal mission. Available at http://jnnurm.nic.in/wp-content/uploads/2010/12/CDP_Kanpur.pdf.

JNNURM and Environmental Sustainability | IGC. Accessed on June 7, 2015. Available at http://www.theigc.org/blog/jnnurm-and-environmental-sustainability/.

Kaur, R, SP Wani, AK Singh and K Lal (2012). Wastewater Production, Treatment and Use in India. Paper Presented at the Second Regional Workshop of the Project Safe Use of Wastewater in Agriculture, 16–18 May 2012, New Delhi. Available at http://www.ais.unwater.org/ais/mod/page/view.php?id=61. Accessed on May 15, 2015.

Kong, DQ and BS He (1999). Analysis of Key Factors of BOT Project Financing. *China Harbour Engineering*, 5, pp. 57–61 (in Chinese).

Marin, P (2009). Public-Private Partnerships for Urban Water Utilities: A Review of Experiences in Developing Countries, Vol. 8. World Bank Publications.

MOEF (2009). State of Environment Report, India 2009. New Delhi: Ministry of environment and forests, Govt of India.

Namami Gange. *Down To Earth*. Accessed on June 8, 2015. Available at http://www.downtoearth.org.in/content/namami-gange.

Narain, S (2014). Ganga: The River, Its Pollution and What We Can Do To Clean It. New Delhi; Centre for Science and Environment. Available at http://www.cseindia.org/userfiles/ganga-the-river-pollution.pdf. Accessed May 1, 2017.

Narain, SP and P Pandey (2012). *Excreta matters: State of India's environment: A Citizens' Report 7*. New Delhi: Centre for Science and Environment.

NMCG (2015). Clean Ganga | Market Conference. *National Mission for Clean Ganga*. Accessed on June 29, 2015. Available at https://nmcg.nic.in/writereaddata/fileupload/8_CleanGanga.pdf.

Rodrik, D (2008). A Practical Approach to Formulating Growth Strategies. *From the Washington Consensus towards a New Global Governance*.

Seth, BL (2013). Status Paper for River Ganga | Past Failures and Current Challenges. CSE India. Available at http://www.cseindia.org/userfiles/Status%20Paper%20Ganga%202013.pdf.

Tare, V, G Roy and P Bose (2013). Ganga River Basin Management Plan: Interim Report. Kanpur: Indian Institute of Technology.

The Scientific American. Which Nations Consume the Most Water?, June 1, 2012. http://www.scientificamerican.com/article/water-in-water-out/.

The World Bank (2011). National Ganga River Basin Project FAQ. *The World Bank: Working for a World Free of Poverty*. Available at http://www.worldbank.org/en/news/feature/2011/05/27/india-ganga-river-basin-faq.

OECD (2005). Encouraging Public-Private Partnerships in the Utilities Sector: The Role of Development Assistance. In *International Investment Perspectives 2005*, OECD Publishing, Paris. DOI: http://dx.doi.org/10.1787/iip-2005-4-en.

TNN (2014). Lucknow Nagar Nigam Discusses Development Plan with Various Departments. *The Times of India*. Available at http://timesofindia.indiatimes.com/city/lucknow/Lucknow-Nagar-Nigam-discusses-development-plan-with-various-departments/articleshow/28488802.cms.

Upadhyay, P (2014). National Mission for Clean Ganga. Presented at the Address to LKYSPP/ NUS MPA students, New Delhi.

——— (2015). Re: Request for Information in Relation to PPP Projects on Ganga Cleanup Program 'Namami Gange,'.

Uttar Pradesh Pollution Control Board (2015). River Water Quality in Uttar Pradesh; Averges Jan-Dec 2013. *UP Pollution Control Board*, n.d. Accessed on February 13, 2015.

World Bank (2000). *Initiating and Sustaining Water Sector Reforms (South Asia Development Series)*. South Asia Rural Development Series. World Bank Publications.

Wu, X, RS House and R Peri (2016). Public-private partnerships (PPPs) in water and sanitation in India: Lessons from China. *Water Policy*, 18(S1), 153–176.

Xun, W, R Schuyler House and R Peri (2015). Public-Private Partnerships in Water and Sanitation in India: Lessons from China. *TBA*, n.d. Accessed on June 29, 2015.

Zayed, TM and L-M Chang (2002). Prototype Model for Build-Operate-Transfer Risk Assessment. *Journal of Management in Engineering* 18(1), 7–16.

Zerah, M-H and G Harrison (2000). The Cancellation of the Pune Water Supply and Sewage Project — Challenges in Private Sector Participation. The World Bank, December 31. Available at http://documents.worldbank.org/curated/en/2000/12/1742970/cancellation-pune-water-supply-sewage-project-challenges-private-sector-participation.

Chapter 7

PRINCIPLES OF WATERFRONT DEVELOPMENT — LESSONS FOR VARANASI

Kok Hoong Chan, Thi Thanh Huyen Do, Lay Ling Lim, Salai Aung Myint and Shikha Thaman

1. Introduction

1.1. *Urban design and waterfront development*

Water is a critical resource for human survival. Since the beginning of time, most human settlements have developed near natural bodies of water for ease of harnessing the water for drinking, irrigation, transportation and other purposes. Over time, rapid population growth and urbanisation resulted in tremendous pressure on these water bodies. Waterfront development, as one element of urban design and planning, allows cities to better manage competing needs to ensure greater sustainability.

The phenomenon of waterfront development and regeneration has spread since the 1960s, triggered by successful transformations and creation of public spaces in waterfront areas in cities of North America (Smith and Ferrari, 2012). Since then, various studies have been carried out to examine and identify the essential factors that facilitate waterfront development ranging from conceptual criteria to process and procedural factors. For example, Bruttomesso (2001) lists three conceptual factors that contribute to the complexity of waterfront development, whereas Millspaugh (2001) identifies several sets of lessons such as a masterplan, consensus and support from the community and public-private partnership as being crucial to the success of waterfront development, while Eckstut (1986) breaks down the approach of solving complex urban design problems on the waterfront to key principles such as "think small; learn from what exists; design streets, not buildings".

This chapter begins by analysing why waterfront development in Varanasi is important before presenting a vision for the waterfront in Varanasi. This is followed by a discussion of the key challenges to waterfront development in Varanasi. Thereafter, lessons on some good practices for processes and governance from other successful waterfront development projects are highlighted and a discussion of how these lessons can be applied to Varanasi is presented. These lessons can broadly be classified into (a) taking a strategic approach, (b) bringing the institutions together and (c) examining the process. A set of principles that can be used to guide the planning, development and management of Varanasi's waterfront is then derived, taking into consideration these lessons and the context of Varanasi. Finally, the chapter ends with some governance challenges that may hinder waterfront development in Varanasi.

2. Waterfront Development in Varanasi

2.1. *Why waterfront development in Varanasi?*

Varanasi is one of the oldest cities in India that has been mentioned in various ancient texts dating back to 800 B.C. (Singh, 1993). The city has a total of 84 ghats (steps and landings leading to the Ganga which flows through the city) that cover a length of 6.8 kilometres along the crescent shaped bank of the Ganga from the Assi-Ganga confluence in the south to the Varuna–Ganga confluence in the north (Singh, 1993). Built in the 18th and 19th centuries, these ghats are spectacular works of architecture that form one of the most magnificent waterfronts in the world (Jalais, 2008).

Varanasi is closely associated with the Ganga, which is considered to be the most holy river for Hindus. The Ganga generally follows a southward path from the Himalayas towards the Bay of Bengal, except in Varanasi where it suddenly changes its course and flows from south to north (Singh, 1993). To Hindus, this is important as it symbolises the life cycle from death to life (Singh, 1993). Varanasi assumes special religious significance because of this (Singh, 1993) and attracts close to 1 million pilgrims every year (Green Pilgrimage Network, 2014). This, in turn, supports a wide range of professions which are dependent on the waterfront (and the Ganga) in Varanasi. Pandas (brahmin priests who officiate religious rituals), ghatias (brahmins who oversee ghat rituals), and pujaris (brahmin priests in charge of worship in the temples) offer their services to pilgrims, vendors sell everything from flowers to oil lamps for making offerings, while cold drink hawkers help to quench the thirst of pilgrims.

Despite its importance, the urban development of Varanasi had not been well planned. In 1826, the city of Varanasi had only one main sewer carrying all the sewage of the city and depositing it near Rajghat (Singh, 1955). It was only in 1917 that the city had a sewage system (Jalais, 2008). However, by then, the sewage system that was constructed was unable to meet the needs of the rapidly growing city. Sewage leaking from the pipelines flows directly into the Ganga (Jalais, 2008). Much later, sewer pumping stations were constructed along the waterfront to pump out sewers draining into the Ganga and bring the sewage to Sewage Treatment Plants (Jalais, 2008). However, due to an unreliable supply of electricity in Varanasi, these sewer pumping stations are often not switched on (Jalais, 2008). As a result, the sewage in the drains empties into the Ganga at the waterfront.

Besides untreated sewage, cremation is another prominent source of pollution in Varanasi. The Manikarnika ghat and the Harishchandra ghat are considered to be great cremation grounds for Hindus. Around 35,000 cremations are performed each year, mostly on wood pyres that do not completely consume the body (Hays, 2011). After cremation, the remains are deposited into the Ganga at the waterfront. Along with these remains, bodies of those who cannot afford cremation are simply thrown into the Ganga (Nina, 2011).

Not surprisingly, the Ganga is particularly polluted in Varanasi. According to a study conducted by the Central Pollution Control Board of India in 2008–2009, Varanasi had the highest level of faecal coliform and the second highest biochemical oxygen demand (BOD) level among a number of cities along the Ganga. Separately, the scientists of the Sankat Mochan Foundation, also reported that the BOD level is 20 to 50 milligram per litre at the junction of the Varuna–Ganga confluence (in Varanasi), whereas the level deemed to be safe for bathing is less than 3 milligram per litre (Sankat Mochan Foundation, 2015). The faecal coliform count there is also very high at 1 to 2 million per 100 millilitres — an indication of pollution contributed primarily by the city of Varanasi (Sankat Mochan Foundation, 2015).

A closer examination reveals that rapid population growth of the city is one of the factors leading to increased pollution of the Ganga in Varanasi. The large increase in the population over the years exerted immense pressure on the city's infrastructure and the Ganga. From 1991 to 2001, the population of Varanasi grew by about 25% (Census2011.co.in, 2015). In the following decade from 2001 to 2011, the population grew by another 17% (Census2011.co.in, 2015). In 2011, the population of Varanasi was 3,676,841 (Census2011.co.in, 2015). According to data from Census 2011, the density of the city of Varanasi could be as high as 2,395 persons per square kilometres (Census2011.co.in, 2015). The *Times of India* also reported that in 2014, more than 17% of the city's land was covered by slums, with a total of 227 slums housing 37% of the population of Varanasi (Dikshit, 2014b). Livelihoods of people dependent on the Ganga and the waterfront have been adversely affected by the large increase in the city's population. For example, some washermen had problems washing clothes in the Ganga as the ashes from the traditional cremation areas settle on the clothes that they were washing (Rogers, 2013).

While waterfront development of Varanasi has been on-going for years (as can be seen from the development of the ghats and other buildings along the waterfront), this process seems to be somewhat haphazard. In fact, one may even question the presence of any advance planning given the uncoordinated developments along the waterfront as well as the half burnt bodies, unburnt dead bodies and solid waste that can be observed floating in the waters of the Ganga. Proper planning, development and management of the waterfront will help to address these and other issues, especially in relation to the last mile connectivity of the sewage drains with the Sewage Treatment Plants, unchecked religious practices and haphazard encroachments into the waterfront area. Furthermore, this will also help to conserve the religious and historical aspects of the city of Varanasi.

3. A Vision for Varanasi's Waterfront

3.1. *A clean, vibrant and holy waterfront*

Given the religious significance of Varanasi and the Ganga, Varanasi's waterfront is naturally holy since it is the interface between the city and the river. It will continue to thrive as one of the most venerated pilgrimage sites for Hindus in the future. The cleanliness and beauty of the waterfront will be another aspect that appeals to visitors. In particular, the allure of the magnificent and well conserved heritage buildings and ghats will attract a wide range of visitors. Even as the waterfront continues to attract religious pilgrims, the number of international tourists who visit the waterfront will increase as Varanasi's waterfront gains a reputation as one of the most unique and stunning waterfronts in the world.

As mentioned earlier, proper planning, development and management of Varanasi's waterfront is essential in order for this vision to be achieved. Furthermore, public awareness of the importance of the waterfront will also be crucial. Finally, the public needs to be mobilised in this endeavour so that the citizens are involved and feel a sense of ownership.

3.2. *Key challenges to waterfront development*

According to Samperi (1986), there are six key challenges to waterfront development. They are land acquisition, environmental issues, construction complexities, transportation access, master plan development, and community/business acceptance. However, due to the uniqueness of each waterfront, not all these challenges may be applicable in every instance. Furthermore, there could also be other specific key challenges that different waterfront development projects face due to the city's socio-economic and political environment. In the context of Varanasi, four key challenges could be identified. They are practical difficulties in acquisition of land near the waterfront by the government, ensuring environmental sustainability in waterfront development, institutional coordination among government agencies, and enabling public participation (which is related to two of Samperi's (1986) challenges i.e. master plan development and community/business acceptance).

Practical Difficulties in Acquisition of Land near the Waterfront by the Government

Land acquisition near the waterfront in Varanasi could be challenging even though the new Land Acquisition Bill (2013) had been passed by Lok Sabha on 29 August 2013 (PTI, 2013). Difficulties arise primarily due to private ownership of the land and the lack of political will. Much of the land near the waterfront is privately owned by religious institutions, trusts and influential families. It would be a politically unpopular move for the government to proceed with land acquisition even though it has full legal authority to do so. The exercise of the government's legal rights in land acquisition is best illustrated in the case of the Sabarmati waterfront development in Gujarat, where the current Prime Minister Narendra Modi was then Chief Minister of Gujarat. At that time, the government acquired private land for waterfront development and paid nominal amounts for it (TNN, 2015b). The key issue concerning practical difficulties in acquisition of land near the waterfront by the government, therefore, lies in the lack of political will.

Ensuring Environmental Sustainability in Waterfront Development

In many developing countries like India, protecting the environment is often not a high priority for the people or the government as there are usually more immediate and urgent concerns such as elimination of poverty and economic growth (World Wildlife Fund, 2015). This is unfortunately also true in Varanasi where religious practices take precedence over environmental sustainability of the waterfront. Changing the mindsets of the government and the community to place greater emphasis on environmental sustainability of the waterfront will be an uphill task, especially if certain religious practices need to be changed or stopped.

Institutional Coordination among Government Agencies

Governmental institutions in India can generally be classified into three groups — (i) those under the central government such as the Ministry of Environment and Forest, and the Ministry of Water Resources, (ii) those under the state government such as state ministries for urban development, water resources and irrigation, power, and State Pollution Control Boards, and (iii) urban local bodies such as municipal corporations. However, it is observed that there is no single institution among these that is overall responsible for waterfront development. Issues concerning ownership and authority in relation to waterfront development are therefore common. It is also not surprising to observe coordination problems among agencies responsible for various aspects of a waterfront development project.

The Constitution of India classifies water as a state subject (Iyer, 2011). However, the River Boards Act 1956 empowers the central government to create and regulate boards for interstate rivers such as the Narmada Control Authority and the Brahmaputra Board (Ministry of Water Resources, River Development and Ganga Rejuvenation, 2014). Due to political differences between the Samajawadi Party that controls the Uttar Pradesh state government and the Bharatiya Janata Party that controls the central government, there is often little or no coordination among agencies under the Uttar Pradesh state government and those under the central government (Singh, 2015). For instance, the National Mission for Clean Ganga (NMCG),[1] which comes under the central government, depends on the Varanasi Development Authority (VDA), which comes under the Uttar Pradesh state government, to implement its action plans. However, political quibbling and lack of coordination have often resulted in inaction.

Similarly, there is a general lack of coordination among agencies under the Uttar Pradesh state government and the Varanasi local urban bodies. For example, while VDA is responsible for urban planning,

[1] The National Mission for Clean Ganga, formed on 12 August 2011, is the national coordinating body for activities relating to the clean-up and conservation of the Ganga (National Mission for Clean Ganga, 2015)

development and management, the Varanasi Municipal Corporation (VMC), a local urban body in Varanasi, is responsible for solid waste management. Varanasi Development Authority is, therefore, dependent on VMC to ensure that no solid waste from the city flows directly into drains which flow into the Ganga at the waterfront. In 2009, VMC had engaged a private company A2Z for waste management (TNN, 2015a). However, due to a longstanding dispute between VMC and A2Z, solid waste management in the city was disrupted for close to two years, resulting in uncollected garbage in many parts of the city (TNN, 2015a). Rainwater washed the garbage into open drains which flow directly into the Ganga. As VDA and VMC did not have a strong working relationship, there was little that VDA could do about the situation even as the Ganga continued to be polluted by solid waste from the city.

Enabling Public Participation

In India, the mode of governance employed by the government can generally be described as traditional public administration, where the government is viewed as the trustee and the citizens are the voters. Public participation in policy making and implementation is generally quite limited. This can probably be attributed to the political economy of the country. According to Midgley (1986), there are four forms of state responses to public involvement: (i) the anti-participatory mode, (ii) the manipulative participatory mode, (iii) the incremental mode of participation, and (iv) the participatory mode. In the Ganga clean-up movement, the government of India used both the manipulative and incremental approaches which indicate that the state welcomes participation when it chooses to do it or in other words when it is in the interest of the state (such as to gain votes or political support) (Ahmed, 1994). This illustrates how public participation is used by politicians in India.

Even when effort is made to engage the public, the extent of the public's involvement is fairly constrained. For example, during the development of the Varanasi Master Plan 2011, the draft master plan, which had already been approved by VDA, was displayed at only eight public sites for about one month for public viewing and comments (TOI, 2013). Aside from the issue of unequal access to public participation (due to the small number of sites where the document could be viewed), the intended purpose of public participation is also unclear given that the plan had already been approved by VDA. It is doubtful that the government truly wanted to seek the citizens' input through this exercise.

The lack of regular public participation has also led to obsolete schemes in India. A development programme for urban areas called Jawaharlal Nehru National Urban Renewal Mission (JNNURM) under the Ministry of Urban Development/Ministry of Urban Employment and Poverty Alleviation was launched with the primary objective to create productive, efficient, equitable and responsive cities (Kumar, 2014). However, in implementing the urban housing initiative for the underprivileged in Varanasi, there were some glaring gaps because the data used for the exercise had not been updated since 2006. The result was that an insufficient number of houses were built as there was inadequate land set aside to meet the actual demand. More than 17% of the city's land thus remained as slums (Dikshit, 2014b).

Interviews conducted with the locals during the fieldwork in Varanasi on 4th December 2014 yield further evidence to substantiate the proposition that public participation is indeed a challenge in Varanasi. More details are provided in Box 1.

From a broader perspective, it should be noted that the particular circumstances of the country itself pose some formidable challenges to the use of public participation in public policy. Firstly, India has a huge population of 1.21 billion people (Mail Online India, 2013) with a deeply ingrained caste system. Even though the caste system has now been outlawed, divisions based on caste remain firmly etched in the people and can easily be observed in routine activities in the society especially in Varanasi (Deshpande, 2010). Getting meaningful and representative participation from a large number of people divided across lines of caste is definitely difficult for the Indian government. Secondly, India is a developing country with a low income per capita. According to a Gallup survey in 2013, India's annual median per capita income is about US$616, below that of other comparable developing countries such as Brazil, Russia, China and South Africa (*Business Standard*, 2013). Earning a living would likely override participation in public forums and townhall meetings for most people in the country. All these factors further complicate any effort by the government to involve the public in planning and implementation of public policy.

Box 1: Interviews with Locals during Fieldwork in Varanasi on 4th December 2014

During the fieldwork in Varanasi on 4th December 2014, informal interviews were conducted with some locals. A total of 10 locals, comprising boatmen, washermen, weavers and sweepers were interviewed. The interviews with the boatmen and washermen were particularly informative as these professions are closely linked to the waterfront. Furthermore, these communities are also good witnesses of the changing landscape of the city as many generations have been residing and working at the ghats along the waterfront. According to the boatmen and the washermen, the root cause of pollution in the Ganga in Varanasi is the lack of proper urban planning. This is exacerbated by the rapid increase in the population of the city over the years.

Although the boatmen do not earn much, they are generally reluctant to leave the profession as they feel that the particular ghat which serves as their base is their property. Even though motor boats are not allowed in the Ganga by the government, many of the boatmen continue to operate motor boats as the government generally does not enforce this law. As a community, the boatmen feel weak and powerless, with no influence or voice in government decision making. As for the washermen, they continue to wash clothes along the Ganga even though this has been banned by the government. The main reason for this is because the government had not made alternative arrangements for the washermen when they implemented the ban.

In general, a lot of angst and disappointment was detected during the interaction with the different communities. The main message communicated by most of them is that they have not been involved in any government policy decision even though those decisions affect their livelihood directly.

4. Good Practices for Processes and Governance from Other Successful Waterfront Development Projects

4.1. *Waterfront development master plan for Varanasi*

There is generally no lack of plans and schemes to redevelop the city of Varanasi. However, most of them have limited focus on development of the waterfront. Given the importance of and potential for waterfront development in Varanasi, the Uttar Pradesh state government should place more emphasis specifically on developing the waterfront. Having a holistic in-depth waterfront development master plan that is separate from the Varanasi Master Plan would benefit the city and also support the central government's agenda for a clean Ganga.

Furthermore, the Varanasi Master Plan has its fair share of challenges. The *Times of India* had reported that nothing had been done to 26 roads of the old city area that were supposed to have been widened under the 2011 Master Plan (TNN, 2014b). Furthermore, the land supposedly reserved for industrial estate, warehouses, city centre and institutional hubs, tourist services facilities, training institutes, hospital, horticulture, fishery, poultry, dairy and other uses under the 2011 Master Plan has instead been converted into residential zones (TNN, 2014b). The plans on improving the transportation network by moving the transport terminal outside the city and constructing a highway that circles the city seem unlikely to materialise given the state government's unwillingness to pay the central authority to do the work (Dikshit, 2013). Despite not having any budget allocated to it, the 2031 Master Plan has been given the go-ahead by the government (TNN, 2014a). This master plan makes no explicit mention of decongestion and development of the waterfront. Therefore, it would be prudent to develop a separate waterfront development master plan for Varanasi instead of relying on the Varanasi Master Plan.

4.2. *Taking a strategic approach*

Typically, waterfront development projects are massive projects that take years for implementation. With the passage of time, there will likely be changes to the physical environment and organisations in Varanasi. In order to prevent mission creep during the later stages of development and implementation, it would be prudent to establish a clear set of high level guidelines in relation to waterfront development in Varanasi at the outset. This helps to ensure a sense of coherence and continuity in implementing the waterfront development master plan.

Besides this, Varanasi could also consider formulating a well-defined strategy to implement its waterfront development master plan. This facilitates decision making and reduces haphazardness in the implementation process thus increasing the chances of success. Box 2 describes how the

Box 2: Setting Guidelines to Steer Future Actions to Implement the Upper River Master Plan in Minneapolis

The Mississippi River is one of the longest rivers that flow across many states in the United States. The city of Minneapolis in the state of Minnesota is one of these cities. Over the years, the riverfront in Minneapolis had become built up with heavy industry, while adjacent parcels of land struggle to attract new investment. The city of Minneapolis thus decided to embark on an exercise to redevelop the riverfront. The Upper River Master Plan was formulated in 1999 to steer the redevelopment efforts.

A critical element of the Upper River Master Plan was to establish a set of high level guidelines to steer future actions. This was essential because the scale of the whole exercise to develop the entire riverfront in Minneapolis into a continuous park system was a humongous task that would realistically take years to complete. To scope the master plan as a single project to be completed within a few years is simply a guaranteed recipe for failure. The city of Minneapolis thus developed a set of high level guidelines under the Upper River Master Plan as given below:

a. Acquire properties that are contiguous with existing parks.
b. Seek stand-alone projects that can be completed in specific timeframes.
c. Pursue strategic acquisitions to connect parcels in public ownership.
d. Utilise public lands as catalyst for assembling larger redevelopment areas.
e. Hold tax-forfeit parcels along the river and in redevelopment areas.
f. Work with industries that can benefit from relocation.
g. Create trail loop projects across the river.
h. Connect local streets to the riverfront.

Besides this, the city of Minneapolis also developed a clear strategy to implement the master plan. The strategy involves setting up the necessary organisations (such as the Upper River Development Corporation to oversee the implementation of the master plan), organising the work areas into different phases and projects, identifying stand-alone projects, understanding the relationship between projects and establishing priorities for projects. All these contributed to the success of the exercise. shows the implementation phases and projects of the master plan.

Source: *City of Minneapolis — Above The Falls — A Master Plan for the Upper River in Minneapolis.*

city of Minneapolis in the United States used this approach to develop their Upper River Master Plan.

At present, Varanasi's waterfront is largely used for religious purposes. One of the strategies that the Uttar Pradesh state government may wish to consider when developing the waterfront could be to establish more varied uses for the waterfront in Varanasi. The state government could set aside more land near the waterfront for tourist facilities, recreation (for the local residents), and other economic activities. This model will be more economically sustainable in the long term as the commercial activities will contribute more to the economy. Some of this money could be used to fund efforts to maintain the waterfront and clean up the river. Box 3 describes the success of the Manado waterfront in Indonesia which adopted this strategy.

Box 3: Mixed Use Waterfront for Tourism and Recreation in Manado, Indonesia

Manado is the capital city of the province of North Sulawesi in Indonesia. When the Manado government decided to redevelop the Manado waterfront, it made a deliberate effort to design the waterfront to serve multiple uses. In particular, the waterfront was designed as a tourist destination as well as a place of recreation for local residents. By attracting tourists to the waterfront through the provision of various facilities and services targeted at them, the waterfront generated economic value for the city. Local residents also visited the waterfront due to the presence of restaurants, cafes and public spaces (where performances were sometimes staged). This strategy of having a mixed use waterfront worked well as it helps to maintain the overall patronage of the waterfront through a more diversified base i.e. tourists and local residents. Furthermore, this also enabled the local residents to better appreciate the socio-economic value of the waterfront which might indirectly encourage them to upkeep and maintain the waterfront e.g. by supporting the government's efforts to keep the water clean. Overall, the project was considered to be a success. The quality of life of the local community improved and the economy also boosted as a result of the redeveloped waterfront.

Source: *Lagarense, BES — Evaluating Waterfront Uses for Tourism and Recreation with Acceptance to Changes: The Case of Manado Waterfront Development.*

When developing a waterfront development master plan for Varanasi, it would be good for the Uttar Pradesh state government to place more emphasis on anticipating potential issues and planning ahead to address them (as far as possible). By doing this, the number of problems encountered will likely be reduced, leading to a greater likelihood of success for the project. In order to develop this ability, relevant officers in charge of this area of work would need to be trained in techniques such as systems thinking and scenario planning. Box 4 describes how the Singapore

Box 4: Anticipating Issues and Planning Ahead to Address Them during the Clean-up of Singapore River

The Singapore River, with a length of 2.95 kilometres, was the main trade artery of Singapore in the late 18th century when the city-state was under the British colonial rule. However, rapid growth in its entrepot trade, population and industrial activities in the 19th century led to overcrowding of the area near the Singapore River and also polluted the river. Stench pervaded the area and organic and inorganic waste were perennially found floating in the waters. Although the colonial government made plans to clean up the river, the plans were not implemented due to financial constraints. It was only in February 1977 when the first Prime Minister of Singapore Lee Kuan Yew issued a directive to the Ministry of Environment to clean up the Singapore River within a ten-year timeline that the initiative gathered momentum and eventually yielded results. Within a period of ten years, the Singapore River was transformed into a clean river with the return of aquatic life.

The redevelopment and clean-up of the heavily polluted Singapore River in the late 1970s was a massive exercise. The main sources of river pollution were domestic and industrial waste, as well as trash thrown into the river (National Archives, 1969). The persons responsible for the pollution were those living along the river or in the river catchment areas such as squatters, farmers, boat-builders and hawkers. After identifying the root causes of pollution, action plans were drawn up to remove, resettle, relocate and phase out some of these activities. The resettlement of the population and the relocation and phasing out of industrial and commercial activities were the main challenges in the implementation of the clean-up operation. In 1977, there were

(*Continued*)

Box 4: (*Continued*)

still some 44,000 squatters living in unsanitary conditions (primarily unsewered premises around the river), and around 610 pig farms and 500 duck farms still discharging untreated waste into the river. Majority of the squatters used night soil buckets or pit latrines, while others used overhanging latrines that discharge waste directly into the river. There were also no pollution control facilities for the different industries along the banks of the river.

The number of unsewered premises that discharged waste into the river was reduced from 3,961 in 1977 to 36 by September 1981. Sand-washing was under control as the activity was centralised under a public holding company and monitored closely while private sand quarries were phased out. By 1987, all of the 11,847 night soil latrines and 621 overhanging latrines were phased out. Squatters and businesses were resettled under a resettlement plan that was introduced in the 1960s. Under the resettlement plan, the affected individuals and businesses were compensated and relocated to facilities of better quality than before. As a part of the resettlement plan, private land was acquired under the Land Acquisition Act before it was cleared for subsequent development. Eviction notices were issued by the Land Office, Resettlement Department or Jurong Town Corporation depending on who owned the land. Table 1 gives an indication of the success, scale and speed of the clearance of squatters. Despite

Table 1: Squatters Cleared in the Singapore River and Kallang Basin

		By 1979		By 1981		By 1983		By 1985	
Catchment Area	Target Numbers to be Cleared	Squatters Cleared	%	Squatters Cleared	%	Squatters Cleared	%	Squatters Cleared	%
Singapore River	3,959	1,097	27.7	1,921	48.5	3,213	81.2	3,744	94.5
Kallang Basin	42,228	9,657	22.9	24,781	58.7	34,596	86.9	40,830	96.7
Total	46,187	10,754	23.3	26,702	57.8	37,809	81.9	44,574	96.5

Source: Joshi, YK, C Tortajada, and AK Biswas — Cleaning of the Singapore River and Kallang Basin in Singapore: Economic, Social and Environment Dimensions.

(*Continued*)

Box 4: (*Continued*)

their initial resistance to the new sites offered, all the polluting industries were relocated by 1985 while all the squatters were resettled by 1986 (Joshi *et al.*, 2013).

To facilitate the whole resettlement process, the government ensured that new and suitable facilities are constructed and completed in time for the relocation of the affected people and businesses. The planning and construction of these new facilities is the crucial factor behind the success of the resettlement exercise. For example, high rise public housing estates were constructed within a radius of five miles from the city centre given that the activities of most people were generally confined to the city centre. Roads were built and public transport was also catered for. The high rise public housing flats were also equipped with piped water supply, gas and electricity — a vast improvement over the old facilities. The higher quality of these new facilities and their accessibility made them more acceptable to the affected individuals and businesses.

Source: Joshi, YK, C Tortajada, and AK Biswas — Cleaning of the Singapore River and Kallang Basin in Singapore: Economic, Social and Environment Dimensions.

government solved the problem of pollution in the Singapore River through an implementation plan with the use of forward planning in resettling the affected persons.

There are many obstacles in real life that hinders the development and implementation of a comprehensive waterfront development master plan. It is often better to take a more pragmatic approach of focusing on a few specific areas that have a greater chance of success than to persist with the ideal of trying to produce a comprehensive master plan. A relentless pursuit of the latter might lead to inaction due to the inability to reconcile the varied interests of different stakeholders. However, efforts must nevertheless be made to ensure that the broad directions and intentions of the master plan are intact so that the entire exercise remains meaningful.

This is very relevant to waterfront development in Varanasi as there are a sizeable number of stakeholders involved in the process. These stakeholders include the federal government, VDA, VMC, non-governmental organisations and civil society groups such as Sankat Mochan Foundation and INTACH Varanasi chapter, religious organisations, and communities dependent on the Ganga for their livelihood such as the boatmen and the washermen. Achieving a consensus on many issues pertaining to waterfront development might be a challenge for the Uttar Pradesh state government. The state government may, therefore, wish to identify the more critical issues and focus on them first in the waterfront development master plan. These issues will likely include direct discharge of untreated sewage into the Ganga, religious practices at the ghats, livelihoods of persons directly dependent on the Ganga and heritage conservation. Other issues might have to be accorded lower priority. The state government can also consider working with private landowners who are more amenable to development first, and focusing on a certain stretch of the waterfront instead of the entire waterfront. The development of Copenhagen's waterfront illustrates how this was put into practice by the Danish government (refer to details in Box 5).

Box 5: Taking a Focused Approach towards Waterfront Development in Copenhagen

In the late 1980s, Copenhagen became the 'growth locomotive' for Denmark in a bid by the Danish government to enable the country to compete internationally. Large infrastructure projects were started to attract investors to Copenhagen. At the same time, there was a revival of interest in waterfronts in Denmark which had been neglected due to economic restructuring and technological advances. Copenhagen's waterfront thus attracted the government's attention and efforts to rejuvenate the waterfront began in 1988.

Although all relevant aspects such as land use, environment and urban planning should ideally be considered in waterfront development so that a complete assessment can be made and subsequently a comprehensive solution could be developed, this is often difficult to achieve in reality. The Danish government, thus, decided to adopt a strategy of focusing on specific areas of Copenhagen's waterfront instead of trying to develop a comprehensive plan.

(*Continued*)

Box 5: (*Continued*)

The challenge in this case was to reconcile the differences between three major departments (the Finance Directorate, Building and Construction Administration, and the Department of Environment) under the Port of Copenhagen, all of which are responsible for urban planning. Each of these departments was managed by a politician and all three politicians happened to be from different political parties with distinctly different political objectives such that the Port of Copenhagen was somewhat dysfunctional. As a result of this, it was impossible to develop a comprehensive plan that is acceptable to all the parties concerned. The Port of Copenhagen thus decided on a more focused approach, placing greater emphasis on four areas: housing, offices, culture and entertainment. This enabled the waterfront development plan to proceed as there was generally less resistance among the different departments on issues relating to these areas.

Source: *Desfor, G and J Jorgensen — Flexible Urban Governance: The Case of Copenhagen's Recent Waterfront Development*

A study conducted by Smith and Ferrari (2012) revealed three typical modes of land use planning and urban development in countries situated near the North Sea. The first mode is a predominantly public sector led approach where land is acquired and subsequently developed by the local government. This mode tends to be the norm in Denmark, the Netherlands and Sweden. The second mode is a mixed approach where the public sector promotes development and coordinates acquisition of land often through joint ventures and partnerships. This mode is commonly used in France and Belgium. The last mode is a predominantly private sector led approach where land is privately owned and funding for development comes primarily from the private sector. The urban regeneration schemes for specific areas in the United Kingdom use this mode.

Given that most of the land near the waterfront in Varanasi is privately owned as well as the practical difficulties in land acquisition by the government, the private sector led approach may be more appropriate for waterfront development in Varanasi. The Uttar Pradesh state government can play the role of a facilitator to engage and persuade the religious institutions (and other parties) that own the land near the waterfront to develop the land in a way that is aligned to the waterfront development master plan. One of the greatest advantages in doing this is that it enables the government to garner greater support and buy-in for the master plan. In fact, VDA could even engage the religious institutions to develop guidelines on redevelopment and conservation of heritage buildings and temples (mostly owned by the religious institutions). An illustration of how a strong grassroots movement can drive waterfront development can be found in the development of the New Bedford waterfront in Massachusetts, United States (described in Box 6).

Box 6: The Role of Active Grass roots in Saving New Bedford's Waterfront

The city of New Bedford in Massachusetts, United States was the whaling capital of the world in the 19th century. With the decline of whaling activities over time, the waterfront of New Bedford also gradually lost its importance. In the 1980s, some members of the Waterfront Historic Area League together with the New Bedford Historical Society attempted to conserve the historic waterfront by developing a master plan for it. However, state funding for the project dropped just as the master plan was developed, leaving just a few thousand dollars for implementing the project. This project would have failed at this point due to lack of funds if not for the good fortune of a donation of a $500,000 building by Fleet Bank, a commercial lending institution. With this building, the conservation groups were able to continue with the project and eventually managed to convince Congress to designate the area as a National Historical Park, which, in turn, led to allocation of funds by the federal government for development purposes. From then on, the revitalization of the waterfront gained momentum. A $124 million aquarium was developed to serve as a tourist attraction. There were also plans to build a conference centre, a hotel, a parking garage and an IMAX theatre. In this case, it is the strong and active grassroots actions that enabled this waterfront development project to become a success.

Source: *Mullin, J, Z Kotval and C Balsas — Historic Preservation in Waterfront Communities in Portugal and the USA.*

4.3. *Bringing the institutions together*

After a waterfront development master plan has been developed for Varanasi, the next big task is to implement it. Projects have been known to fail during this critical stage where ideas and concepts are translated into reality. One of the ways to increase the chances of successfully implementing the waterfront development master plan for Varanasi is to create a new agency for this specific purpose. Although it is possible to assign this role to an existing government agency, this often does not produce the desired results. The government agency that has been assigned this role would still have its existing responsibilities and functions and thus often not dedicate sufficient attention and resources to the waterfront development project. As a result, there is a substantial risk of the waterfront development project not being implemented properly, according to timeline or at all.

In terms of structure, the new agency could be set up as a private, non-profit corporation instead of a governmental agency. This will give it greater flexibility in raising funds through private donors and international agencies. In terms of operations, it will also be more nimble as it can potentially side-step some of the red tape that is typically associated with government agencies. The government could lend its support for this agency e.g. by providing some start-up capital for the agency. Another possibility is to nominate some senior government officials to sit on the agency's Board (as Directors). This has the added advantage of facilitating interactions between the agency and other government agencies which will be involved in implementing the waterfront development master plan. An illustration of such an approach can be found in Box 7 which describes how the city of Minneapolis set up a separate agency to oversee implementation of the Upper River Master Plan.

Besides this, another area that deserves more attention is institutional coordination in implementing the waterfront development master plan (especially since this is one of the key challenges to waterfront development in Varanasi as highlighted earlier). Given the large number of agencies that have a direct or indirect role in waterfront development, the extent of coordination among these agencies could mean the difference between success and failure of the project. Agencies that would likely be

Box 7: Creating the Upper River Development Corporation to Oversee Implementation of the Upper River Master Plan in Minneapolis

A group of national advisors in the areas of real estate, conservation and waterfront development, who were invited to critique the concept plan alternatives during the development of the concept plan for the Upper River Master Plan, recommended that a dedicated agency be created to facilitate implementation of the master plan. With a committed staff focused solely on implementing the master plan, this ensures that the master plan will not be forgotten or accorded low priority as is likely to happen if a government agency whose core responsibilities lie elsewhere were to be tasked with this job.

In the case of Minneapolis, it was recommended that a private, non-profit corporation, the Upper River Development Corporation (URDC), be created to promote and lead the implementation of the master plan. There are a number of advantages in creating a private, non-profit corporation for this purpose. For example, private, non-profit agencies tend to be more successful than government agencies in lobbying for grants from private foundations. Furthermore, they would also not be perceived as rival agencies by other government agencies thus enabling them to cultivate better relationships with these government agencies. An outline of the main features of URDC together with the advantages and disadvantages of instituting it as a private, non-profit corporation can be found in Table 2.

URDC would be the advocate for the redevelopment effort and also help to build support in the community by engaging the people, businesses and private landowners in the neighbourhood. URDC would also facilitate inter-agency coordination and discussion with regards to implementation of the master plan. The main responsibilities of URDC staff would be to implement decisions made by its Board of Directors and to recommend actions that promote implementation of the master plan. In particular, they would be involved in identifying priorities in the implementation process, advising elected officials and commissions, writing grant requests to potential donors, communicating with the citizens and the media, and fundraising.

Source: *City of Minneapolis — Above The Falls — A Master Plan for the Upper River in Minneapolis*

(*Continued*)

Box 7: (*Continued*)

Table 2: Proposed Upper River Development Corporation as a Private, Non-Profit Corporation

Characteristics	Examples	Advantages	Disadvantages
• More development-oriented than a multi-agency management council. • Board is independent of local government. • Functions include: negotiating agreements with developers, coordinating public and private development activities, contracting for design and maintenance of public improvements, and fundraising.	• Charles Center Inner Harbor Management, Baltimore • St. Paul Riverfront Development Corporation • Riverfront Recapture, Hartford • Sheyboygan Development Corporation	• Free from certain constraints typical of public bodies. • Enjoys privacy in negotiation and financial decisions. • Partially sheltered from political pressures. • May receive private or charitable donations. • May take on less profitable ventures than would a for-profit company.	• No bonding, taxing or eminent domain authority. • Less willing to subsidise risky ventures than a purely public body. • May be somewhat less responsive to public opinion.

Source: *City of Minneapolis — Above The Falls — A Master Plan for the Upper River in Minneapolis.*

involved in waterfront development in Varanasi include VDA, VMC, NMCG and others. The clean-up of the Singapore River gives a good illustration of effective institutional coordination in practice (refer to Box 8 for details).

Since water is a state subject in India, the Uttar Pradesh state government will likely need to take a leadership role in initiating and managing institutional coordination among all the agencies involved in waterfront

Box 8: Coordination Among Institutions Involved in Singapore River Clean-up Project

The Ministry of Environment was tasked to achieve the goal of cleaning up the Singapore River and was the key coordinator behind the operation which involved various other agencies. Drainage engineers from the Public Works Department (PWD) and water engineers from the Public Utility Board (PUB) had to identify and solve the environmental issues on river pollution. Phasing out of farming activities was overseen by the Primary Production Department while the resettlement exercise itself required close collaboration between many agencies including the Land Office, Resettlement Department in the Housing and Development Board (HDB), Jurong Town Corporation (JTC), Ministry of Environment, the Urban Redevelopment Authority (URA) and the Port of Singapore Authority (PSA). Hawkers were relocated to new food centres built by HDB and URA. The cottage industries and backyard trades were moved to more appropriate facilities in industrial estates built by JTC and HDB. Lighterage activities and lighters were moved to new mooring and upgraded facilities provided by PSA. PUB, HDB and JTC also had to work closely to ensure that the new housing and industrial developments had the necessary infrastructure network to connect to direct utilities services and for sewage drainage. All these agencies worked very well together with a common goal and helped to make the project a great success.

Source: *Joshi, YK, C Tortajada and AK Biswas — Cleaning of the Singapore River and Kallang Basin in Singapore: Economic, Social and Environment Dimensions.*

development in Varanasi. An advantage of this arrangement is that the state government might be in a better position to gather resources (especially money) to support the waterfront development project than other agencies. Although it is advocated that the state government plays a central role in waterfront development in Varanasi, it is nevertheless possible to integrate this with a more decentralised approach for other aspects of the city's management and development. For example, a highly centralised approach to waterfront development whereby the state government maintains an oversight of the development of the entire waterfront is fully compatible with a decentralised solid waste management system in Varanasi as proposed in Chapter 7.

One of the ways to facilitate institutional coordination in a project is to set up an empowered high level steering committee with representatives from the agencies that are more heavily involved in the project. In order to ensure that they have sufficient authority to make critical decisions during the steering committee meetings, the representatives should preferably be the highest ranking executives of their respective agencies such as Chief Executive Officer, Chairman, and Managing Director. This high powered steering committee that has a substantial degree of flexibility in carrying out its work provides a forum where these senior executives could interact and discuss freely with one another in relation to the waterfront development project thus enabling greater cooperation among the agencies. This was practised by the Danish government during the development of Copenhagen's waterfront (refer to Box 9 for details). For Varanasi, a similar practice could be adopted whereby the top executives of VDA, VMC, NMCG and other relevant agencies would represent their agencies at the steering committee for the waterfront development project.

Box 9: Flexible Urban Governance in Copenhagen's Waterfront Development

As a result of a recommendation in the Copenhagen Harbor Committee's report, a "Vision" group was formed to ensure that waterfront development in Copenhagen proceed in a reasonable and timely manner. The Vision group or the "flexible urban governance" group comprised of high-level powerful politicians and bureaucrats who have the authority and autonomy to make decisions on behalf of their agencies and who can influence other members of their agencies. As conflicts often arise between different stakeholders in developing and implementing a waterfront development master plan due to different interests, the Vision group provided a platform for these representatives to discuss and resolve contentious issues pertaining to the implementation of the master plan.

Source: *Desfor, G and J Jorgensen — Flexible Urban Governance: The Case of Copenhagen's Recent Waterfront Development.*

4.4. *Examining the process*

One of the common mistakes made in waterfront development is that much effort is spent on developing the master plan but insufficient attention is paid to the process through which this is achieved. It is therefore worthwhile to examine the process in greater detail. One of the good practices that the Uttar Pradesh state government could consider is to provide more options for the waterfront development master plan during the preliminary phase of its development. By providing more choices, there is greater room for creativity and this allows better and more innovative designs and solutions to be developed. In addition, better decisions can also be made as the decision makers have the opportunity to compare different features of the alternatives before deciding on a particular option.

Besides this, the state government could also think of ways to encourage/enable greater public participation in the process of developing the waterfront development master plan. As mentioned earlier, enabling public participation is one of the key challenges to waterfront development in Varanasi. The advantages of public participation have been acknowledged by various scholars in the field of public administration. For example, Irvin and Stansbury (2004) have argued that citizens gain from public participation through learning from and informing government representatives, persuading and enlightening the government, and gaining skills for activist citizenship while the government gains from public participation through learning from and informing the citizens, persuading and building trust with the citizens, establishing greater legitimacy for decisions made, and building strategic alliances with the citizens.

In addition, Ahmed (1994) also noted that the people's participation is often the main factor contributing to the success and sustainability of development projects. Moreover, public participation should not be seen as simply involvement of citizens. It should be carried out in a meaningful manner as "meaningful participation is concerned with achieving power, that is, the power to influence the decisions that affect one's livelihoods" (Oakley and Marsden, 1984). People are more likely to see the value of a project if they are a part of it, their voices are heard, and their needs and concerns are noted.

In the context of Varanasi, it would be good to seek the input of stakeholders such as VDA, VMC, the federal government, religious authorities, boatmen, washermen, residents, tourist agencies, non-governmental organisations and the general public when developing the master plan. For a start, the Uttar Pradesh state government can first develop awareness among these stakeholders so that participation is more informed. Thereafter, these stakeholders could even be given the opportunity to critique the preliminary proposals for the waterfront development master plan. Undoubtedly, managing all these stakeholders would not be easy given that they may have different self-interests. Nevertheless, the benefits of engaging and involving them in the process would likely outweigh the costs of doing so. The sense of empowerment and the higher level of buy-in that are created would significantly increase the likelihood of the project's success. Box 10 describes how the city of Minneapolis made use of these strategies in developing their Upper River Master Plan and also reinforces the viability of having varied uses for a waterfront.

Box 10: Developing a Concept Plan for the Upper River Master Plan in Minneapolis

After a careful analysis of the existing conditions and based on the stated planning objectives of the Upper River Master Plan, three concept plan alternatives were created. They were "Heavy Industries and Parks", "Parks and Light Industry" and "Parks and Residential". Comments on these concept plan alternatives were then solicited during a series of public meetings, small group discussions, presentations to elected officials, and regular meetings with Park Board, City Planning, Minneapolis Community Development Authority, and County staff sitting as the Upper River Master Plan technical advisory committee. In addition, a group of national advisors in the areas of real estate, conservation and waterfront development were also invited to critique the plans. A sample of the "Parks and Residential" concept plan is shown in Figure 1. Besides a description of the concept plan and its main features, a critique of the plan and implementation issues in relation to the plan were also included based on the comments received from various stakeholders

(*Continued*)

Box 10: (*Continued*)

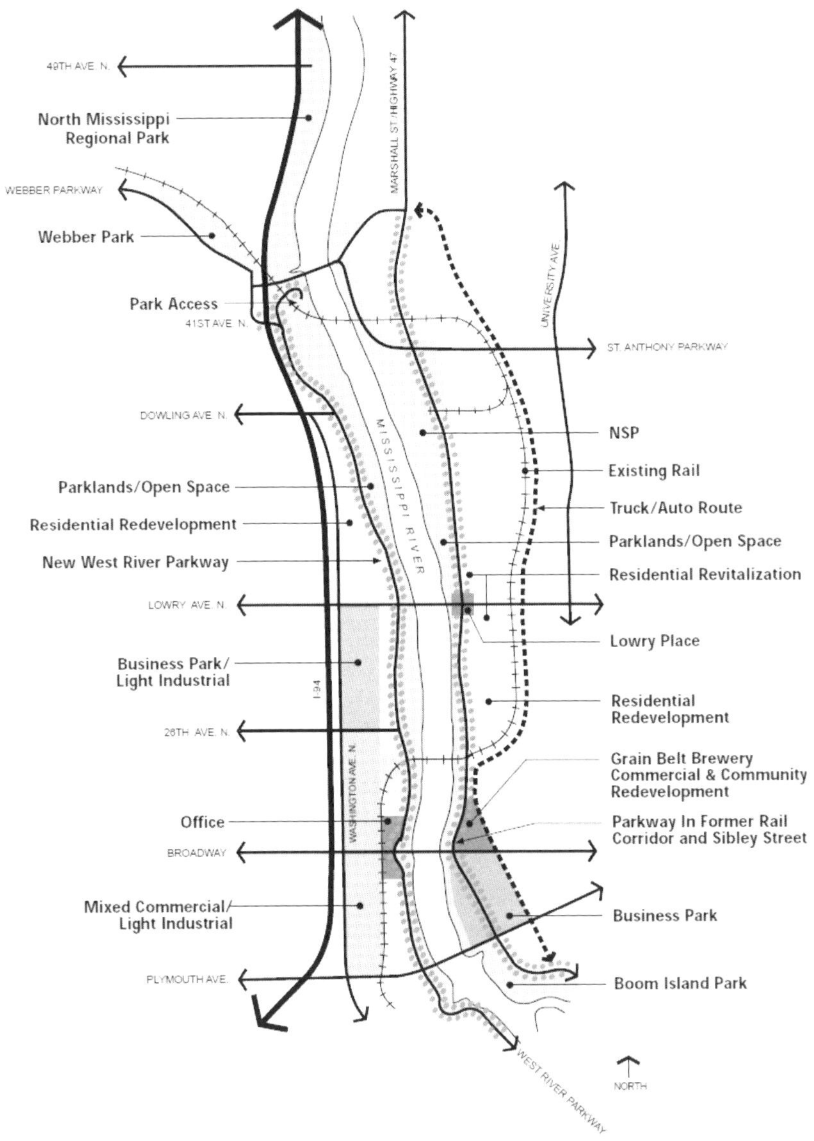

Figure 1: Parks and Residential Concept Plan "The River Green"

Source: *City of Minneapolis — Above The Falls — A Master Plan for the Upper River in Minneapolis.*

(*Continued*)

Box 10: (*Continued*)

(as described earlier). This provides a comprehensive overview of the entire concept plan and facilitates the evaluation process that comes thereafter.

A set of assessment criteria was developed to enable evaluation of the three concept plan alternatives. After the evaluation was completed, a "Preferred Plan" concept was synthesized by adopting recommendations from the most suitable concept plan alternative (which was "Parks and Residential") with some further refinements. This thorough and well thought out process contributed to the successful development of a concept plan that is widely accepted by all the stakeholders.

Source: *City of Minneapolis — Above The Falls — A Master Plan for the Upper River in Minneapolis*

Another aspect that the Uttar Pradesh state government may wish to place greater emphasis on is enforcement of property rights with regards to public land near the waterfront. As early as 2000 and 2001, the state government had issued orders number 320/9-A-3-2000-127 of 5 February 2000, and 840/9-A-3-2001 of 11 April 2001 that forbid any development activity from taking place within 200 metres from the riverbank for all the towns situated along the Ganga in the state (TNN, 2010). However, encroachments and housing clusters are still easily spotted along the riverbanks in Varanasi despite these orders. At places where VDA had taken action to demolish these illegal developments, little has been done to restore the affected buildings or defaced ghats (Dikshit, 2014a). In view of the current situation, the state government should make more effort to take firm punitive actions against the offenders. To facilitate this, the state government could set more explicit goals for VDA in relation to this issue, monitor VDA's work more closely to ensure that they achieve the goals set for them (e.g. enforcement of the regulations, demolition of illegal developments and redevelopment of 'reclaimed' public land), or even be directly involved in the exercise to demolish the illegal developments and restore the buildings thereafter.

Finally, the state government should have a greater focus on acquisition of privately owned land near the waterfront to enable a more coherent design and use of the waterfront. This is also one of the key challenges to waterfront development in Varanasi as mentioned earlier. Although a private sector led approach was advocated for waterfront development in Varanasi earlier in this chapter, the government may still need to acquire some parcels of land near the waterfront to facilitate the implementation of the waterfront development master plan. A possible scenario for this is when a particular land parcel is critical to the waterfront development master plan and the private landowner does not want to or agree with the government's suggestions on development of the land parcel. It may then be inevitable for the government to acquire the land parcel in order for the waterfront development master plan to proceed.

However, although the Uttar Pradesh state government has legal powers to acquire land under the Land Acquisition Act, it is generally reluctant to do this in practice as it is a politically unpopular exercise. On the other hand, the state government may be severely constrained in terms of what it can do along the waterfront so long as critical land parcels are under private ownership and the owners do not cooperate with the government (despite its efforts to engage them). Therefore, if the state government is serious about developing the waterfront in Varanasi, it must have the political will to carry out this unpopular task if and when the situation arises. It may need to engage the public and work with the media to communicate the rationale of land acquisition so that there is greater public understanding of and support for this exercise. Once the government is able to secure the cooperation of landowners of the more critical land parcels or has acquired control of these land parcels, it will be in a better position to implement its waterfront development master plan.

5. Principles for Waterfront Development in Varanasi

After analysing the specific context of Varanasi and understanding some good practices for processes and governance from other successful waterfront development projects, a set of principles that help to ensure the long term environmental and economic sustainability of the waterfront is derived. These principles are given below.

1. Diversify uses of the waterfront to avoid excessive burden on any particular aspect of the waterfront
2. Enable a reasonable level of public participation in a meaningful way to secure the success of the project
3. Adopt a long term perspective in the planning of waterfront development with the understanding that revitalisation is an on-going process

6. Conclusion

According to Kumar *et al.* (2010), there is no proper land use planning in Varanasi. This has resulted in land degradation as well as devastation of the valuable ecosystem, with serious repercussions for Varanasi especially in the long term. The current study has drawn lessons on some good practices for processes and governance from successful waterfront development projects around the world and shown how they can be applied to Varanasi. A summary of these lessons is provided below:

Taking a Strategic Approach

- Establish a clear set of high level guidelines in relation to waterfront development in Varanasi at the outset
- Formulate a well-defined strategy to implement the waterfront development master plan
- Establish more varied uses for the waterfront in Varanasi
- Anticipate potential issues and plan ahead to address them during the development and implementation of the waterfront development master plan
- Identify the more critical issues and focus on them first in the waterfront development master plan
- Engage and persuade owners of land near the waterfront to develop the land in a way that is aligned to the waterfront development master plan

Bringing the Institutions Together

- Create a new agency in the form of a private, non-profit corporation to oversee the implementation of the waterfront development master plan
- The Uttar Pradesh state government will need to take a leadership role in initiating and managing institutional coordination among all the agencies involved in waterfront development in Varanasi
- Set up an empowered high level steering committee with representatives from the agencies that are more heavily involved in the waterfront development project

Examining the Process

- Provide more options for the waterfront development master plan during the preliminary phase of its development
- Actively involve the stakeholders when developing and implementing the master plan
- Enhance enforcement of property rights with regards to public land near the waterfront
- Cultivate the necessary political will to carry out land acquisition near the waterfront where necessary

By adopting the strategies mentioned above, the Uttar Pradesh state government would be able to have better planning, development and management of the waterfront in Varanasi. This will, in turn, help to address the pollution of the Ganga in Varanasi since the waterfront is the point at which untreated sewage and other pollutants such as half burnt bodies enter the Ganga. In addition, this will also facilitate the conservation of the heritage buildings and ghats that make up the magnificent waterfront in Varanasi.

A set of principles for waterfront development in Varanasi is then derived based on these lessons and an understanding of Varanasi's context. These principles (given below) help to ensure the long term environmental and economic sustainability of the waterfront.

1. Diversify uses of the waterfront to avoid excessive burden on any particular aspect of the waterfront
2. Enable a reasonable level of public participation in a meaningful way to secure the success of the project

3. Adopt a long term perspective in the planning of waterfront development with the understanding that revitalisation is an on-going process

It should however be pointed out that there are various governance challenges that may hinder waterfront development in Varanasi. Firstly, the central government and the state government are under the control of different political parties — the Bharatiya Janata Party and the Samajawadi Party respectively. Differing political ideologies, views and interests might result in conflict and non-cooperation between these two levels of government (Singh, 2015). This would impede the development and implementation of the waterfront development master plan.

Secondly, the bureaucracy in India tends to be entangled in red tape. In the World Bank Ease of Doing Business Index 2014, India ranked 142 out of 189 countries in the world (World Bank, 2015). The pace of developing and implementing the waterfront development master plan would foreseeably be slow. There is also a substantial risk that the master plan would not even be developed or implemented due to the obstacles posed by the bureaucracy.

Thirdly, corruption is still a challenge in India. In the Corruption Perceptions Index 2014, India ranked 85 out of 175 countries (Transparency International, 2015). The "I Paid a Bribe.com" website gives an illustration of the pervasiveness of corruption in the country (IPaidABribe.com, 2015). It would be difficult to ensure that funds and other resources dedicated for the waterfront development project are not siphoned off given the large scale of the project.

Finally, given the religious significance of Varanasi and the Ganga, there may be some deeply ingrained religious beliefs and practices that are hard to change. For example, cremation of dead bodies and spreading of the ashes into the Ganga in Varanasi, placing religious offerings into the Ganga and other practices are still firmly adhered to by many Hindus. These may be formidable obstacles to development of the waterfront.

Despite these challenges, efforts should be made to pay more attention to waterfront development in Varanasi as the waterfront is probably one of the most important aspects of the city. Neglecting the waterfront will be a huge costly mistake that could have serious adverse repercussions for this ancient city.

References

Ahmed, S (1994). The Rhetoric of Participation Re-examined: The State, NGOs, and Water Users at Varanasi, Uttar Pradesh, India. *Environmentalist*, 14 (1), 3–16.

Bruttomesso, R (2001). Complexity on the Urban Waterfront. In *Waterfronts in Post-Industrial Cities*, Marshall, R (ed.). London and New York.

Business Standard (2013). India's Median Per Capita Income Lowest Among BRICS: Gallup. *Business Standard*, 16 December. Accessed on 15 June 2015. Available at http://www.business-standard.com/article/economy-policy/india-s-median-per-capita-income-lowest-among-brics-gallup-113121600968_1.html.

Census2011.co.in (2015). Varanasi (Varansi) District: Census 2011 Data. Accessed on 22 May 2015. Available at http://www.census2011.co.in/census/district/568-varanasi.html.

Central Pollution Control Board (2014). Presentation to NUS LKYSPP students during India Study Trip, 1 December.

City of Minneapolis (2011). Above The Falls: A Master Plan for the Upper River in Minneapolis. Accessed on 6 March 2015. Available at http://www.minneapolismn.gov/cped/planning/master-plans_above-falls_index.

Deshpande, MS (2010). History of the Indian Caste System and Its Impact on India Today. Research Proposal, California Polytechnic State University.

Desfor, G and J Jorgensen (2004). Flexible Urban Governance: The Case of Copenhagen's Recent Waterfront Development. *European Planning Studies*, 12 (4), 479–496.

Dikshit, R (2013). Masterplan-2031 Exposes Varanasi Development Authority's Failure. *The Times of India*, 14 April. Accessed on 30 April 2015. Available at http://timesofindia.indiatimes.com/city/varanasi/Masterplan-2031-exposes-Varanasi-Development-Authoritys-failure/articleshow/19546611.cms.

Dikshit, R (2014a). Demolished Structures Deface Varanasi Riverfront. *The Times of India*, 7 September. Accessed on 12 May 2015. Availabe at http://timesofindia.indiatimes.com/city/varanasi/Demolished-structures-deface-Varanasi-riverfront/articleshow/41894549.cms.

Dikshit, R (2014b). City Houses 227 Slums on 17% of Total Area. *The Times of India*, 7 Feb. Accessed on 23 May 2015. Available at http://timesofindia.indiatimes.com/city/varanasi/City-houses-227-slums-on-17-of-total-area/articleshow/29981477.cms.

Eckstut, S (1986). Solving Complex Urban Design Problems. Paper presented at the Symposium on Waterfront Development at the ASCE Boston Convention, 28–30 October.

Green Pilgrimage Network (2014). Pilgrim Numbers. Accessed on 10 May 2015. Available at http://greenpilgrimage.net/resources/pilgrim-numbers/.

Hays, J (2011). Hindu Funerals. Accessed on 10 May 2015. Available at http://factsanddetails.com/world/cat55/sub354/item1343.html.

IPaidABribe.com. (2015). I Paid A Bribe. Accessed on 2 June 2015. Available at http://www.ipaidabribe.com/bribe-trends#gsc.tab=0.

Irvin, RA and J Stansbury (2004). Citizen Participation in Decision Making: Is it Worth the Effort? *Public Administration Review*, 64 (1), 55–65.

Iyer, RR (2011). Should Water Be Moved to Concurrent List? *The Hindu*, 23 June. Accessed on 10 May 2015. Available at http://www.thehindu.com/opinion/lead/should-water-be-moved-to-concurrent-list/article2113384.ece

Jalais, S (2008). The River Front of Benares: between 'Sacred' Waters and Sewage Water. Paper presented at the 8th World Wide Workshop for Young Environmental Scientists WWW-YES-2008 — Urban Waters: Resource or Risk?, Creteil, France, 13–16 May.

Joshi, YK, C Tortajada and AK Biswas (2012). Cleaning of the Singapore River and Kallang Basin in Singapore: Economic, Social and Environment Dimensions. *International Journal of Water Resources Department*, 28 (4), 647–658.

Joshi, YK, C Tortajada and AK Biswas (2013). *The Singapore Water Story: Sustainable Development in an Urban City-State*. Singapore: Routledge.

Kumar, M, N Mukherjee, GP Sharma and AS Raghubanshi (2010). Land Use Patterns and Urbanization in the Holy City of Varanasi, India: a Scenario. *Environ Monit Assess* 167 (1–4), 417–422.

Kumar, M (2014). JNNURM Failed — UPA's Urban Housing Scheme Ineffective. *Niti Central*, 22 July. Accessed on 23 May 2015. Available at http://www.niticentral.com/2014/07/22/jnnurm-failed-upas-urban-housing-scheme-ineffective-234424.html.

Lagarense, BES (2013). Evaluating Waterfront Uses for Tourism and Recreation with Acceptance to Changes: The Case of Manado Waterfront Development. *GSTF Journal on Business Review*, 3 (1), 114–122.

Mail Online India (2013). India's Literacy Rate Rises to 73 Per Cent As Population Growth Dips. *Mail Online India*, 1 May. Accessed on 15 June 2015. Available at http://www.dailymail.co.uk/indiahome/indianews/article-2317341/Indias-literacy-rate-rises-73-cent-population-growth-dips.html.

Midgley, J (1986). Community Participation: Histories, Concepts and Controversies. In *Community Participation, Social Development and the State*, Midgley, J *et al*. (ed.). London: Methuen.

Millspaugh, ML (2001). Waterfronts as Catalysts for City Renewal. In *Waterfronts in Post-Industrial Cities*, Marshall, R (ed.). London and New York.

Ministry of Water Resources, River Development and Ganga Rejuvenation. (2014). River Boards Act 1956. Accessed on 10 May 2015. Available at http://wrmin.nic.in/forms/list.aspx?lid=439

Mullin, J, Z Kotval and C Balsas (2000). Historic Preservation in Waterfront Communities in Portugal and the USA. *Landscape Architecture and Regional Planning Faculty Publication Series* Paper 13. Scholarworks@UMass Amherst. Available at http://scholarworks.umass.edu/larp_faculty_pubs/.

National Archives (1969). Interim Report on Improvement to Rivers and Canals in Singapore. *PWD/223/53,* Vol. IV, 181A.

National Mission for Clean Ganga (2015). About NMCG. Accessed on 8 June 2015. Available at https://nmcg.nic.in/about_nmcg.aspx.

Nina (2011). The Burning Ghats of Varanasi, Uttar Pradesh. *Water & Megacities*, 12 October. Accessed on 10 May 2015. Available at http://www.waterand megacities.org/the-burning-ghats-of-varanasi-uttar-pradesh-3/

Oakley, P and D Marsden (1984). *Approaches to Participation in Rural Development*. Geneva: International Labour Office.

PTI (2013). Lok Sabha Passes Land Acquisition Bill. *The Economic Times*, 29 August. Accessed on 10 May 2015. Available at http://articles.economic-times.indiatimes.com/2013-08-29/news/41582220_1_land-acquisition-bill-resettlement-bill-wakf-land.

Rogers, J (2013). India's Polluted Ganges River Threatens People's Livelihoods. *DW*, 21 November. Accessed on 10 May 2015. Available at http://www.dw.de/indias-polluted-ganges-river-threatens-peoples-liveli hoods/a-17237276.

Samperi, SJ (1986). Waterfront Development: Getting It Started — The Public Sector Role. Paper presented at the Symposium on Waterfront Development at the ASCE Boston Convention, 28–30 October.

Sankat Mochan Foundation (2015). Pollution in Ganga & Ganga Action Plan Failures. Accessed on 10 May 2015. Available at http://www.sankatmochan foundationonline.org/PollutionofGanga.html.

Singh, RPB (1955). *Banaras: A Study in Urban Geography*. Varanasi: Banaras Hindu University, Nand Kishore and Bros.

Singh, RPB (1993). Varanasi as Heritage City on the Scale the UNESCO World Heritage List: From Contestation to Conservation. In *Banaras (Varanasi): Cosmic Order, Sacred City, Hindu Traditions*, Singh, RPB (ed.). Varanasi: Tara Book Agency.

Singh, R (2015). Varanasi's Development Seems to Have Stuck in Red-Tapism. *DNA*, 20 February. Accessed on 8 June 2015. Available at http://www.

dnaindia.com/india/report-varanasi-s-development-seems-to-have-stuck-in-red-tapism-2062555

Smith, H and MSG Ferrari (2012). *Waterfront Regeneration: Experiences in City-Building*. London: Routledge.

TNN (2010). Ganga Continues to Be Exploited. *The Times of India*, 9 January. Accessed on 12 February 2015. Available at http://timesofindia.indiatimes.com/city/varanasi/Ganga-continues-to-be-exploited/articleshow/5425476.cms.

TNN (2014a). Varanasi City Master Plan Gets Going Without Budget. *The Times of India*, 23 July. Accessed on 22 May 2015. Available at http://timesofindia.indiatimes.com/india/Varanasi-city-master-plan-gets-going-without-budget/articleshow/38913719.cms.

TNN (2014b). Hopes Soar Again for Masterplan 2031. *The Times of India*, 30 June. Accessed on 23 May 2015. Available at http://timesofindia.indiatimes.com/city/varanasi/Hopes-soar-again-for-Masterplan-2031/articleshow/37510681.cms

TNN (2015a). Kyoto Varsity Team in Holy City Today. *The Times of India*, 26 March. Accessed on 23 May 2015. Available at http://timesofindia.indiatimes.com/city/varanasi/Kyoto-varsity-team-in-holy-city-today/articleshow/46698256.cms.

TNN (2015b). New Policy to Quicken Riverfront Land Acquisition. *The Times of India*, 18 March. Accessed on 23 May 2015. Available at http://timesofindia.indiatimes.com/city/ahmedabad/New-policy-to-quicken-riverfront-land-acquisition/articleshow/46603587.cms.

TOI (2013). Varanasi Development Authority Makes Masterplan-2011 Public at Eight Sites. *The Times of India*, 26 April. Accessed on 23 May 2015. Available at http://timesofindia.indiatimes.com/city/varanasi/Varanasi-Development-Authority-makes-Masterplan-2011-public-at-eight-sites/articleshow/19732265.cms.

Transparency International (2015). Corruption Perceptions Index 2014: Results. Accessed on 2 June 2015. Available at http://www.transparency.org/cpi2014/results#myAnchor1.

World Bank (2015). Doing Business: Measuring Business Regulations. Accessed on 2 June 2015. Available at http://www.doingbusiness.org/rankings.

World Wildlife Fund (2015). Environmental Problems in India. Accessed on 25 June 2015. Available at http://wwf.panda.org/who_we_are/wwf_offices/india/india_environmental_problems/.

Chapter 8

DATA GOVERNANCE IN FOSTERING POLICY COHERENCE AND COLLABORATION FOR CLEANING THE RIVER GANGA

Doris Maharlika H Dizon, Prudente S Parcia, Rakesh Kumar, Wai Min Kwok and Maitreyee Mukherjee

1. Introduction

Efforts to clean up the Ganges River can be traced back to as early as 1985, through the Ganga Action Plan (GAP). Since then three decades of attempts to address acute pollution issue in the river basin have yielded no satisfactory results. With renewed thrust to rejuvenate River Ganga, coming directly from the current Indian Prime Minister Mr. Narendra Modi, various stakeholders including general public have got a fresh impetus to connect to the noble cause. As a result, availability of accurate data has gained utmost importance for undertaking right decisions towards desired goals. The multiplicity of agencies involved, along with limited governance capacity for managing the river's resources, have created substantial lapses in data collection and dissemination of information regarding river pollution, environmental flow, water quality and other parameters. Moreover, due to frequent overlapping of agency jurisdictions and their lack of authority, concrete steps towards policy formulation, implementation and subsequent evaluations still remains unclear.

According to World Bank India's Country Director Onno Ruhl and Senior Water Resource Specialist Genevieve Connors, consistent and accurate measurements should be considered critical as plans to rejuvenate the river are being pursued (Ruhl and Genevieve, 2014). Water quality bears a complex range of indicators to describe its physical, chemical and biological parameters. In addition, existence of multiple guidelines at global, national as well as local levels makes it a difficult task to set proper compliance standards. The challenges to effectively implement data governance in consistent monitoring of water quality changes resulting from industrial disposals, urban sewage and agricultural run-offs; are still significant in India due to existence of weak policies and inadequate technological equipment. To complicate matters, in the case of River Ganga, there is an uphill task of collating the existing fragmented data, identifying their authenticity, setting baseline standards and then defining future plans. Hence, efficient data governance forms an integral part of formulating an effective Ganga River basin management plan.

The lack of authoritative data and concrete evidence regarding impacts of Ganga Action Plan has led to polarised views and disparate arguments for its unsatisfactory results (Markandya and Murty, 2000). Although Comptroller and Auditor General (CAG Report, 2000) assessed that GAP could only accomplish 39% of its sewage treatment targets between 1993–2000, some researchers argued that conditions could have been worse had there been no initiatives at all. The GAP was a static plan which failed to take into account the phenomenal population increase and escalating economic activities the country would face in coming years. Moreover, coordination between Central and state organizations were very poor. One such example is the partial failure of state pollution control boards to collect requisite monitoring data due to lack of capacity or resources. In an official evaluation of the Central Pollution Control Board (CPCB) and State Pollution Control Boards (SPCBs) by the Indian

Institute of Management, Lucknow, it was observed that the inventory tracking of polluting sources had not been completed in spite of its inception since 34 years ago (IIM, Lucknow, 2010).

This chapter will, therefore, focus on the significance of data governance in the basin-wide management of the Ganga River. As numerous agencies are involved in collecting, storing and disseminating data to various stakeholders, it is essential to assess their responsibilities with reference to the kind of data they own, its accuracy and easy availability. By introducing the concept of data governance through the four pillars of policy, people, process and technology, we aim to apprise policy makers an in-depth analysis on how data governance can play a crucial role in fostering policy cohesion while enhancing collaboration on the numerous efforts to help clean the Ganga River. Availability of competitive data in public domain will also promote increased transparency and accountability of respective authorities. Lastly, efficient data governance can significantly benefit the general public as well; ensuring greater knowledge dissemination, awareness and their inclusive participation in this massive river-basin rejuvenation drive.

2. Relevant Literature and Policy Development on Data Governance

Data informs and empowers action. Without accurate, consistent, complete and reliable data, policymakers are handicapped. It may lead them in giving misleading, conflicting, or even potentially crippling policy directions. While data governance is an emerging area, it has proved to be relevant and impactful with broad applicability for managing large river basins. Data is that which helps to resolve uncertainty (Berson and Dubov, 2011). By learning from successful cases like that of the Mekong River and the European Union's (EU's) river basin management, it is evident that data governance is highly instrumental in operational management of river basins as well as pollution in water bodies. Effective data governance will result in harmonisation of all important records in measuring, monitoring and assessing the health of riverine ecosystem.

Data governance shares the characteristics of any other strategic governance process. One theoretical definition treats data governance as a set of processes, treating it as a strategic area within the organisation (Otto, 2011). The Master Data Management Institute, a research institute based in the United States, has defined data governance as the formal orchestration of people, processes and technology to enable an organisation to leverage data as an enterprise asset (Berson and Dubov, 2011). Yet another definition states data governance as "the organization and implementation of policies, procedures, structure, roles, and responsibilities which outline and enforce rules of engagement, decision rights, and accountabilities for the effective management of information assets" (Ladley, 2012). In a nutshell, effective data governance indeed is the adaptive use of authority, people and resources combined with appropriate processes and technologies to ensure the good governance of data.

Very often, official datasets tend to present more optimistic views on ground situation, which may be due to data errors or definitional problems. The questions of ownership and responsibility of providing reliable data are fundamental issues to be addressed first. At the same time, setting proper indicators, or not setting them, is often a political tactic to divert public attention from urgent issues (Tortajada and Chan, 2013).

The emphasis on data and its governance has grown exponentially in the last two decades due to advances made in data collection, processing and analytical technologies (Coleman *et al.*, 2009; McGilvray, 2006). The recent recognition of availability of government data in public domain in order to meet the rights of individuals, business and civic organisations to access and use government information has also enhanced the scope of data governance (UN E-Government Survey, 2014). India is no exception to this trend. The opportunities of e-governance is entering a new episode with the ambitious introduction of "Digital India" project. It is aimed to make India "a digitally empowered society and knowledge-based economy" (DNA India, 2015), encouraging government agencies to advance beyond e-governance to "m-governance", or mobile governance. The nation's first attempt to have Open Government Data was launched in 2012 as part of its Open Government Initiative (data.gov.in website). Although the country has been at the forefront in embracing e-governance in facilitating citizens' access to numerous online services, similar initiatives in water governance sectors are yet to be witnessed.

3. Data Governance — A Framework Approach

Since data governance requires cooperation and collaboration of various government agencies as well as participating stakeholders, it is essential to have a clear strategic vision and a strong political leadership. Issues concerning data authenticity, integrity and quality, as well as data sharing and its re-use standards are important, along with those related to data privacy and its protection against misuse (UN E-Government Survey, 2014). Data governance is a complex interaction of policies, people, and processes along with the effective use of technologies as catalysts or enablers. It is, therefore, necessary to first analyse the current state of data governance in Ganga River management, and then to craft a path forward to support evidence-based policy options. Simultaneously it is also crucial to formulate a data governance framework that is fundamental for both central and state agencies as a reference base.

As derived through the extended definition of the Master Data Management Institute, data governance can be seen as an association of four pillars — policy, people, processes and technologies (Berson and Dubov, 2011; TDWI website). While 'policy' forms an integral pillar to ensure legitimacy and institutionalisation of government actions; 'people' comprises various governmental agencies involved in data collection and monitoring. The third component 'processes' enable data sharing along with facilitating public participation; and finally, the adaptive use of latest 'technologies' supports effective data collection, monitoring and governance.

Figure 1 illustrates the adapted four-pillar framework to be used to evaluate the current state of data governance in the context of Ganga river basin management. Various key principles such as accountability, transparency, inclusiveness, efficiency, data sharing, public participation and use of modern technologies, as seen in recent debates of data governance evaluation, have been considered while applying thc above framework in following sections.

In evaluating the policy component, it is important to determine if there is sufficient **transparency and accountability**, along with **equity of access** to reliable data. The subsequent discussion shares some insight on the importance of data visibility as part of transparency in enabling evidence-based policy decision making. A well-defined policy for data

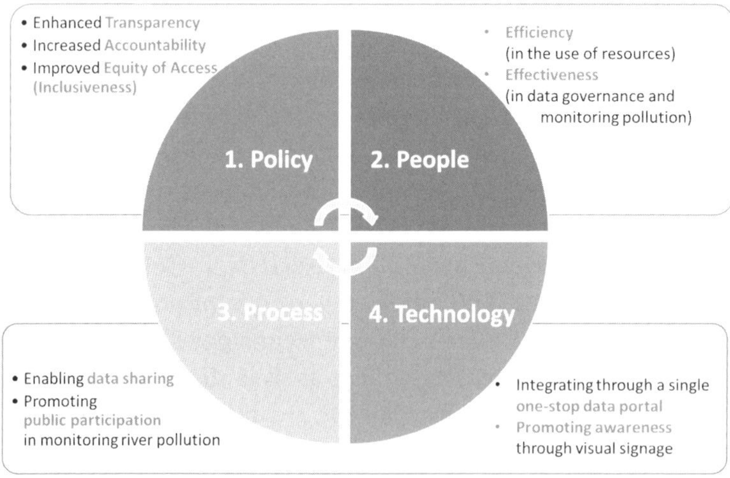

Figure 1: The 4-Pillar Data Governance Framework for Ganga River Basin

Source: Author's Compilation.

related to river basin management will increase transparency and accountability on behalf of respective authorities. It would also aid funding agencies as well as international organisations to assess the river management plans and their implementation feasibility before sanctioning fresh funds. Regarding equity of access to data, debate is still ongoing over whether to release data as open access, registered access or restricted access in public domain. This is linked to several concerns related to national security, policies over data classification and associated risks of having sensitive data in open public domain.

In assessing the second pillar that is people; we are looking at **effectiveness in use of resources as well as in monitoring and governance**. There is need to address responsibility, pro-activeness and efficiency of organisations in delivering the overarching objective of pollution abatement. Very often intrinsic bureaucracies among government agencies, along with overlapping jurisdictions results in obscuring institutional boundaries and respective responsibilities. Effective data governance needs to empower agencies to fuel better outcomes from mitigation efforts. Data quality involves accuracy, span, adequacy, timeliness and relevance of the given information. Both quality and quantity are equally important, in addition there is also a requirement to avoid data redundancy, resource burden and duplication of costs.

In appraising the third pillar that is processes; the essential consideration should be how to enhance data governance coverage through **data sharing, public participation**, public debate and anti-corruption efforts. While government agencies at various levels collect a wide range of data, they do not always share these data in ways that are easily accessible, operational or even comprehensible to general public. Access to clean, coherent data would allow stakeholders at different levels to develop new innovative insights to help mitigate the Ganga river pollution in a better way. However, it would indeed take a lot of time and resources to organise existing data and release them in form of a high-quality complete data-set. It is also important to have elaborate consultations among government agencies along with public participation, in order to identify which data to prioritise for release or improvement.

Finally, the full potential of technologies (the fourth pillar) in enabling data governance, is unlocked only when technologies are used within the context of well-defined governance processes supported by policies and institutions through an integrated approach across the chain of central and state government agencies. Technology should be used to enhance **public awareness** through **one-stop data portal** containing extensive, coherent and updated information. There is need of a clear strategy outlining the roles and responsibilities of institutions and other stakeholders as well as a comprehensive plan to guide the use of these tools in support of specific data governance goals. Equally important is the requirement of clarity on how these tools can help data governors or data stewards to deliver their mandates and at the same time provide compliance within the overarching service delivery framework at both central and state levels.

The subsequent sections in this chapter will provide detailed insights on each of the above mentioned four pillars of data governance, elaborating on the current state in context to river Ganga. This will then lead to proposed policy recommendations on the way forward, in harnessing the power and potential of data through its effective governance for abatement of pollution and long term sustainable management for the Ganga River.

3.1. *Policy as the founding step for laying data governance framework*

An effective data governance framework should be supported by clear and coherent policies. Data governance policies and their efficient implementation will ensure integrity and consistency of data towards competent river basin management. Data policies related to river water quality may include adherence to data rules, enforcing access rights to data, monitoring procedures and compliance with laws and regulations. These policies should also define the precise criteria, specifications, creation, usage and storage. Policies should guide governance decisions as to what data and indicator measurements are to be collected, stored and shared; what quality controls should be in place to assure their validity, and who should have access (Bhansali, 2014). Strong data policies, at the macro level, are fundamental towards realising the political vision and legal framework on various aspects of Ganga river basin governance.

Table 1 shows a list of the main legislations and policies relevant to water pollution and its monitoring in India.

As seen in Table 1, there are a number of legislations related to water resource management, including collecting, monitoring and sharing of data. Nonetheless, there is no specific policy on data governance that internalizes principles of enhancing transparency, promoting accountability or ensuring equity of access. Due to a lack of integrated policy approach towards comprehensive data governance, the functional co-ordination among agencies in the management of the Ganga River is highly unsatisfactory, mitigation efforts have failed to meet expectations, resulting in increasing water pollution crisis over the years.

Without interdisciplinary data co-ordination, different water resources are being managed separately and are not obligated to function in a comprehensive or integrated manner. This fragmented approach has affected policy making, including those at resource and data levels, therefore degrading both transparency and accountability (Reddy, 2009).

As seen in Table 1, prevention and control of water pollution in India started with the Water Act enacted in 1974 and Central and State pollution control boards were established under Ministry of Environment, Forest and Climate Change (MoEFCC); to implement the provisions of the Act.

Table 1: List of Legislations and Policies Related to Water Quality in India

Legislations/ Policies	Year	Brief Description
Indian Constitution (on water)	1949	Water is basically a state subject and the Centre has role only in dispute cases of inter-state river waters. List II in the 7th Schedule of constitution, dealing with subjects on which states have jurisdiction, has Entry 17 which says : "Water, that is to say, water supplies, irrigation and canals, drainage and embankments, water storage and water power subject to the provisions of Entry 56 of List I. Entry 56 of List I (Union list) reads as follows: "Regulation and development of inter-state rivers and river valleys to the extent to which such regulation and development under the control of the Union, is declared by Parliament by law to be expedient in the public interest"[1] (MOWR website, a).
Water (Prevention and Control of Pollution) Act[2]	1974	This Act provides for the prevention and control of water pollution and maintaining the integrity of water bodies. To achieve this objective, the Act provides for establishing the Central Pollution Control Board (CPCB) and respective State Pollution Control Boards (SPCBs) to be responsible for implementation of the provisions of the Act. Water pollution is broadly defined to include contamination or change in physical, chemical and biological properties of water or such discharge of sewage, industrial effluents or any kind of solid or gaseous matter into water which can render water harmful or pose as threat to domestic, commercial, industrial or agricultural use.
Water (Prevention and Control of Pollution) CESS Act	1977	This Act empowers SPCBs to levy and collect a cess from local authorities and specified industries on consumption of water for certain specified activities. This cess is meant to augment the funds required by SPCBs for their effective functioning in discharge of duties under the Water Act, 1974. Thus, it provides a market-based instrument for controlling pollution.

(*Continued*)

[1] http://wrmin.nic.in/forms/list.aspx?lid=298. Accessed on 28 May 2015.

[2] http://www.moef.nic.in/legis/water/wat1c1.html. Accessed on 28 May 2015).

Table 1: (*Continued*)

Legislations/ Policies	Year	Brief Description
Environment (Protection) Act	1986	This Act provides an umbrella framework for the coordination of central and state authorities established under Water Act. The Act authorizes the central government to protect and improve the environment and deal with all matters relating to pollution control.
National Water Policy[3]	2012	The National Water Policy lays down rules and regulations for planning and management of water resources through an integrated approach. Such an approach would take care of local, regional and national contexts and will be environmentally sound. The policy *inter alia* states that all hydrological data, except classified data should be placed in public domain. However, periodic review is required to further declassify data. It also calls for integration of all water related data with well-defined procedures and formats so that communication and information technology can be used for development of database for informed decision making.
National Data Sharing and Accessibility Policy (NDSAP)	2012	The aim of the NDSAP is to share and govern the data, generated by different government agencies, in digital format through a national data portal. This policy extended the mandate of Right to Information Act of 2005 by proactive disclosure of government data. The Open Government Data Platform has been established to collect and share open government data.
National Hydro-meteorological Dissemination Policy	2013	The policy is the result of the recommendations of the National Water Policy of 2012 for placing all hydrological data, other than those classified on national security consideration, in the public domain in open and transparent manner. According to this policy, all water data maintained by the Central Water Commission (CWC) and the Central Ground Water Board (CGWB) will be made available in the public domain on India-Water Resource Information System (India-WRIS).

[3] http://wrmin.nic.in/writereaddata/NationalWaterPolicy/NWP2012Eng6495132651.pdf. Accessed on 28 May 2015.

While Central Pollution Control Board (CPCB) is responsible for establishing or modifying the standards of water quality parameters to be maintained; it is the state level Pollution Control Boards (SPCB) that are in charge of enforcement and compliance of the Water Act to prevent and control pollution of water bodies at regional level. The CPCB receives data related to monitoring of water quality from SPCBs in different states. It collects and publish technical as well as statistical data pertaining to water pollution, along with disseminating relevant information to stakeholders. Thus, the main government agency with respect to the monitoring of the water pollution in the Ganga river basin is the CPCB along with its state counterparts.

Another government agency involved in the Ganga River Basin management is the National River Conservation Directorate (NRCD), also under MoEFCC. The NRCD is tasked upon to monitor water quality of River Ganga from Rishikesh in Uttaranchal to Uluberia in West Bengal through accessory institutions such as Pollution Control Research Institute (Haridwar), CPCB Zonal Office (Lucknow), Indian Institute of Technology (Kanpur), Patna University, Bharat Heavy Electricals Limited (BHEL), and Bidhan Chandra Krishi Vishwavidyalaya, Kalyani.

For hydro-meteorological data, Central Water Commission (CWC) is responsible for collecting relevant data as per Hydro-Meteorological Data Dissemination Policy, 2013. Similarly, for groundwater related matters, Central Ground Water Board (CGWB) collects the data for water quality and water level. In addition, Central Pollution Control Board also monitors groundwater quality through 807 water monitoring stations. Besides Water Act of 1974, there are fourteen other policy and legal provisions related to prevention of groundwater pollution (Garduno *et al.*, 2011). Moreover, there are multiple dedicated organisations like CGWB, CPCB, National Environmental Engineering Research Institute (NEERI), SPCBs at state level, for groundwater quality monitoring (Reddy, 2009).

Other central ministries involved in the water quality management are the Ministry of Urban Development, which works through various Urban Bodies in monitoring drinking water quality in the urban agglomerations; and the Ministry of Drinking Water and Sanitation, working through State Public Health Engineering Departments, is also monitoring drinking water quality in rural areas of the country (WFR website).

Hence we see that under current situation, there are numerous policies, legislations and institutional arrangements, authorising a variety of government agencies to deal with data pertaining to hydrological data as well as water quality information of water bodies all over the country. Each individual organisation may be maintaining large amount of data pertaining to its respective jurisdiction but this is not available in an integrated manner so as to provide a holistic view on the status of the water resources nation-wide, including that of Ganga river basin. Therefore, by simply making data available in public domain does not provide transparency or comprehensiveness, instead, it should be made available in a non-fragmented way such that the relevant data is more consistent and applicable to water resources issues from any perspectives.

The National Water Policy, 2012, calls for integration of all water related data within well-defined procedures and formats for development of an extensive database. This has resulted in formulation of hydro-meteorological data policy which is currently maintained by CWC and CGWB. But to enhance its scope, such hydro-meteorological information should be consolidated with other water pollution related records so that there is wholesomeness to the data linking it with the water quality parameters. It is to be noted that as part of the overall hydrological cycle, the Ganga River basin constitutes of its major riverine system, tributaries, surface water storages, groundwater resources and associated water catchment areas (GRBMP Interim Report, 2013). Thus, integration of all possible data related to hydrological as well as water quality parameters would present a holistic and reliable base for enactment of further action plans. Moreover, there should be one single agency responsible for collating all water related data to assure better process implementation and accountability through an existing data governance legislative framework arrangement.

Despite clear mandates given to CPCB and SPCBs, there are failures from some SPCBs to collect requisite monitoring data at regular intervals; on grounds of lack of capacity or resources (IIM, Lucknow, 2010). As per Comptroller and Auditor General of India Report No. 21 (CAG Audit report, 2011), on water pollution in India, though CPCB had established Environmental Data Bank but no data was received from Water Quality Monitoring agencies other than that from a handful of SPCBs. The data established under Environmental Data Bank is in public domain and can be accessed from the website of CPCB (http://cpcbedb.nic.in/).

As part of the government's drive for data availability CPCB is the main agency for maintaining an overall set of the water quality data through its network of manual as well as 10 real-time monitoring stations across the country. While it publishes reports on the pollution assessment of river Ganga in the form of periodic reports, the results of 10 real-time monitoring stations are being made available at its website (http://cpcb.rtwqms.com/). At the same time, CWC is also running real-time monitoring stations at three locations at Yamuna and two tributaries of Ganga (Ramganga and Gomti), publishing its collection at the following website http://cwc.rtwqms.com. Similarly, NRCD too monitors water quality parameters through its monitoring stations, some of which are available as part of the Annual Report of MoEFCC but it provides limited information from only a few monitoring stations only. Hence, there is a dire need for a central co-ordinated approach to integrate the entire network of real-time water quality monitoring stations in river Ganga for better coherence and effectiveness.

It might be worthwhile to point out here that inland water quality monitoring programs in India generally measure only basic parameters relating to pH, BoD, DO, temperature, nitrite, nitrate etc. On contrary, in other countries like the United States, it is required to prepare water quality inventory every two years for documenting the status of water bodies in order to identify those that has been adversely affected by over 65 classes of toxic compounds. This class of toxic parameters include both conventional (such as BoD, total suspended solids, fecal coliform) and non-conventional (such as ammonia, chlorine and iron) pollutants (Rajaram and Das, 2008). Such an inventory indeed helps to update stakeholders at regular intervals, regarding current status of water bodies, while initiating timely necessary actions on part of authorities.

The amount of flow in the river is another important factor in determining water quality and interpretation thereof. The potential effects of pollutants on the different kinds of water use and aquatic life depend largely on the environmental flow of a river. Furthermore, if standard levels are not reached by natural dilution of pollutants, then adequate treatment of discharge water is required to reduce pollutants' concentration. Less water flow in the river decreases its assimilative capacity to self-clean. In addition, the cities along the Ganga basin have gone through rapid urbanisation over time, without adequate planning or development of waste treatment infra-structure. So to prepare long-term policies, current as well as projected demographic data also need to be considered. Moreover, all information related to proposed initiatives (like barrages, transport routes and river-front development plans) which would have effect on overall river health should also be published in clear transparent manner to enable informed public discussion.

According to the Hydro-meteorological Data Dissemination Policy (NDDP, 2013), the data pertaining to the Ganga-Brahamputra-Meghna river basin is labelled as classified. Such data need to be requested for access through specific government channels by stating clearly the purpose of requirement, coupled by a secrecy oath. Nevertheless, NDDP aims to place all hydrological data, other than those classified on national security consideration, in the public domain in an open, transparent manner. There is need for review of the classified data so that gradually more of it can be put in public domain. Further the hydro-meteorological data that is being collected by other government organisations like Indian Meteorological Department (IMD), state governments, National Environmental Engineering Research Institute (NEERI) should also be made part of the overall available data.

Without extensive data in public domain, researchers and other non-governmental stakeholders have limited capacity to contribute to the process of Ganga rejuvenation. For example, Sankat Mochan Foundation, a non-governmental organisation, had been collecting water samples to monitor water quality from various Ghats of the Varanasi city, and maintaining their own catalogue. They have even contested government authorities and their fallacious claims by using their own datasets (Alley and drew, 2012; Upadhyay, 2009). But such limited, regional data would be of no use to the broader cause unless they are united with similar fragmented but contemporary data.

3.2. *People as an integral component towards realizing the goals of data governance*

Data governance is an organisational function distinct from data management. Ideally there should be a separation of duties between those who manage and those who govern data. The roles of various stakeholders should be well defined, such as data stewards or caretakers; advocates of data assets; data owners, responsible for creation and maintenance of data; and lastly data users. The enforcement of equal sharing, dissemination and reuse by users who may be governmental or non-governmental agencies, at the same time should not be overlooked.

Although in India, government agencies are key to agenda setting, creation of policy alternatives and implementations, they lack the mechanism for an integrated institutional approach, whereby coordinated actions may lead to multiplier effect and result in a coherent network of data collection, analysis, delivery and access. Let us first understand the individual agencies involved in this process and their respective functions, responsibilities and jurisdiction in the overall government structure. Table 2 summarises the same.

As evident from Table 2, at present, the monitoring of water quality in the Ganga river basin is being supervised by various government agencies and ministries. While CWC, CGWB, and NMCG are under the control of Ministry of Water Resources; CPCB, NGRBA and NRCD function under Ministry of Environment Forestry and Climate Change. Again, the individual SPCBs are managed by respective State Governments, whereas the Union Territories (UTs) have Pollution Control Committees (PCCs) as their regional monitoring agencies. Thus, with involvement of so many agencies and their authorities, it is important to establish a roadmap for achieving efficiency in use of resources as well as effectiveness in pollution monitoring.

In an attempt to introduce an uniform nation-wide procedure for sampling, analysis, data storage and reporting of water quality parameters, the Water Quality Assessment Authority (WQAA), in exercise of its powers conferred by the Environment (Protection) Act, 1986, had issued the 'Uniform Protocol on Water Quality Monitoring Order' in 2005 for generation of reliable data in water related sectors. This protocol is applicable to all organisations, agencies and any other body monitoring surface and groundwater quality in the country. The WQAA has also facilitated the creation of state level Water Quality Review Committees (WQRCs) to review and interpret state water quality data (Garduño *et al.*, 2011). However these state level Review Committees are yet to be formed in all states and wherever they have been formed, their functional capacities are

Table 2: List of Government Agencies and Their Brief Mandates and Scope of Work in Data Collection and Monitoring of River Water Pollution

Minstry/Agency	Brief Description	Data Collection and Monitoring
Ministry of Water Resources, River Development and Ganga Rejuvenation (MoWR)	The MoWR is the main body for the formulation and administration of policies related to the development as well as regulation of the water resources in India. Its functions include overall planning, policy formulation, coordination and guidance.[4] In July 2014, the Ministry was renamed to "Ministry of Water Resources, River Development & Ganga Rejuvenation", making it officially responsible for the conservation, development, management and abatement of pollution in River Ganges and its tributaries.[5]	Done by sub-ordinate agencies like NGRBA, NMCG, CWC, and CGWB.
Ministry of Environment, Forest and Climate Change (MoEFCC)	The MoEFCC is the nodal agency for planning, promotion, co-ordination and overseeing the implementation of India's environmental and forestry policy programs.[6] The Ministry is chiefly responsible for the conservation of natural resources such as lakes, rivers, forests, their biodiversity and wildlife, along with the task of pollution control.	Responsible agencies are CPCB and NRCD.

[4] http://wrmin.nic.in/forms/list.aspx?lid=239. Accessed on 28 May 2015.

[5] http://wrmin.nic.in/forms/list.aspx?lid=277. Accessed on 28 May 2015.

[6] http://envfor.nic.in/about-ministry/about-ministry. Accessed on 28 May 2015.

Central Water Commission (CWC)	The CWC is a technical organisation under MoWR. Its main functions include to initiate, coordinate and conduct consultation of the respective State Governments. They oversee schemes or projects for flood control, irrigation, navigation, drinking water supply and water power development.[7]	• The CWC collects meteorological data as mandated by the Hydro-Meteorological Data Dissemination Policy (2013).[8] • The CWC is running a Real Time Water Quality Monitoring System (RTWQMS[9]) on tributaries of Ganga namely Gomti and Ramganga, and Yamuna at three locations in Uttar Pradesh. • Results of real-time monitoring are being made available on their website (http://cwc.rtwqms.com).
Central Ground Water Board (CGWB)	The CGWB too is under jurisdiction of the MoWR as a multi-disciplinary scientific organisation, responsible for providing scientific inputs for management, exploration, monitoring, assessment, augmentation and regulation of ground water resources.[10]	• The CGWB collects groundwater data as mandated by the Hydro-Meteorological Data Dissemination Policy (2013)[11]

(*Continued*)

[7] http://www.cwc.nic.in/. Accessed on 28 May 2015.
[8] http://www.cwc.nic.in/main/downloads/hddp2013.pdf. Accessed on 30 May 2015.
[9] https://pibindia.wordpress.com/. Accessed on 30 May 2015.
[10] http://www.cgwb.gov.in/. Accessed on 30 May 2015.
[11] http://www.cwc.nic.in/main/downloads/hddp2013.pdf. Accessed on 28 May 2015.

Table 2: *(Continued)*

Minstry/Agency	Brief Description	Data Collection and Monitoring
National Ganga River Basin Authority (NGRBA)	As an initiative in river conservation mission NGRBA was constituted in 2009, under the chairmanship of the Prime Minister as a planning, financing, monitoring and coordinating authority for the Ganga River as per Environment (Protection) Act, 1986. The NGRBA is both multi-disciplinary and multi-sectoral agency.[12] Program activities are both broad-based as well as locally targeted through stakeholder communications and community participation campaigns, and research focusing on critical knowledge gaps.[13]	• A proposal is now under study for real time monitoring at 113 locations for pollution abatement along the main stem of river Ganga and its selected tributaries covering the States of Uttarakhand (8), Uttar Pradesh (57), Bihar (13) and West Bengal (35). • Of these 113 locations, 30 locations will be common between real time and manual monitoring network to help CPCB in maintaining past and current database of parameters not covered under real time monitoring. • Though the real-time system can monitor 21 parameters, only the 18 parameters[14] have been selected for the proposed system.[15]
National Mission for Clean Ganga (NMCG)	The NMCG is placed under MoWR. It is supported by State Level Program Management Groups (SPMGs) of Uttar Pradesh, Uttarakhand, Bihar and West Bengal.[16] Its functions are to coordinate and oversee the implementation of projects under NGRBA.	• CPCB acts as the nodal point of NMCG's proposal for real-time monitoring at 113 stations.

[12] http://www.moef.nic.in/sites/default/files/ngrba/ngrbaread.html. Accessed on 31 May 2015.

[13] http://www.moef.nic.in/sites/default/files/ngrba/ngrbaread.html. Accessed on 31 May 2015.

[14] Note: The 18 parameters are Biochemical Oxygen Demand (BOD); Benzene, Toluene, and Xylenes (BTX); Chloride; Chemical Oxygen Demand (COD); Dissolved Oxygen (DO); Dissolved Organic Carbon (DOC); Electric Conductivity (EC); Fluoride; Ammonia; Nitrate; pH; Potassium; Temperature; Total Organic Carbon (TOC); Total Suspended Solids (TSS); Turbidity; Water level; and Colour.

[15] http://www.moef.nic.in/sites/default/files/ngrba/Appraisal%20note%20for%20WQM.pdf. Accessed on 31 May 2015.

[16] https://nmcg.nic.in/about_nmcg.aspx. Accessed on 31 May 2015.

Central Pollution and Control Board (CPCB)	The CPCB serves as a field consultant and also provides technical services to the Central Government under the provisions of the Environment (Protection) Act, 1986. Its mandates are governed by the Water (Prevention and Control of Pollution) Act (1974).[17]	• Under the National Water Quality Monitoring Programme (NWMP),[18] the CPCB has set-up 57 water quality monitoring stations at crucial points of the river in association with SPCBs of Uttarakhand, Uttar Pradesh, Bihar, Jharkhand and West Bengal (CPCB Report, 2013). • Like CWC, the CPCB is currently running parallel RTWQMSs[19] which monitors water parameters, on river Ganga, Yamuna and their tributaries at 10 locations in Uttarakhand (1), Uttar Pradesh (5), Bihar (1), West Bengal (1) and Delhi (2). • The results of the 10 real-time monitoring stations are being made available online.[20]
State Pollution Control Boards (SPCBs) and Pollution Control Committees (PCCs)	The SPCBs and PCCs have generally the same function, the difference being SPCBs act for the state while the PCCs act for the Union Territories (UT). These agencies plan and advise the state and UT; collect and disseminate information; encourage, conduct and participate in investigations and research; collaborate with the central board in organising the training; to inspect sewage or trade effluents, works and plants, etc.[21]	SPCBs and PCCs are responsible for collection, dissemination of information and implementation of the provisions of the Acts regarding water pollution.

(Continued)

[17] http://cpcb.nic.in/Functions.php. Accessed on 31 May 2015.

[18] http://www.cpcb.nic.in/divisionsofheadoffice/pams/NWMP.pdf. Accessed on 31 May 2015.

[19] http://pib.nic.in/newsite/PrintRelease.aspx?relid=106835. Accessed on 30 September 2015.

[20] http://cpcb.rtwqms.com/. Accessed on 15 May 2015.

[21] http://hspcb.gov.in/annex-a-3.htm. Accessed on 31 May 2015.

Table 2: *(Continued)*

Minstry/Agency	Brief Description	Data Collection and Monitoring
National River Conservation Directorate (NRCD)	NRCD placed under MoEFCC, looks after the implementation of action plans to improve water quality in rivers under National River Conservation Plan.[22]	• NRCD operates a water quality monitoring network on river Ganga in collaboration with universities and other research institutions. There are 21 common stations between the NRCD and CPCB water quality monitoring networks. Overall, 98 stations are operated exclusively by NRCD and CPCB operates 36 independent stations[23]. • Pollution monitoring information is not available through online or offline reports.
Water Quality Assessment Authority (WQAA)	The mandate of WQAA is to direct agencies to standardise water quality monitoring methods, ensure proper treatment of wastewater to restore the water quality of surface and ground waters, take up R&D related activity for water quality management and promote recycling and reuse of treated wastewater.	• It has issued the 'Uniform Protocol on Water Quality Monitoring Order' in 2005 to ensure availability of reliable uniform data.

[22] http://envfor.nic.in/division/national-river-conservation-directorate-nrcd. Accessed on 29 June 2015.

[23] http://www.moef.nic.in/sites/default/files/ngrba/ESC%20Note%20for%20WQM.pdf. Accessed on 30 September 2015.

still redundant (data.gov.in website). The insufficient data received in the Environmental Data Bank is currently in public domain and can be accessed through CPCB's website (CAG Report, 2011; Environmental Data Bank website). Similar works carried out by different ministries and agencies have resulted in different datasets, unintentionally challenging one another on both authenticity and sources. Hence there is absolute necessity for a cohesive approach to the governance and harmonisation of all available datasets related to river Ganga.

In addition, there are duplications of efforts, mostly due to overlapping of institutional jurisdictions. Overlapping and duplication are helpful if they serve the purpose of checks and balances, but without such functionalities data creation is a simple wastage of time and resources. As new knowledge and complexity of the operational environment becomes intricate with more datasets, specialization along with multi-disciplinary collaboration becomes more of a requirement for organisations. Data collection and creation not only require a clear-cut institutional arrangement and specified process path, but also requires substantial funds for implementation.

Although each organisation is managing data in accordance with its mandate but the effect tends to be diluted due to an enormous scope of the Ganga River basin's challenges. There is a need to do it through more specificity and collaboration. Organisations involved in data governance must revisit their capacity and resources, and compare them with other concerned organisations identifying the areas of common interest, resource pool, competency and scope of collaboration. The challenge, now, is to have a platform where organisations involved can collaborate objectively. These challenges can become more complex when organisations and their responsibilities are reshuffled according to their specialisation and competence.

3.3. *Processes to define an effective implementation path for data governance*

Collecting, organising and disseminating data or observations can be both time and resource intensive processes. Challenges in ensuring data accuracy, consistency, integration and usability need to be addressed systematically.

The process on how government agencies share data is very much dependent on their mandates and functions. In general, communication among government agencies' channel up and down through the bureaucratic hierarchy. The problem with this set-up is that it requires agencies to seek permissions from the higher hierarchies for information that is processed elsewhere in other agencies. There are insufficient mechanisms for lateral coordination of information transfer and dissemination to other agencies since it is not specified in the mandate and functions of these agencies. Good data governance requires government agencies to share data on a regular basis irrespective of their bureaucratic framework. Some agencies resort to data sharing memorandum of agreements to bypass institutional limitations but it can be cumbersome due to complex process of liaising different agencies working at different interfaces. It does not provide enough flexibility as compared to an integrated yet open forum where data is available while preserving its security as well.

The challenge of data governance, therefore, is to transmit data and information without compromising security concerns. The task of separating what is critical information for different categories of use will require an agency dedicated specifically in the management, transfer and dissemination of data and information. Presently, for river Ganga, numerous agencies manage their data and information individually leading to silo effects. Compartmentalisation of works in the Ganga River management plan will lead to reduced effectiveness of the broader project. While it is necessary for regional municipalities and urban settlements of the participating states to manage and develop their waste treatment facilities; a central nodal organisation is needed for supervising and monitoring the functioning and progress of individual states.

Although the Ministry of Water Resources does not have a data sharing and dissemination mandate as part of its functions but its sub-agencies like the CGWB and the CWC publishes scientific reports based on the data generated through various investigations (MOWR website). The CGWB reports the State and District hydrogeological data, ground water year books and Atlases, ground water user maps and guides/manuals/pamphlets on various aspects of ground water management (CGWB website). The CWC performs functions such as to collect, coordinate, publish and analyse the data relating to river tides, rainfall, surface runoff, water temperature, silting of reservoirs, behaviour of hydraulic structures, and other environmental aspects. It also promotes mass awareness regarding the progress and achievement made in water resources development, use and conservation (CWC website).

The CPCB, on the other hand has four functions relating to data and information, namely: to organise a comprehensive mass awareness program; to collect, compile and publish technical and statistical data relating to water pollution and the measures devised; to prepare manuals, codes and guidelines relating to treatment and disposal of sewage and trade effluents; and to disseminate information in respect of matters relating to water pollution and their prevention and control (CPCB website). For the SPCBs and the PCCs, they are generally regarded as regulatory bodies in the control or abatement of pollution.

Thus, without collaboration among these agencies that have separate responsibilities and different capacities, inefficiency in streamlining data outputs is inevitable. Enabling data sharing amongst agencies encourages in-depth research inquiry, promotes innovation and draws potential new data users. Likewise, stakeholders will have easier access to data needed for their intended project or research. Various non-government organisations such as Sanckat Mochan Foundation, Eco-Friends, also have been collecting data pertaining to the River Ganga on their own. All these lead to numerous varying conclusions creating multiple source information, while affecting decision making regarding the Ganges Management Plan.

The process of public participation in Ganga river basin management also requires alignment of processes following agreed protocols and standards. As seen in recent development, disparate views between governments and civic groups, are often characterised by unclear objectives, absence of or limited data, and differing opinions (Alley and Drew, 2012; Upadhyay, 2009). These observations, in turn, demand greater public participation and public access to data to close the gaps, in order to promote public trust.

Obviously, there is untapped potential of engaging people such as the boatman and other river community members to act as new "sensors" since they are well-informed and connected stakeholders. Innovative community management is another key to successful overall performance. Evidence shows that projects with public participation are more successful in achieving intended results with long-lasting effects (Prasad, 2015).

Various countries such as Australia and the United States are partnering with academic institutions to engage students to monitor river water quality through standardised water test kits. These test kits are sourced through recommended supply vendors, which can be centralised or decentralised. In the case of Australia, one can also borrow monitoring kits from the local Catchment Management Authority or Water-watch program coordinator (Waterwatch Victoria website). Even though such a plan has been included in the project document of NMCG as part of its future project plan, this initiative could not be pushed forward to be launched earlier given the urgency of the situation, supported by available technologies and funding resources.

Enhancing public participation ensures success in government policies because it manages perception change and most importantly results in positive change in behavior amongst ground level stakeholders. The more people are made aware of the magnitude of the problem, the more they are receptive to extend help in finding a solution to the problem. By properly educating the people in Ganga River basin about repercussions of the river's pollution threats, they will become more eager to better understand the need for themselves to be engaged in setting the rules, determining sanctions, and other issues related to the management of the river (Ahsan, 2012).

3.4. *Technology as an indispensible tool to achieve competent data governance*

Advances in information communication technologies such as the Geographic Information System (GIS), the Internet, big data and cloud computing, have brought in immense opportunities and innovation to data governance. These include methodologies in monitoring river pollution as well as innovations in facilitating data sharing to establish a **one-stop data platform** that enhances **public awareness and action**.

Technologies play critical roles as enabling instruments towards policy coherence and collaboration, reducing duplication work while delivering accurate, consistent and reliable data. Technologies on sharing and re-use will also help in connecting researchers to the civil society, raising awareness among public and supporting transparency in water governance.

The Government of India formulated a National Data Sharing and Accessibility Policy in 2012, with the objective to facilitate access to government-owned shareable data and information proactively, while updating existing ones periodically (NDSAP-12). The Hydro-Meteorological Data Dissemination Policy (2013) echoes the same stating that "data collected with the deployment of public funds should be made more readily available to all, for facilitating rational debate, better decision making and in meeting society's needs" (Hydro-Meteorological Data Dissemination Policy, 2013).

The aim of good data governance for consensus building will be accomplished only if full and transparent data information is available to both government and non-governmental stakeholders. This will first require a central, integrated and accessible database for Ganga river basin that reflects the hydrological, meteorological, geomorphological, ecological, sociological and economic parameters with up-to-date datasets. At present, however, there is a distinct segmentation in data representation for the Ganga River basin project. As a result, duplications and mutual conflicts among datasets are unavoidable. Table 6 below enlists those government websites that separately maintains and put up data related to their respective jurisdiction.

3.5. *Examples of good data governance from international experiences*

Ideally, all data should reside in one single window or platform as a single one-stop platform for all users. The Mekong River Commission Data Portal (see Figure 2) is one such example that extends beyond nation-wide to transboundary sharing and dissemination of river basin data. Some of its features may be replicated for Ganga river basin in an adaptive manner to meet the needs of both central as well as state governments.

The Mekong River Commission (MRC) was formed on 5 April 1995 as a river basin commission with the goal of creating regional cooperation at a basin scale. Four countries, namely Lao PDR, Cambodia, Thailand and Vietnam entered into the "Agreement on the Cooperation for Sustainable Development of the Mekong River Basin". These countries border what is considered as the "Lower" Mekong River Basin.

Table 3: Various Websites and Portals with Datasets on Ganga River Basin

Portals[24]	Owner	Description	Datasets Related to River Ganga
Environment Data Bank http://cpcbedb.nic.in/	CPCB	To provide data and information on various environmental parameters that are often required by various user groups, including Pollution Control Boards, researchers, students and the general public.	Information on DO, BOD, pH, FC, TC and conductivity,
CPCB's Real Time Water Quality Monitoring System (RTWQMS — CPCB) http://cpcb.rtwqms.com/	CPCB	Real-time monitoring of river quality along Ganga, Yamuna and tributaries of Ganga (10 locations).	Website is non-operational at time of checking.
CWC's Real Time Water Quality Monitoring System (RTWQMS-CWC) http://cwc.rtwqms.com/	CWC	Real-time monitoring of river quality along Yamuna three tributaries of Ganga (three locations).	Website is non-operational at time of checking.

[24] Note: Last accessed on 25 May 2015.

Water Resources Information System of India (India-WRIS), http://www.india-wris.nrsc.gov.in/	CWC and Indian Space Research Organisation (ISRO)	The India-WRIS aims as a 'Single Window' solution for "comprehensive, authoritative and consistent data and information of India's water resources along with allied natural resources in a standardised national GIS framework with tools to search, access, visualize, understand and analyse the data for assessment, monitoring, planning, development and finally Integrated Water Resources Management (IWRM)".[25]	Hydro-meteorological data of river Ganga, including water level, river discharge (or flow) sediment flow and water quality.
Ground Water Information System http://gis2.nic.in/cgwb	CGWB	Ground water quality monitoring.	Groundwater level, fluoride content, quantity of nitrogen, and acidity.
India Open Government Portal,[26] https://data.gov.in/	Ministry of Communications & Information Technology	A joint initiative between the Government of India and Government of the United States, with the intent to be used by Indian Ministries, Departments and their organisations to publish datasets, documents, services, tools and applications collected by them for public use. It aims to increase transparency in the functioning of Government and also open avenues for many more innovative uses of Government Data to give different perspective.	Four catalogues are found on river Ganga, with data dated 2008, 2011 and 2012; limited to DO, BOD, pH levels, nitrates and TC.

(Continued)

[25] http://www.cwc.nic.in/main/downloads/hddp2013.pdf. Accessed on 1 October 2015.
[26] https://data.gov.in/. Accessed on 1 October 2015.

Table 3: *(Continued)*

Portals[24]	Owner	Description	Datasets Related to River Ganga
Ganga Knowledge Center (GKC)	NMCG	A knowledge base centre as an integral part of the NMCG setup, given its key function in supporting NGRBA. The mandate includes conducting modelling and targeted research to facilitate basin-scale decision-making, mapping the entire basin through a GIS-based mapping system, as well as disseminating knowledge, including through a high-quality web portal.[27] Proposed outputs[28] include: Digitisation ad integration of legacy records; comprehensive spatial database; system water quality model; e-flows model; water quality investment decision support system; prioritised research and innovation needs; state-of-art Ganga website; online portal of Ganga basin monitoring data; crowd-sourcing interface for community monitoring	Under development; implementation period: 8 years.
West Bengal Water Quality Information System http://www.wbpcb.gov.in/	West Bengal Pollution Control Board (State)	Online system generating PDF reports	Available report for downloading on BOD, conductivity, faecal coliform, nitrate, pH, temperature, total coliform.

[27] http://www.worldbank.org/en/news/feature/2015/03/23/india-the-national-ganga-river-basin-project. Accessed on 1 October 2015.

[28] http://www.moef.nic.in/sites/default/files/ngrba/ESC%20Note%20_GKC.pdf. Accessed on 1 October 2015.

Figure 2: Mekong Basin Data and Information Services Portal[29]

The MRC Data and Information Services Portal enables search, discovery, and download of information or statistics held by the MRC, including spatial data, time-series, non-spatial data, and technical documents. The key feature of the Portal is the Master Catalogue, which enables users to search its catalogue using keywords, geographic location, and time period for relevant material. The materials on MRC Portal include information on water quality monitoring; water Level monitoring; hydro-meteorology; hydro-power database; irrigation; climate change and adaptation; geo-spatial information, river-bio monitoring and many more. Modest data fees have also been set to cover operational costs.

The Portal in addition provides an overview of some of MRC's databases and information services, atlases, interactive maps and multimedia.

[29] http://portal.mrcmekong.org/cms/information-services. Accessed on 28 May 2015.

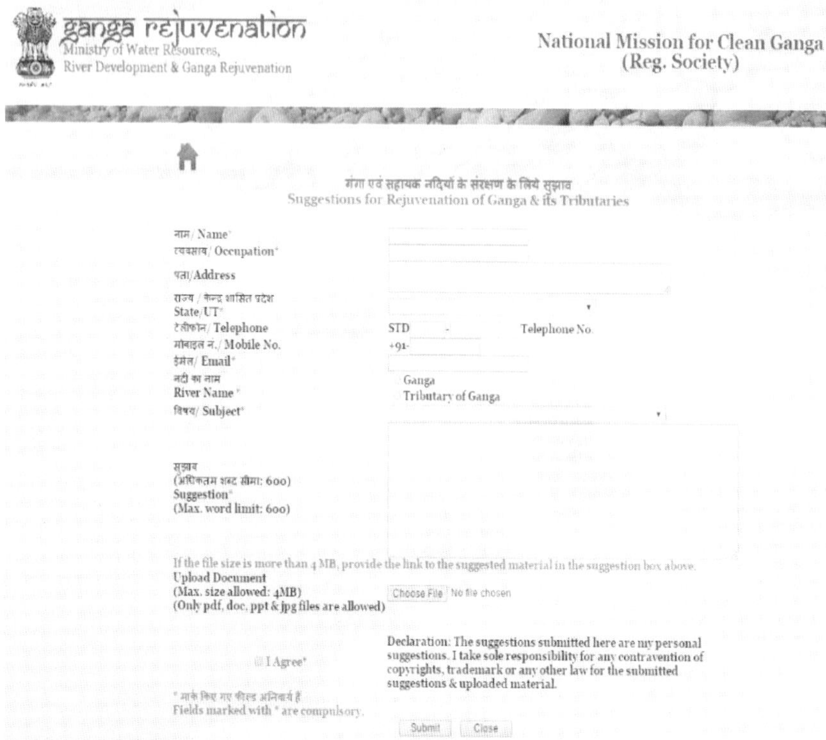

ganga rejuvenation
Ministry of Water Resources,
River Development & Ganga Rejuvenation

National Mission for Clean Ganga
(Reg. Society)

गंगा एवं सहायक नदियों के संरक्षण के लिये सुझाव
Suggestions for Rejuvenation of Ganga & its Tributaries

नाम / Name*
व्यवसाय / Occupation*
पता / Address
राज्य / केन्द्र शासित प्रदेश
State/UT*
टेलीफोन / Telephone — STD - Telephone No.
मोबाइल नं. / Mobile No. — +91-
ईमेल / Email*
नदी का नाम
River Name* — Ganga / Tributary of Ganga
विषय / Subject*
सुझाव
(अधिकतम शब्द सीमा: 600)
Suggestion*
(Max. word limit: 600)

If the file size is more than 4 MB, provide the link to the suggested material in the suggestion box above.
Upload Document
(Max. size allowed: 4MB)
(Only pdf, doc, ppt & jpg files are allowed) — Choose File No file chosen

I Agree* — Declaration: The suggestions submitted here are my personal suggestions. I take sole responsibility for any contravention of copyrights, trademark or any other law for the submitted suggestions & uploaded material.

* मार्क किए गए फील्ड अनिवार्य हैं
Fields marked with * are compulsory.

Submit Close

Figure 3: Current online form requesting for "Suggestions for Rejuvenation of Ganga & its Tributaries[30]

It also provides an overview of the MRC Toolbox that is used for modelling of development scenarios. This way the Mekong River Commission seeks to promote sustainable development by engaging non-governmental stakeholders in data governance.

The government of India has also launched a similar platform under its '*Namami Gange*' or 'Ganga Rejuvenation' programme. However, except for a simple online web-based form (Figure 3) and a fund donation page (Clean Ganga Fund website), no concrete plans on engaging the public has been presented so far. Hence we suggest that the MRC portal

[30] https://nmcg.nic.in/suggestions.aspx. Accessed on 10 May 2015.

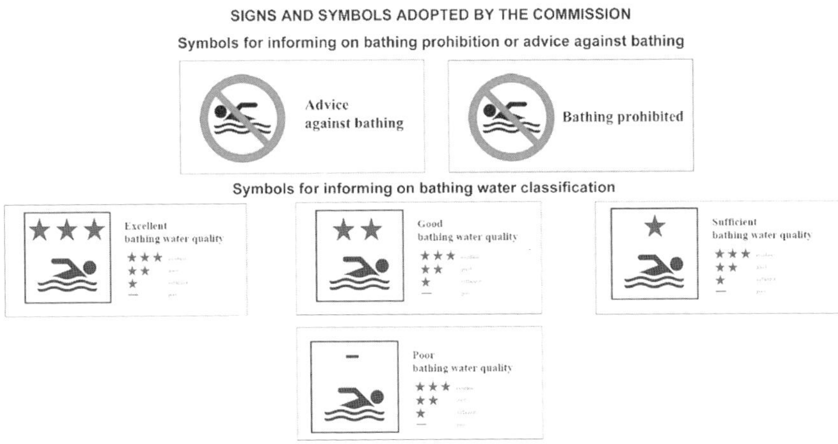

Figure 4: Signs and Symbols Adopted by the European Commission on Bathing[31]

might serve as a reference point to learn key features on how to evolve the '*Namami Gange*' website into a better equipped one-stop data platform.

Apart from informing stakeholders on quantitative parameters of river health, alerting the river users on water quality through visual indicators and data visualisation is another innovative way to make people aware of daily pollution levels. The European Commission adopted an initiative ensuring timely water quality information to public during bathing season, with an obligation on Member States to disseminate information actively. The European Union's Bathing Water Quality data viewer provides users a quick check on locations and statistics on the quality of coastal and freshwater at bathing sites. One can easily search by country, region, and province through interactive maps to find out the status of waterbodies in their local area. The visual also provide trends on how the water quality has changed over past years. Similar easily-identifiable notices banning or encouraging bathing can also be adopted for River Ganga, with standardised symbols of the water classification system (Official Journal of the European Union, 2011) (See Figures 4).

---

[31] http://ec.europa.eu/environment/water/water-bathing/signs.htm. Accessed on 29 September 2015.

4. Conclusion — Possible Policy Recommendations

The complexity and dynamics of Ganga River pollution abatement problem calls for policymakers, researchers, civil society, private sector and the common people to join hands in addressing its long neglected causal issues, their consequences to find long term sustainable solutions. Ganga River's rejuvenation project requires a holistic approach with involvement of multi-disciplinary stakeholders from various walks of life and expertise. The benefits of identifying a credible and reliable data source would generate actions that would serve as introductory step towards a meaningful and productive collaboration with such stakeholders. This chapter, therefore, summarises the following recommendations in accordance to the already discussed four-pillar framework of data governance.

First and foremost, competent **policy** formulation holds key to a more efficient Ganga River basin governance. As already identified before, we have seen that multiple agencies are involved in collecting data on surface and groundwater related parameters. This fragmented approach results in duplication and ambiguity of functions, while hindering comprehensive data collection and analysis. The Comptroller and Auditor General (CAG), India, Report 2011–12 observes that water pollution has not been adequately addressed in any national or regional policy yet, as a result government efforts in this area fails to receive its required emphasis and thrust (Water Pollution in India, Report No. 21, 2011–12).

Though there has been an increase in the number of monitoring stations for measuring the water quality status at different locations of the river over the years, still, their number is not sufficient to extensively monitor the overall river condition (Rajaram and Das, 2008). In addition, repeat monitoring should be done more frequently, as opposed to a current trend of performing these tests only on a half-yearly basis. On top, only 32% of the monitoring agencies have been found to do the monitoring exercises twice a year. (Status of Water Quality in India, 2012). Although in recent times, GIS-based real-time, near-real-time monitoring as well as manual monitoring of water quality has improved a lot, further development is still needed to cover all strategic locations and hotspots to provide sufficient data for enforcement of planning and regulation. Moreover, the number of parameters being monitored are only the basic quality parameters and are not sufficient to develop a good quality inventory (Rajaram and Das, 2008).

A review of the Water Act of 1974 is critical to solve the inter-linked problems such as over-lapping roles, performance of duties and implementation jurisdiction of various agencies under its mandate. Currently, the CPCB and SPCBs work autonomous of each other. As per the scheme of the Water Act, while CPCB is under the administrative control of the MoEFCC, SPCBs are administratively under their respective State governments but work within the overall policy frame work of CPCB leading to accountability problems. New provisions need to be incorporated within the Act, so that it can allow CPCB to exercise sole authority over all data related to water quality monitoring, along with supervising roles over concerned agencies. It would also help CPCB to deliver its duty more effectively by engaging both public as well as private stakeholders simultaneously to share data and at the same time, impose strict policies for governing well-organised storage of data.

In addition, a revision of the Hydro-Meteorological Policy also need to be initiated to declassify parts of the hydro-meteorological data. Currently any hydro-meteorological data related to Ganga–Brahamputra–Meghna basin is treated as classified due to national security concerns Declassification of parts of these information may enhance a better understanding to manage problems of water pollution, water allocation and development in the Ganga basin.

The **institutional arrangement (people)** is important in the overall data governance framework. Having a proper institutional structure will help in optimal collection of data and its subsequent management. In this framework, efficiency of organisation is critical and duplication of monitoring efforts should be avoided to optimise the use of resources. Changes in functioning of organisations may be needed in order to harmonise the efforts of government agencies and eliminate duplication. Simultaneously, institutions would need to establish a strict code for undertaking stringent monitoring programs and storage of resulting data.

The **processes** pertaining to data governance in the water sector may be improved by introducing a standard uniformed basis defining clearer water parameters and associated methodologies for data collection. All these will not only help identify and indicate the current state of water along the river stretches, but also expand its assessments into categories such as if the water is fit for bathing, drinking or any other human uses. More coherent efforts should be made to promote awareness of water quality through use of visual signage, both online and offline. Clear signs need to be installed at physical locations, for example, Ghats or ferry points, to display water suitability for bathing.

Moreover, by encouraging public participation a more productive collaboration between universities, research institutes, as well as civic society groups such as the Sankat Mochan Foundation may be established. Datasets can be relied as both complementary and comparable, when they are also found usable by other stakeholders. In addition, promoting portable monitoring kits in local universities and schools may be coordinated by regional SPCBs. Public participation should also be stimulated through incident reporting, for example, launching of hotline, mobile apps for smart phones, text messages (SMS) and lastly through social media.

Lastly, **technologies** also confer a crucial contribution in formation of effective data governance. A single one-stop integrated portal is recommended to host and share all datasets related to river Ganga. Open standard formats such as XML or other machine-readable formats, as well as standardised data collection and monitoring methodologies should be adopted and publicised through this portal. A set of data governance policies could be made available to lay out the roles and responsibilities of various stakeholders, both data owners and users, with harmonisation of rules, processes and standards. In order to encourage more usage and sustain public participation, government agencies, may consider publishing more data proactively, or "open by default", rather than reactively respond to data or information request (UN E-Government Survey, 2014).

In addition, different communication channels should be explored such as web interfaces, smartphones applications and SMS-enabled access for feature phones, thus effectively reaching out to more audience through multiple ways. For example, in the United States, water quality data on current conditions of rivers and streams across the country, can be accessed through smartphone, SMS text or email, via WaterNow, an integrated system of centralised database with more than 16,000 monitoring stream gauges across the country (USGS website). In California, there is a new approach through social media (Twitter) to monitor water quality in the Sacramento-San Joaquin river system, which contains two-thirds of the state's fresh water, via GIS-enable monitoring robots that float and tweet findings concurrently, providing a dynamic and detailed understanding of changing water quality parameters (Davies, 2012).

Furthermore, an integrated portal would act as a gateway for more coherent efforts in promoting awareness of water quality by introducing the use of visual signage, both online and offline. Other forms of visual images with help of color coding or infographics could likewise be incorporated as a more effective communication tool for users at the online data portal. The general public should be encouraged to participate in reporting incidents on river pollution. In this regard, citations may be drawn from the government Digital Service of the United Kingdom that publishes as well as maintains a 24-hour hotline at their Environment Agency to facilitate all environment related incident reporting (Gov.UK website). To better serve the need of the Ganga community, such a hotline would indeed be helpful for concerned stakeholders.

Finally, modern technologies such as GIS and remote sensing also holds enormous potential for progress in monitoring water quality. Technological advances in methodologies are lowering the costs of monitoring significantly (Tortajada and Chan, 2013). Hence the quantity and quality of data can be easily improved over time with effective use of such technologies. If necessary, a substantial part of the data collection and collaboration job may be outsourced to private agencies as well. This might serve to provide punctuality and accountability of data produced, as government agencies will then have the authority to scrutinize their performance carefully and only then sanction payments.

Such a public-private initiated integrated-basin approach (OECD, 2003), coordinated at the central level by MoWR along with CWC and CPCB as governing bodies, will definitely be able to leverage and harness data governance as an enterprise asset (Berson and Dubov, 2011). We sincerely hope that our appraisal of inter-national case-studies and prospective recommendations would find some consideration, while future policy making in the Ganga Basin management by the concerned Indian authorities. Indeed, we firmly believe that a strong data governance platform, on the first hand, would serve as one of the founding steps towards sustainable management and development of one of the most polluted and complicated river basin systems in the world.

References

Ahsan, S (2012). Cleaning Up the River Ganga. Available at: http://www.fairobserver.com/region/central_south_asia/cleaning-river-ganga-examining-role-public-participation-ganga-action-plan. Accessed on 1 October 2015.

Alley, KD and G Drew (2012). *Ganga.* Oxford University Press.

Berson, A and L Dubov (2011). *Master Data Management and Data Governance,* 2nd edn. McGraw-Hill, New York.

Bhansali, N (2014). *Data Governance: Creating Value from Information Assets.* CRC Press, Boca Raton, Fla.

CAG Report of 2000/Union Government. Retrieved from: http://www.saiindia.gov.in/english/home/public_folder/training/Structure%20Training%20Module/Enviornmental%20Auditing%20RTI,%20Mumbai/Day%202/Ganga%20Action%20Plan%20_Box_.pdf. Accessed on 9 September 2015.

CAG Audit Report (2011). Available at http://www.saiindia.gov.in/english/home/public/In%20_Media/PR21_2011_12.pdf. Accessed on 30 September 2015.

Central Ground Water Board (CGWB) website. Available at http://www.cgwb.gov.in/. Accessed on 30 September 2015.

Central pollution Control Board (CPCB) website. Available at http://cpcb.nic.in/Functions.php. Accessed on 30 September 2015.

Central Water Commission (CWC) website. Available at http://www.cwc.nic.in/main/webpages/rti/rti_item1.html. Accessed on 30 September 2015.

Clean Ganga Fund website. Available at http://www.cleangangafund.com/. Accessed on 1 October 2015.

Coleman, DW, A Hughes and WD Perry (2009). The Role of Data Governance to Relieve Information Sharing Impairments in the Federal Government. In *Computer Science and Information Engineering, 2009 WRI World Congress on* (Vol. 4, pp. 267-271). IEEE.

CPCB Report (2013). Pollution Assessment: River Ganga. Available at http://cpcb.nic.in/upload/NewItems/NewItem_203_Ganga_report.pdf. Accessed on 30 September 2015.

Data.gov.in website. Available from: https://data.gov.in/. Accessed on 30 September 2015.

Davies, A (2012). Tweeting Robots Monitor Water Quality in California Rivers. Available at http://www.treehugger.com/clean-technology/swimming-robots-hit-california-rivers.html. Accessed on 1 October 2015.

DNA India (2015). Narendra Modi addresses National Conference on E-governance through Twitter. Available at http://www.dnaindia.com/india/report-narendra-modi-addresses-national-conference-on-e-governance-through-twitter-2056756. Accessed on 30 September 2015.

Environmental Data Bank, CPCB website. Available at http://cpcbedb.nic.in/. Accessed on 30 September 2015.

Ganga River Basin Management Plan (GRBMP) Interim Report (2013). Available from https://nmcg.nic.in/writereaddata/fileupload/25_GRBMPInterim_Rep.pdf. Accessed on 30 September 2015.

Garduño, H, S Romani, B Sengupta, A Tuinhof and R Davis (2011). India groundwater governance case study, (15). Gov.UK website. Report an environmental incident. Available at https://www.gov.uk/report-an-environmental-incident. Accessed on 1 October 2015.

Ministry of Water Resources (2013). *Hydro-Meteorological Data Dissemination Policy (2013)*. Report by Government of India. Available at Central Water Commission website http://www.cwc.nic.in/main/downloads/hddp2013.pdf. Accessed on 1 October 2015.

Ladley, J (2012). Data Governance: How to Design Deploy, and Sustain an Effective Data Governance Program. Boston, MA: Morgan Kaufmann Publishers.

Indian Institute of Management (IIM), Lucknow (2010). Available at http://cpcb.nic.in/IIMLko.pdf. Accessed on 30 September 2015.

Markandya, A and MN Murty (2000). *Cleaning-up the Ganges: A Cost-Benefit Analysis of the Ganga Action Plan*. New Delhi: Oxford University Press.

McGilvray, D (2006). Data Governance: A Necessity in an Integrated Information World. *Information Management*, 16(12), p. 24.

Ministry of Environment, Forest and Climate Change (MoEFCC) website (1974). Water (Prevention and Control of Pollution) Act. Available at http://www.moef.nic.in/legis/water/wat1c1.html. Accessed on 28 May 2015.

Ministry of Water Resources, River Development and Ganga Rejuvenation (MOWR) website. Available at http://wrmin.nic.in/forms/list.aspx?lid=239. Accessed on 28 May, 2015.

NDSAP-2012 or National Data Sharing and Accessibility Policy (2012). Avaialable at http://ogpl.gov.in/NDSAP/NDSAP-30Jan2012.pdf. Accessed on 1 October 2015.

National River Conservation directorate (2009). Status Paper on River Ganga. Available at http://www.moef.nic.in/sites/default/files/Status%20Paper%20-Ganga_2.pdf ; 19. Accessed on 30 September 2015.

Official Journal of the European Union, Commission Implementing Decision (2011). Available at: http://eur-lex.europa.eu/legal-content/EN/TXT/?uri=CELEX:32011D0321. Accessed on 1 October 2015.

OECD (2003). *Improving Water Management: Recent OECD Experience,* OECD Publishing.

Onno Ruhl and C Genevieve (2004). Five-point plan to clean Ganga. World Bank India. Available at http://www.worldbank.org/en/news/opinion/2014/06/19/five-point-plan-to-clean-the-ganga. Accessed on 29 September 2015.

Otto, B (2011). Data Governance. *Business & Information Systems Engineering.* 3(4), 241–244.

Prasad, K (2015). *Water in the Coming Decades: Policy and Governance Issues in India.* Foundation Books: New Delhi.

Rajaram, T and A Das (2008). Water Pollution by Industrial Effluents in India: Discharge Scenarios and Case for Participatory Ecosystem Specific Local Regulation. *Futures.* 40(1), 56–69.

Ratna, RV (2009). *Water Security and Management: Ecological Imperatives and Policy Options.* Academic Foundation, in association with Centre for Economic and Social Studies, Hyderabad, New Delhi. India (239).

Status of Water Quality in India (2012). Available at http://www.cpcb.nic.in/WQ_Status_Report2012.pdf. Accessed on 1 October 2015.

TDWI (The Data Warehouse Institute) website. Available at https://tdwi.org/articles/2008/05/27/data-governance-strategies-helping-your-organization-comply-transform-and-integrate.aspx. Accessed on 16 October 2015.

Tortajada, C and NW Chan (2013). *Water Quality Policy and Management in Asia.* London: Routledge.

UN E-Government Survey 2014. Available at http://unpan3.un.org/egovkb/Reports/UN-E-Government-Survey-2014. Accessed on 30 September 2015.

Upadhyay, V (2009). Ganga at Varanasi: Lessons from Environmental Abuse. *Economic and Political Weekly*, 44(37), 64–66.

USGS (United States Geological Survey) website. USGS Water data for the Nation. Available at http://waterdata.usgs.gov/nwis. Accessed on 1 October 2015.

Water Freedom Revolution (WFR) Worldwide Website. Available at http://waterfreedomrevolution.com/operations-and-projects/commercial-project/role-of-different. Accessed on 30 September 2015.

Water Pollution in India, Report No. 21, 2011–12. Available at http://www.environmental-auditing.org/Portals/0/AuditFiles/India_f_eng_Water-Pollution-in-India.pdf. Accessed on 1 October 2015.

Waterwatch Victoria website. Available at http://www.vic.waterwatch.org.au/. Accessed on 1 October 2015.

Chapter 9

CREATING A KNOWLEDGE MANAGEMENT CULTURE FOR GANGA RIVER: AN ESSENTIAL TOOL FOR SUCCESS

Thanathorn Asanarong, Sowon Jeon, Ren Yuanlin, and Christopher Yeo

1. Introduction

Since Prime Minister Narendra Modi announced his plans to clean up the Ganga River, there has been a much renewed effort to tackle the problem of river pollution. The problems of river pollution and their proposed solutions are neither new nor lacking, nor is there a lack of reasons why these solutions have not been implemented. Meanwhile, river pollution continues unabated. What is needed is a renewed effort to look at the fundamental and structural issues that prevent policies and solutions from working. This chapter focuses on one such aspect that is both glaringly obvious, but yet is both within reach of current government capacity and promises great reward — the field of knowledge management (KM). In addition, this chapter aims to enhance policy makers' understanding of KM in the

Ganga Rejuvenation Project which includes the analysis of the current KM landscape, the identification of the knowledge gaps, and possible solutions to guide the future direction of KM in the area of river pollution management.

Firstly, we conduct an overall problem analysis and find that Ganga River pollution knowledge is often missing, unreliable, or not shared effectively. Without good knowledge generation, validation, and dissemination, there will be basic lack of usable pollution data. Without good pollution data, the sources and causes of pollution cannot be determined correctly, and resources will be misdirected to ineffective measures. Effective governance and pollution abatement policies cannot be achieved as poor policies will be designed and implemented, and there is no accurate data to evaluate policies and adjust them. Pollution laws also cannot be enforced if pollution data cannot hold up in the court of law, and polluting companies and even cities and states have little incentive to reduce pollution, if pollution data is poor and cannot be linked directly to them.

In the area of knowledge dissemination, ineffective knowledge sharing firstly deprives other parties from using knowledge and generating further knowledge, and can create mistrust and misperception. If reliable pollution data is not available to the public, the public is also often misled by inaccurately high and sensationalist figures and this leads to low public trust and cooperation in the effort to reduce river pollution.

In the area of governance and policy making, a lack of emphasis on tacit knowledge and people as knowledge assets leads to missed opportunities to make use of local knowledge and expert groups. There is also a low emphasis on policy making and implementation experience. Effective governance requires a recognition of people having the right knowledge comprising of skills and experience to make and implement solutions to reduce pollution.

In our second part, the theoretical framework of KM is fleshed out. Knowledge management is not just about information technology or computer databases. At its core, KM is a conceptual toolset that allows users to recognise intellectual capital as knowledge assets. If one can learn to properly identify and use these knowledge assets, or improve their utility and value through validation and sharing, KM can prove to be a critical tool in the success of the Ganga Rejuvenation Project. Knowledge management recognizes the difference between raw *data*, *information* that is useful, *knowledge* that is practical, and *wisdom* that is the understanding coupled with analysis of knowledge. This is known as the DIKW chain or pyramid, which is representation and relationship between data, information, knowledge, and wisdom. Knowledge management also recognizes the difference between *explicit knowledge* that is written down and *tacit knowledge* that is not written down but resides in our brains — and importantly includes experiences. By enabling the use of conceptual framework to identify knowledge gaps, and to increase the value of existing knowledge, KM needs to be seen as valuable tool and a focus for development by all parties involved in Ganga Rejuvenation Project.

In the final section, we will look at specific KM focus areas where we have identified knowledge gaps, and propose possible solutions to address those gaps. Areas include the generation, validation, and dissemination of pollution data, and how KM can be applied towards reduction of industrial pollution and non-point source pollution. In the area of governance and policy making, we identify the need to recognise tacit knowledge and people as knowledge assets. Also, critical knowledge is required at several junctures of policy making, and a KM perspective highlights the need for good knowledge in successful policy formulation and implementation. Finally, there is a need for a national framework for knowledge dissemination, and for government to take up a leadership role in facilitating knowledge sharing and champion the principles of KM. We have used publically available data and plans published by the Indian government, and from academic journals to support our analysis.

We believe that KM is a fundamental area critical to the success of the Ganga Rejuvenation Project that is underappreciated. We hope that by addressing these gaps from a KM angle, effective use of KM can bring about real solutions to the problem of Ganga River pollution.

2. Problem Analysis

2.1 *KM and the causes and effects of river pollution*

Four main sources or causes of river pollution and their effects are generally well articulated and researched in academic literature.[1] Untreated

[1]Murty *et al.* (2011). 285

sewage, industrial waste, and non-point source pollution from agricultural run-off and religious practices have variously been fingered as culprits. In addition, seasonal water flow levels and water diversion complicate the analysis by making pollution levels more concentrated and dangerous even with the same base level of polluting sources. Table 1 illustrates the causes and the effects of pollution in the Ganga River.

Point-source pollution such as municipal sewage and industrial effluence constitute up to 95% of pollution, while non-point sources such as open defecation, disposal of dead bodies, and mass bathing accounts only for the remaining 5% of the river pollution.[2] Currently, the Ganga River is highly polluted and had crossed the safe level thresholds issued by the WHO. The level of coliform bacteria is over 2800 times[3] higher than the level considered as safe. In the river's mainstream of the Ganga River, the faecal coliform levels remained above the acceptable level across some 2,500 km from Gangotri to Diamond Harbour.[4] The major cause of the increase in faecal coliform is contributed by the municipal sewage.

Another point source, or industrial waste, accounts for an estimated 260 MLD of pollutants into the Ganga River. In the CPCB report, there are an estimated 687 industries situated in UP dumping industrial wastes into the Ganga River. Of these industries, tannery industries near Kanpur region within UP are considered to be major pollutant producers, followed by paper & pulp, chemical, sugar and textile industries.[5]

Non-point sources are another contributing factor to Ganga River pollution. Non-point sources pollutants include agriculture run-off, open defecation, dumping of animal carcasses, disposal of dead bodies, flower offering, solid waste, agricultural run-off from rural settlements, and mass bathing. Majority of the non-point sources stem from religious faith and social practices with little, or no, support to increase public awareness and knowledge of the negative side-effects of pollution to the river. For instance, even though mass bathing is widely practice during religious

[2] Sewage Pollution of the River Ganga (2015).

[3] Ganges River Pollution (2015).

[4] Ganga Water handbook (2015).

[5] Sewage Pollution of the River Ganga (2015).

Table 1: The Causes and Effects of Pollution in the Ganga River

Problem Tree Analysis of Pollution in Ganga River				
Higher Level Effects	Social Increased healthcare and water costs.	Political Voter dissatisfaction	Economic Reduced tourism, agriculture and fish yields	Political International relations with Bangladesh affected
Direct Effects	Increased health problems	Lack of drinking water	Death of river ecology	Polluted water flows downriver
Focal Problem	**SEVERE POLLUTION IN THE GANGA RIVER**			
Direct Causes	Untreated Sewage	Industrial Pollution	Non-point source pollution	Reduced water flow
Root Causes	*Lack of KM?*			

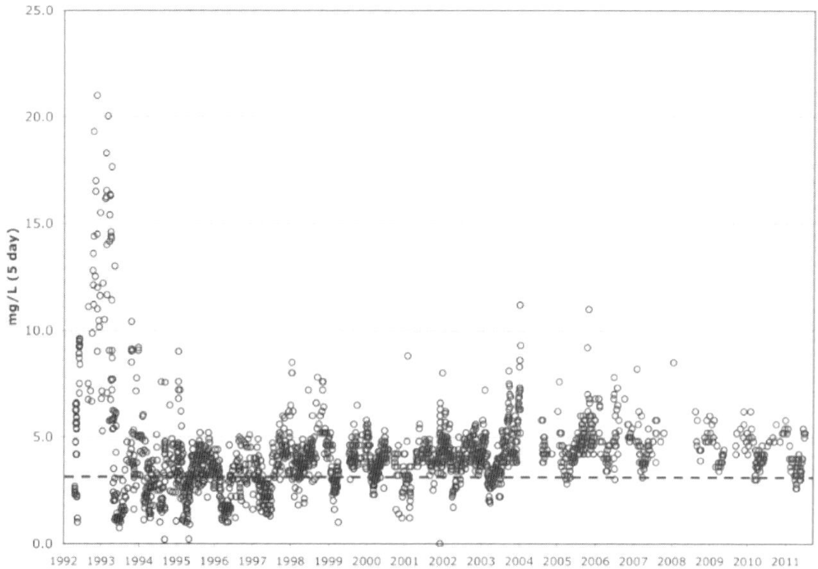

Figure 1: BOD measurements along Ganga Tulsi Ghat September 1992 – November 2011.[6]

festivals, the consequences of mass bathing have harmed the Ganga River through dramatic increase in the biochemical oxygen demand (BOD) when thousands of people bathed by the river at the same time. Figure 1 provides evidence of the increase in BOD along the Ganga River.

The magnitude of the Ganga River pollution is reaching a massive scale where huge amounts of wastes are released into the river (i.e. 1.3 billion liters of sewage, 260 million liters of industrial waste, runoff from 6 million tons of fertilizers and 9000 tons of pesticides used in agriculture and large quantities of solid waste[7]) affecting nearly an estimated population of 500 million, or 40% of India's population across 11 states.[8] In addition, the total annual volume of untreated household, and industrial effluents in the Ganga River amounts to 328.9 million kiloliters.[9] It is

[6]Ibid. 269.

[7]Cleaning-up the Ganges: A Cost-Benefit Analysis of the Ganga Action Plan

[8]Pollution of the Ganges (2015).

[9]Cleaning of the Ganga (2015).

reported that the nearly 3000 MLD of sewage generated in the towns along the river. Moreover, nearly 50% of wastewaters are discharged as untreated, leading researchers to believe that the river is no longer capable in purifying itself.[10]

The pollution of Ganga River had threatened and increased the health related-problems to the local population living along the river. According to an article in The Hindu, it is reported that the pollution levels in the Ganga River had contributed to 9%–12% of total disease related illness in UP; the fourth biggest state in India. The report estimated the total health related-problems are accounted to the water pollution, in the state at an estimated 6.4 million DALY (Disability Adjusted Life Year).[11]

In addition, the municipal and industrial waste released to the Ganga River affects the ecology; especially to the vulnerability of the ground aquifer. According to Umar and Ahmad's studies, traces of water elements from the shallow aquifer showed concentrated toxic metals such as Fe, Mn, Cd, Pb, and Cr+6 well above the permissible limits in parts of central Ganga basins.[12] While they may entail various human health hazards, these concentrated toxic metals are adversely affecting the economic activities along the river including industries and agriculture.

2.2 *Effects of poor knowledge management in pollution abatement*

How is the lack of knowledge generation, validation, and dissemination leading to continued pollution in the Ganga River? We find that continued river pollution is explicitly linked to the lack of proper knowledge management in the areas of pollution knowledge and governance and policy-making knowledge.

Without good knowledge generation, validation, and dissemination, there will be basic lack of usable pollution data. It may be that pollution data is simply not being generated at all or often enough, or that pollution data is generated but is inaccurate, incomplete, inconsistent, or irrelevant

---

[10] Das (2011).

[11] Pollution Levels in Ganga Alarming *et al*. (2004).

[12] Umar *eet al*. (2004).

and therefore unusable. It may also be that data exists, but is not effectively shared and cannot be found or used in a timely manner.

Without good pollution data, the sources and causes of pollution cannot be determined correctly, and resources will be misdirected to ineffective measures. Effective governance and pollution abatement policies cannot be achieved as the wrong policies will be designed and implemented, and there is no accurate data to evaluate policies and adjust them. Without good data, specific pollutants, pollution levels, and their health effects cannot be identified. Pollution laws also cannot be enforced if pollution data cannot hold up in the court of law, and polluting companies and even cities and states have little incentive to reduce pollution, if pollution data is poor and cannot be linked directly to them (see Table 2).

In the area of knowledge dissemination, KM advocates an open sharing of data and knowledge between organisations and the public. Not sharing knowledge firstly deprives other parties from generating further knowledge, and creates mistrust and misperception. For example, if pollution data is shared openly and effectively, researchers and other experts in the private sector will be able to identify specific causes and thus generate further useful data and knowledge. If different government ministries 'own' their data generated and do not share openly, mistrust and 'turf wars' over data will make effective governance harder. Lastly, if reliable pollution data is not shared with the public, the public is apt to refer to inaccurately high and sensationalist figures quoted in the news. This leads to low public trust and cooperation in the effort to reduce river pollution.

In the area of governance and policy making, a lack of emphasis on tacit knowledge and people as knowledge assets leads to missed opportunities to make use of local knowledge and expert groups. Many policy failures can be attributed to the failure to recognise the importance of local knowledge and tacit knowledge. For example, the efforts to build STPs under the Ganga Action Plans were criticized for not using the right STP technologies given the unreliable power supply, and the lack of operational and maintenance expertise of municipal authorities after they were built.[13]

[13] Ganga Action Plan (2015).

Table 2: How Poor Knowledge Management Affects River Pollution

	Pollution Knowledge			Governance and Policy Knowledge	
Specific KM Gaps	Knowledge Generation	Knowledge Validation	Knowledge Dissemination	People as Knowledge Assets	Tacit Knowledge
	Lack of data	Unreliable data Lack of Pollution Monitoring Standard	Lack of knowledge sharing platforms and database	Low recognition of local knowledge and expert groups	Low emphasis on policy know-how and experience
KM Direct effects	• Unable to determine pollution sources and causes	• Unable to determine specific pollution levels and health impact	• Available knowledge not used effectively • Low trust between groups, • Public misperception	• Failure of policies due to local factors • Alternative technologies and solutions missed	• Roles and responsibilities not clear • Human resource and performance management underutilized
			↓Leads to↓		
KM Secondary Effects	Unable to design effective policies, direct resources, track results	Unable to enforce pollution laws	States/Cities have no incentive to reduce pollution	Low public trust and understanding	Low accountability
Outcome			Failed River Pollution Policy Design and Implementation SEVERE POLLUTION IN THE GANGA RIVER		

There is also a low emphasis on policy making and implementation experience. Effective governance requires a recognition of people having the right knowledge comprising of skills and experience to make and implement solutions to reduce pollution. Human resource management and performance management are therefore areas which must be incorporated as part of KM and part of the Ganga Rejuvenation Project's effort.

3. Knowledge Management Framework

3.1 *What is knowledge?*

Before continuing further into the details of KM in Ganga Rejuvenation Project, let us start discussing the fundamental component of KM — *knowledge*. What is knowledge? Albert Einstein once said that "information is not knowledge. Knowledge is experience. Everything else is information."[14] Einstein understood that information is only the ingredients for knowledge whereas real knowledge is created and stored from human experience. From the perspective of the Ganga Rejuvenation Project, there are vast amounts of knowledge compiled from years of private and public projects and experiences on river pollution, urban planning, water and sewage treatment plants, health, and education, etc. However, this knowledge had yet to be effectively collated, stored, managed, and utilized in a meaningful way.

Many academics and researchers have pointed out that the term data, information, and knowledge are often used interchangeably. However, the meaning of knowledge is completely different from the meaning of data, information, or wisdom. Knowledge is not data, information, or wisdom but "is the appropriate collection of information, such that its intent is to be useful"[15] — it is a form of *intellectual capital*. Intellectual capital, or also known as knowledge capital or knowledge asset,[16] is defined as an intangible asset that comprises of the information, skills, and experiences

[14]Zeleny.

[15]Ahsan (2006).

[16]In this chapter, knowledge capital and knowledge asset are used interchangeably with identical meaning.

including but not limited to on-the-job training and learning. Unlike capital used for factors of production such as land, labour and capital, knowledge capital is based on skills that people share with each other in order to improve efficiencies.[17] By supporting and fostering the management of knowledge capital, policy makers are at a better advantage to achieve success with the intended policies in the Ganga Rejuvenation Projects through utilization of information, skills, and experiences.

In KM, knowledge is categorized under two types — explicit and tacit knowledge. Explicit knowledge is set out in tangible form where organizations, private or public, produce and store reports of various projects.[18] This classification of knowledge is codified and disseminated in various forms such as reports, database, internet, e-mail, graphs, charts, etc. However, tacit knowledge is more ambiguous and intangible form where knowledge is made up of intellectual capital including but not limited to skills and experiences — the know-how.[19] For both private and public sectors, tacit knowledge proves to be a major challenge and difficulty to unlocking its' value and create a blueprint for an effective KM framework to generate, validate and share such intellectual capital. Reasons are traced back to the storage of tacit knowledge (i.e. within a person) and willingness to share tacit knowledge in public. On the other hand, explicit knowledge is easier to store and disseminate compared to tacit knowledge.

3.2 *Tacit vs explicit knowledge*

From an academic perspective, it is easier to distinguish and differentiate between tacit and explicit knowledge than in practice where both are not so separate and discrete. The relationship and interaction between these two types of knowledge is important to the understanding of knowledge management. The key differences between tacit versus explicit knowledge lies in the following areas: a) knowledge codifying and transfer, and b) knowledge storage and retrieval.

---

[17] Investopedia

[18] Baporikar (2014).

[19] Intellectual Capital (2015).

Most, if not all, explicit knowledge can be translated into codes and transferred without having to understanding the knowledge subject, while tacit knowledge is more intuitive and requires an understanding of the knowledge subject. Hence, the transfer of explicit knowledge is relatively easy while the transfer of tacit knowledge requires interaction, shared understanding, and trust between people.

It is easier to aggregate and store explicit knowledge in a single location in various forms and formats without the understanding of the knowledge subject. However, tacit knowledge as stated earlier is more intuitive and personal where it may be distributive but not easily aggregated or stored. For example, it is difficult to store a child's experience in learning how to ride a bicycle. Nevertheless, tacit knowledge may be captured when individual joins a group or organization sharing a common experience or objective.

In the Ganga Rejuvenation Project, we find that there is little emphasis or even mention of tacit knowledge. Our research has reviewed and investigated various websites (i.e. National Mission for Clean Ganga, Ganga Knowledge Centre, Central Pollution Control Board, and Gangapedia) where most websites provided little or no tacit knowledge covering Ganga river pollution. Majority of the knowledge are in forms of explicit knowledge such as history of Ganga River, river pollution data, or current and future projects. We have rarely seen websites where tacit knowledge such as uses of lesson learnt, socialization of successful/failed projects, or innovation sharing in the private and public sectors are shared publicly and made available online. Ergo, we believe that the sharing of tacit knowledge in the Ganga Rejuvenation Project is missing and it is important for policy makers to utilize such knowledge to foster a KM culture.

3.3 *The importance of tacit knowledge*

Knowledge starts of as tacit and once recorded, it becomes explicit. However, tacit knowledge is very personal and take shapes in forms of individual's intellectual capital such as memories, past experience, or learning through trial and error. According to researchers, tacit knowledge is sometimes difficult to record or stored as individual's memories or intellectual capital are never owned by any organizations or groups other than the individual.[20] Michael Polyani introduced the term "tacit knowledge" in 1958 and reinforced the notion that tacit knowledge may not always be verbally articulated in order to be recorded and changed into explicit knowledge.[21] Moreover, individuals are often unaware of the importance and value of tacit knowledge they possess and how it is valuable to organizations and others.

Even though it is challenging to capture tacit knowledge, individual can effectively transfer tacit knowledge through interactions and contacts with organization and or other groups of similar interest. A certain level of trust needs to be established between individual and organization such that individual's tacit knowledge is not misused or damaged others. Such capturing of tacit knowledge happens when individual or group of people joins a network, organization, or community working to achieve an objective.[22] Using explicit knowledge alone is not enough to achieve an objective but the use of tacit and explicit knowledge together will ultimately provide a successful outcome. For instance, having explicit knowledge but not knowing how to use such knowledge does not help the individual, group, or organization perform the objective that is required.

Another aspect of tacit knowledge is described as *know-how* which is different from *know-that* or *facts*. In 1945, Gilbert Rye published a paper arguing against the position that knowledge is *knowledge of proposition*, or *knowing the facts*, and instead showed that knowledge should be considered as *know-how*.[23] According to Rye, *know-how* is the knowledge obtained by experts who can skilfully and automatically perform tasks without hesitation.[24] This *know-how* knowledge is ingrained in the individual's human capability which has become instinctive and intrinsic rather than extrinsic.

To summarize, KM defines tacit knowledge as knowledge possessed and owned by an individual where such knowledge can be difficult to communicate to others via verbal or physical means. It is possible for an

---

[20] Baporikar (2014).

[21] Zeleny

[22] Tacit Knowledge — Wikipedia (2015).

[23] Norström (2014).

[24] Norström (2014).

individual to acquire tacit knowledge without having to use verbal or physical communication. For instance, children learn to imitate their parents and acquire the necessary tacit knowledge in their adult life. The most important factor in acquiring tacit knowledge is experience. People acquire experience through years of gaining knowledge and it is difficult to imagine people sharing each other's thinking process without some form of common and shared experience. Hence, tacit knowledge, even though difficult to capture, can be shared amongst people if the shared experience is common.

Ergo, such tacit knowledge should be promoted in the Ganga Rejuvenation Project and as part of the KM culture. Private and public sectors alike (i.e. government and NGO) should work together to generate, validate, and disseminate tacit knowledge such that the experience on Ganga river pollutions are shared amongst people. Sharing of experience will lead to wisdom in creating innovative solutions or ideas to rejuvenate Ganga River.

3.4 *Data, information, knowledge and wisdom chain*

Moving away from the explanation of tacit knowledge but still building on the foundation of knowledge management is the concept of the Data-Information-Knowledge-Wisdom chain (DIKW chain) created by Russell Ackoff.[25] The DIKW chain in Figure 2 illustrates the relationship between data, information, knowledge, and wisdom. The chain shows the progression of an individual's understanding of data, information, knowledge, and wisdom. The pictorial below encapsulates the DIKW chain.

In this pictorial, the connectivity and understanding explains the movement of an individual's understanding of data, information, knowledge, and wisdom across the x and y-axis. Across the x-axis is connectivity, or how each component in the DIKW chain are connected to each other. The y-axis explains an individual understanding of each component as it transition from data to wisdom. This pictorial showing the diagram chain transition is produced by Amy Hodler at Data Centre Advisory. She states "another interesting outcome of looking at this as a process is that

[25] DIKW Pyramid — Wikipedia (2015).

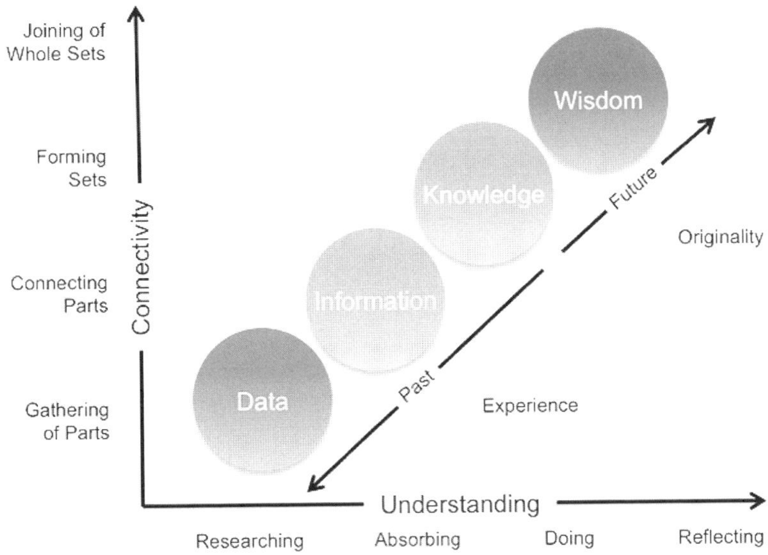

Figure 2: DIKW Chain Transition

it clearly shows that data, information and knowledge are attributable to better understanding the past. We all understand that's a critical basis, but our goals usually depend on insight into what we should do in the future, which requires wisdom."[26] Her statement supports that the explanation of future experience and past experience.

In simple terms, each component of the DIKW chain has a specific description and meaning. Data is usually produced either manually or from an automatic system where its value is added when data is interpreted and converted into useable format, or information. Information is only converted into knowledge once it is applied to a particular situation or use in a meaningful way. Knowledge is achieved when information is used to increase the understanding of the knowledge subject. However, wisdom has a more philosophical meaning where it is intrinsic and typically an accumulation of connections, experiences, interpretations, values, or judgements. In many organisations, whether private or public, data and information are perceived as more important IT tools than knowledge and

[26]Hodler (2015).

wisdom to perform tasks and improve operations. However, knowledge and wisdoms are key concepts and components in in governance and policy-making that cannot be ignored.

3.5 *Data and information vs. knowledge and wisdom*

To better understand the DIKW chain, Figure 3 illustrates the progression with reference to the Ganga Rejuvenation Project.

Figure 3 identifies data as *know-nothing*, information as *know-how*, knowledge as *know-what*, and wisdom as *know-why*. In DIKW chain, each component represents a coordinated action such as identification of objectives and goals, selection of quantitative and qualitative inputs, establishing the progressions order, assigning responsibilities, delegating and performing the tasks, and evaluate the outcomes. The coordination of action requires not just know-how but many other components in the chain.[27] Specifically, coordination of action means the actions that are required to capture data, translate to information, analyse as knowledge, and

For instance, *know-nothing* can be defined as no value add or nothing is known from the data. Raw data by itself does not represent any picture or understanding of the problem as whole and, thus, it is *know-nothing*. In *know-how*, information can be referred to present a picture but an incomplete picture without the understanding of the problem. *Know-how* is also referring to the tacit knowledge such as skills and experiences. *Know-what*, or knowledge, refers to the knowledge and understanding of the problem (i.e. identification of objectives and goals). Such progression involves information previously collected and analysed. In *know-why*, explicability, relationship, or causality is added. Information and knowledge combined can lead to *know-why* as it involves analysis and understanding of the problem.

As each component moves to the next level, its effects are explained as follows:

- Data and muddling through — raw data such as faecal coliform counts do not represent anything by itself and may only manage to provide a part of the puzzle without understanding the problem;

[27]Zeleny

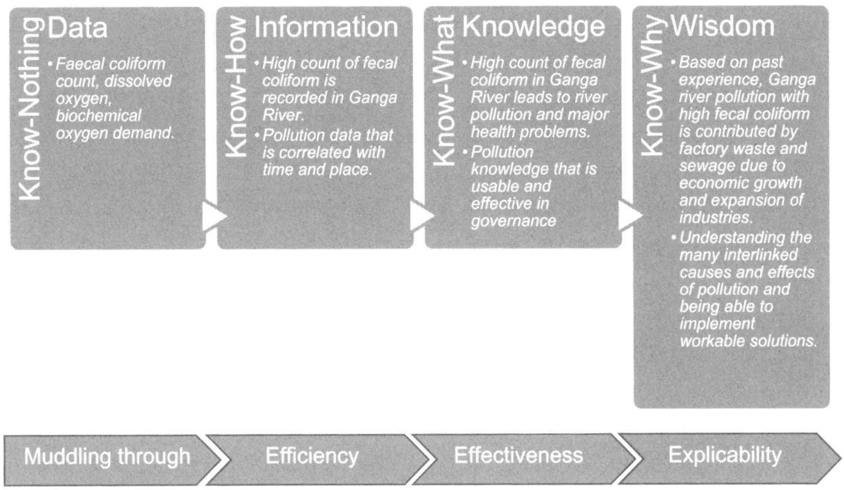

Figure 3: Data & Information vs Knowledge & Wisdom

- Information and efficiency — information is the ingredients such as high count of faecal coliform is recorded in Ganga river which illustrates the problem;
- Knowledge and effectiveness — knowledge is the understanding of the subject, or problem, such as high count of faecal coliform in Ganga River leads to river pollution and major health problems, thus illustrating an analysis of information and understanding the problem causation; and
- Wisdom and explicability — wisdom is the use of past experience, values, or judgements to understanding, predict, and unlock the problem, e.g. Ganga river pollution with high faecal coliform is contributed by factory waste and sewage due to economic growth and expansion of industries.

3.6 *Knowledge gap*

What is a knowledge gap? In 1998/99, the World Bank has published and identified the knowledge gap between and within countries where knowledge is lacking.[28] The World Bank has reported and differentiated two

[28] Evers

types of knowledge that is lacking. One is the knowledge about attributes (i.e. incomplete or garbage data) leading to information problems, and second is knowledge in relation to technology. The knowledge gap in technology is define as little knowledge available in the increasing use of technology. Moreover the World Bank stated "typically, developing countries have less of this know-how than industrial countries, and the poor have less than the non-poor. We call these unequal distributions across and within countries knowledge gaps."[29]

According to Tichenor, knowledge, like any forms of wealth and status, is unevenly distributed throughout the social system. Through the study of mass communication, Tichenor predicted that "as the infusion of mass media information into a social system increases, higher socioeconomic status segments tend to acquire this information faster than lower socioeconomic-status population segments so that the gap in knowledge between the two tends to increase rather than decrease."[30] As stated, a knowledge gap refers to the uneven intensity of knowledge. The uneven intensity, or acquisition, of knowledge is caused by different objectives and perspectives within the organization and whether in private or public sector. Hence, the poor usually will have higher knowledge gap than the rich with little accessibility to the kinds of knowledge in preventing or raising awareness of the Ganga river pollution. From a KM framework, knowledge gap exists within each of the three phases — generation, validation, and dissemination. In order to achieve success in KM, closing the knowledge gap is vital and necessary steps.

3.7 *The principles of KM policy*

As earlier discussed, KM should be considered as a tool in getting the right knowledge to the right people at the right time and, thus, policymakers should design KM policy that enables such actions. In principle, this chapter tries to propose KM policies for Ganga Rejuvenation Project that seeks to enable and support effective achievement of the river clean-up via

---

[29] Evers

[30] Tichenor (1970).

knowledge sharing. Therefore, some key principles to be mindful of are as follows.[31]

- Identification and analysis of available and required knowledge, and the subsequent planning and control of actions to develop knowledge capital so as to fulfil organizational objectives;
- Systematic and organized attempts to store, use, and deliver knowledge assets in a way to perfect the performance of the organization;
- Ensure knowledge assets are identified, available, shared, preserved and managed to maximize effective operation of the relative organizations;
- Provide incentives to promote a learning culture and its role as a key differentiator in achieving objectives of the relative organization; and
- Encourage innovation by creating an open and transparent environment with good practice and documentation managed and shared appropriately.[32]

3.8 *Our KM framework for the Ganga Rejuvenation project*

Our KM framework identifies three distinct phases of the KM lifecycle:

- Generation: data, information, knowledge, and wisdom are acquired, captured or created:
    - On Ganga Rejuvenation: Central problems — what are the knowledge demands and supply? What is missing?
- Validation: data, information, knowledge, and wisdom are validated by refining, verifying, evaluating, organizing, confirming and storing systematically.
    - On Ganga Rejuvenation: Central problem — how to judge and ensure the accuracy, reliability, systematisms and usefulness of knowledge?

[31] CENN Knowledge Management Policy (2015).
[32] CENN Knowledge Management Policy (2011).

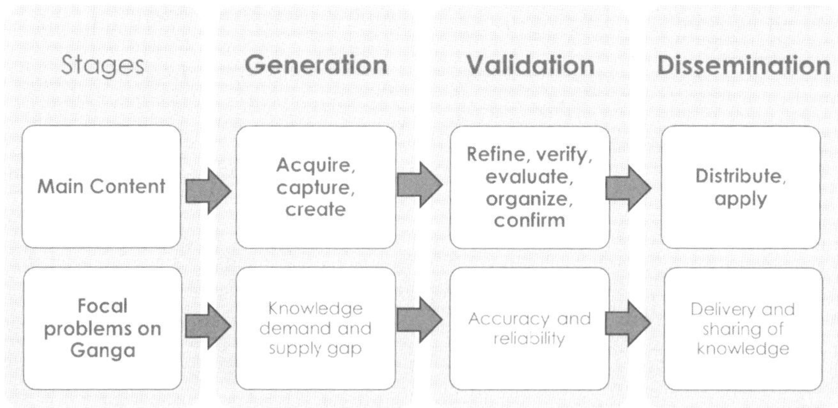

Figure 4: KM Framework

- Dissemination: data, information, knowledge, and wisdom are distributed through education, training programs, automated knowledge based systems, expert networks. Knowledge is applied or leveraged for further learning and innovation via mining of the organizational memory and the application of expert systems such as decision support systems.
    - On Ganga Rejuvenation: Central problem — how to ensure that valid knowledge is delivered and shared with the people who really require it?

Figure 4 illustrates the KM framework used in this chapter.

3.9 *Linking knowledge to knowledge management*

How are the discussions of tacit knowledge and the DIKW chain relevant to knowledge management? Both of these fundamental concepts are the foundation of knowledge management and of the KM framework for the Ganga Rejuvenation Project. Thomas H. Davenport, an expert in KM theories, states that "knowledge management is the process of capturing, distributing, and effectively using knowledge".[33] His definition is shared

[33]Davenport (1998).

amongst many academics and researchers that knowledge management is a discipline for the *generation* (identifying and capturing), *validation*, (validating and evaluating), and *dissemination* (retrieving and sharing) of knowledge. Generation, validation, and dissemination are the three pillars of KM framework that are applied to the Ganga Rejuvenation Project.

In both private and public organizations, KM is an important central strategy with the objective to share and provide access to information and knowledge within the organization and often to the public. To support the central strategy, KM needs to create an environment and culture which information and knowledge are easily accessible to individuals. Moreover, KM should foster and allow individuals to gain, share, use, and learn from the information to develop their own knowledge and ultimately benefit their organization or interest group. From a human resource management and performance management angles, KM is not just information systems, information and communication technology (ICT), and communication system between people and IT systems. Knowledge management is an effective facilitator and promoter of knowledge to the individual and the organization. It should encourage and foster a culture to apply knowledge for the benefit of the organization and in order to achieve its objectives.

From a policy perspective, KM should provide the direction towards policies that recognize the requirements of tacit knowledge from individuals and towards human resource and performance measurement activities to achieve organizational objectives. Typically, many KM policies are created and developed to support the sharing of knowledge within and not outside an organizational. However, the success of KM and Ganga Rejuvenation Projects depends on creating a culture that supports the sharing of knowledge and information inside and outside organizations with no boundaries across the public sector. It has been noted by many academics and researchers that KM theories and tools have been successfully in the private sector. Thus, it is not unreasonable to assume that KM can be successful in the public sector. Many public sector organizations are considered knowledge-based organizations (e.g.. the Ganga Knowledge Centre) and thus, KM could enable better formulation of policies with increased transparency and trust that can deliver effective services by maximizing productivity in terms of process, time and cost.

KM should not be overly focused on the investment of IT technologies but rather an investments in intellectual capital to aid the acquisition of knowledge by individuals. Hence, the investment in intellectual capital creates a KM culture that supports the sharing of knowledge and public participation in policymaking by bringing resources together to solve the Ganga river pollution. Moreover, KM should be considered as a tool in getting the right knowledge to the right people at the right time and helping individuals to turn information into action in ways that promote the rejuvenation of the Ganga River.

Applying KM towards the Ganga Rejuvenation Projects may be a difficult task but our research has shown that KM theories can easily be applied to any private or public organization with enough support from the NGOs or governments. For example, a KM central strategy developed purely for private organization may be utilized to serve the public sector as its objectives are similar (i.e. foster KM culture, knowledge sharing, etc.). Moreover, the KM theories such as tacit knowledge can be translated into policy that enables knowledge sharing in relation to the river pollution and ways to improve or rejuvenate the Ganga River. IT technologies should also be developed along with the use of KM theories in Ganga Rejuvenation Projects to foster a KM culture. Such IT technologies include databases to capture tacit knowledge and networking solution to connect various people and projects to a central knowledge asset directory.

4. KM Issues in Ganga Rejuvenation Project

4.1 *Overview of KM issues*

In our problem analysis, we have shown how a lack of proper knowledge management can lead to failures in policy making and implementation. In this section, we will focus on specific issues of governance and pollution abatement policy, and elaborate on the identified knowledge gaps and show how KM can be implemented to achieve a greater chance of policy success.

In the area of pollution data, developing a Knowledge Asset Directory to inventorize and categorize existing knowledge assets will help identify existing knowledge resources and also missing data needed. This will allow everyone to find and use existing knowledge, while avoiding duplication of resources towards generating knowledge that is already available.

Policy makers can also look towards innovative collection methods and non-traditional sets of data to generate knowledge and address the problem of insufficient pollution data. Developing a national water pollution monitoring standard will also ensure that pollution data generated can be reliable and useful in the fight against pollution.

In the area of industrial pollution, providing technological know-how, pollution data, and environmental programme design knowledge to industries can help industries comply with pollution laws. Knowledge dissemination to local communities and the use of a community and market-based approaches may also prove critical in reducing industrial pollution. Non-point source pollution on the other hand requires a better understanding of the relationship between non-point source pollutants and the ultimate health effects. Public education can then be directed to reduce this source of pollution.

In the area of governance and policy making, there is a need to recognize people and the tacit knowledge they hold as knowledge assets, and to link them up to facilitate knowledge sharing and development. Also, critical types of knowledge are needed at several junctures of policy making. A KM perspective allows us to identify and ensure that important knowledge is available in carrying out successful policy formulation and implementation.

Finally, in the area of knowledge dissemination, we find that developing a nationwide dissemination system or mechanism is key. In addition, the role of a knowledge broker to take the lead in developing knowledge management and sharing is critical. Table 3 provides an overview of the KM problems and recommendation.

4.2 *Knowledge demand and supply*

Knowledge generation is the first phase in KM where knowledge starts in its life cycle explaining how and where knowledge comes from. For the Ganga Rejuvenation Project, the purpose of knowledge generation is to

Table 3: Overview of Pollution and Governance Knowledge Gaps

KM Area	Identified Knowledge Gaps	Proposed KM Solution(s)	Effects of KM Solutions
Knowledge Demand and Supply	Data is not generated in accordance with needs	Develop a Knowledge Asset Directory to identify and categorize existing knowledge assets	More effective use of resources in data generation, identification of missing data needed, and better knowledge sharing
Pollution Data Generation	Inadequacies in number of monitoring stations, frequency of monitoring, types of pollutants, and health effects	Explore new technologies, consider monitoring via local communities, and identify related data	More data, and more cost-effective data
Pollution Data Validation	Lack of national water pollution monitoring standards	To develop a national water pollution monitoring standard	Increased validity of existing data, data from other parties can be used, better understanding of impact of agriculture and household runoff
Industrial Pollution	Lack of pollution data, and inadequate use of community and market based approaches	Develop R&D or provide technological knowledge to industries, provide pollution data and environmental programme design knowledge	Local communities can play an effective role, industries can comply with pollution laws, decreased industrial pollution

Non-point source pollution	Relationship between non-point source pollution and health impact, lack of public awareness	Commission studies on non-point source pollution, increase public awareness	Better understanding of non-point source pollution impact, resource direction towards more polluting practices, reduction in pollution from public
People as Knowledge Assets	No utilization of networks of people and their knowledge assets	Develop a decentralized knowledge asset network	Allows use of experience from people with expert knowledge
Governance and Policy Making	Lack of critical policy making knowledge at important junctures	Identify and ensure sufficient knowledge in policy making	More effective and successful pollution abatement programmes
Pollution Data Dissemination	Lack of standardized method of publishing and sharing data	Develop a consistent and effective way to share pollution data, for e.g. through the GKC	Greater transparency, accountability, and effectiveness in dealing with pollution.
GKC as Knowledge Broker	Lack of impetus and leadership in knowledge sharing Lack of clear roles and responsibilities to manage KM in the Ganga Rejuvenation Project.	For GKC to take on a leadership role in developing KM and knowledge sharing Establish and assign clear roles and responsibilities to organization responsible for implementing KM	Increasing value of existing knowledge, making known information previously unavailable, generating new data and new solutions Greater transparency and accountability of KM in the Ganga Rejuvenation Project.

prepare appropriate knowledge assets for different stakeholders. In this phase, knowledge gaps exist between the knowledge demand and supply for different stakeholders.

To understand what is the current knowledge demand and supply is the starting point of using KM for Ganga Rejuvenation. From the demand side, there are three levels demanders: In the first level, the policy makers and inspectors, mainly from public sectors, who make policies and inspect the implementation of policies. To be more specific, in the federal level, this includes: the Ministry of Water Resources, River Development and Ganga Rejuvenation (MoWR, RD&GR; or MWR), National Ganga River Basin Authority (NGRBA), and its implementation wing- National Mission for Clean Ganga (NMCG); the Ministry of Environment and Forests (MEF) and the Central Pollution Control Board (CPCB); and the supporting department like the Ministry of Finance (MOF), etc. At the state level, the SPMGs (the States level Program Management) of UP (Uttar Pradesh), Bihar, Uttarakhand West Bengal and Nodal Cell in Jharkhand, which covers five mainstream states related to Ganga River. The World Bank is supporting the Government of India in its effort to rejuvenate the Ganga River. The $1 billion of the National Ganga River Basin Project is helping the National Ganga River Basin Authority (NGRBA) build institutional capacity for rejuvenating the river.[34] It is also financing key infrastructure investments in the five mainstream states.

In the second level are the policy executors, who execute and manage the policies. Investments are being implemented by multiple executing agencies, mainly parastatal organizations such as the Uttar Pradesh Jal Nigam, Bihar Urban Infrastructure Development Corporation, Kolkata Metropolitan Development Authority and some Urban Local Bodies (ULBs).[35] This also includes the industrial factories, city sewage discharge and treatment units or individuals, and the educational institutions which are responsible for raising the awareness of stakeholders on cleaning Ganga mission.

In the third level are the general public, including the ordinary people, the research institutions, universities and even media, who are observers and can be approval, opposite or neutral to the policies, and give comments

[34]The National Ganga River Basin Project — The World Bank
[35]Ibid. 159.

or suggestions to current policy. They could be transferred to the stakeholders easily under a certain condition. For example, the ordinary people who live far away from Ganga River may be influenced by the products which made by Ganga water, and are transferred to stakeholders.

From the supply side, the Ganga Rejuvenation Project should be planned, started, guided and monitored by the policy makers and inspectors, so the main knowledge suppliers of this project are focused on this level although the above three level demanders all can supply a certain amount of knowledge.

The current situation of knowledge generation on Ganga Rejuvenation Project is not optimistic. First of all, there is no specific or united administrative institution which is in charge of knowledge generation for Ganga rejuvenation. When there is a requirement for Ganga rejuvenation knowledge, different institutions will generate knowledge assets through their own channels. For example, data needed by MWR is sourced via GKC, MEF data is via CPCB, and some local cities like Varanasi get data via their own research institutions such as Banaras Hindu University. Secondly, the knowledge demand and supply situation for Ganga rejuvenation are not clear. No institutions have done careful investigations on it; no organizations are in charge of this; and no departments know how much knowledge is enough, what kind of knowledge they need, and what is missing. Thirdly, Ganga rejuvenation knowledge assets are scarce. Currently, the useful portal we could find for the public to access Ganga knowledge assets is just the website. However, even for the website portal, the knowledge assets we could find are really limited. The Ganga Knowledge Centre (GKC) website only has introduction information about itself and there are no further resources related to Ganga rejuvenation in this website. Another website portal is Gangapedia, but the knowledge assets here are also limited, lacking systematic structure, completeness, and reliability.

We believe that the functional organization of knowledge generation is more suitable to be put in first level, because it can make a strategic planning, guide and monitor the whole process of knowledge generation. And it also feasible for it to understand the general situation, make clear what is the current situation of knowledge, what are the requirement and supply, and what the knowledge gaps are. From this perspective, the GKC, which is set up under NMCG, is a kind of good practice. GKC is conceptualized as a premiere and autonomous knowledge based institution.[36] The main objectives of GKC are to:

- Create and manage knowledge resources including analysis and modelling of diverse data sets relevant to Ganga River Basin.
- Design and foster research innovation through identification of knowledge gaps, need for new ideas and supporting targeted research.
- Facilitate stakeholder dialogue through public involvement and building partnerships with universities/institutions of national & international repute, public & private entities and NGOs.[37]

The objectives are exactly what the knowledge generation hopes to achieve. So GKC could be the functional and comprehensive organization which is responsible for the generation of Ganga knowledge.

Currently, the required knowledge for Ganga rejuvenation mainly includes the following categories: law and regulations, geography and hydrology, environment/ ecology/biology, demography, health and hygiene, historical lessons, religion/culture, education, cases of river rejuvenation, international relations, financial support, and science and technology.

From the above analysis, we can draw a *Knowledge Asset Directory for Ganga Rejuvenation*.

Table 4 provides a general picture of the knowledge asset demand for Ganga rejuvenation. Policymakers can check what knowledge it has and what is missing based on this table. Then, the next step is to generate relative knowledge assets to support relative institutions, industrial factories and individuals on cleaning or protecting Ganga River.

4.3 *Pollution data generation: addressing the problem of insufficient pollution data*

To generate knowledge assets, there are two basic methods: *collecting and producing*. Knowledge collecting is a very common method of transferring knowledge. It is often the most effective and efficient method, and

[36] NMCG — Ganga Knowledge Center (2015).
[37] Ibid. 160.

Table 4: Knowledge Asset Directory for Ganga Rejuvenation

No.	Type	Main Content	Main Suppliers	Main Demanders
1	Law & regulations	Industrial emission standards, sewage disposal standards, water quality standards, etc.	Public sectors: MWR; MEF; NGRBA;NMCG;CPCB; GKC	Policy makers and inspectors, Policy executors
2	Geography & hydrology	The geographic and hydrological characteristics of Ganga, especially water flow	MWR, GKC	Policy makers and inspectors
3	Environment, ecology & biology	The environmental, ecological and biological influences that the Ganga pollution brought	MWR, MEF, GKC, media & research institutions	Policy makers and inspectors, General Public
4	Demography	The influences of rapid growing population	Ministry of Home & Affairs, GKC	Policy makers and inspectors, General Public
5	Health & hygiene:	The influences of water quality to health	Ministry of Health & Family Welfare, CPCB, GKC, media & research institutions	Policy makers and inspectors, General Public
6	Historical lessons	Historical experiences of cleaning Ganga River	MWR, MEF, GKC & research institutions	Policy makers and inspectors, Policy executors

(Continued)

Table 4: *(Continued)*

No.	Type	Main Content	Main Suppliers	Main Demanders
7	Religious & cultural reasons	The religious and cultural influences for Ganga rejuvenation, such as holy water status of Ganga	Ministry of Minority Affairs, GKC, media & research institutions	All level demanders
8	Education	Raise the public awareness and make deep research on Ganga rejuvenation	The Ministry of Human Resource Development, GKC & research institutions	All level demanders
9	Relative cases	Successful and failed policies, experiences or lessons from other countries	MWR, MEF, GKC, media & research institutions	Policy makers and inspectors
10	International relations	Cooperation among related countries	Ministry of External Affairs, GKC	Policy makers and inspectors
11	Financial support	Financial investment or cost-benefit study	Ministry of Finance, MWR, MEF, World Bank, and GKC	Policy makers and inspectors
12	Science and technology	Use technical methods to solve the technical problem for Ganga rejuvenation	Ministry of Science and Technology, MWR, MEF & GKC	Policy makers and inspectors, Policy executors

also is visible and easy to understand, which enables knowledge to move from one to many regardless of time and space. Collecting knowledge is a difficult task in today's fast-moving virtual organizations.

For explicit knowledge, the traditional library-style collecting methods are useful. An administrative order is another effective way to collect knowledge assets, such as official documents, materials of some government projects related to Ganga rejuvenation and so on.

For tacit knowledge, to elicit the most usable knowledge from an individual related to Ganga rejuvenation, it is better to interview face-to-face rather than a summary of the individual's own experiences. The sharing of knowledge can be "accelerated using various communication media".[38] "Written communication may not be as effective a driver of knowledge transfer as interactive communication methods such as meetings and seminars".[39]

When required knowledge assets are missing, lacking validation or could not be collected, producing knowledge directly by the suppliers is a good choice. Usually this method is useful when the government starts Ganga river policy analysis (like agenda setting, decision making or evaluation) but could not find good enough supporting data. At this time, policy makers may use administrative power, asking subordinated organizations or research institutions to study and produce relative knowledge assets.

The common pitfalls in the implementation of knowledge generation should be avoided. Firstly, avoid generating too much data. Knowledge generation efforts should focus on what the demanders need, and not attempt to generate everything that is known about a particular topic. Do not generate data that isn't used, unless you have identified what the potential demand is. Secondly, do not underestimate the time and effort required. It is a laborious process to harvest knowledge and present it in a manner that people can make sense of it for re-use and adoption.[40] Thirdly, do not assume one size fits all — when it comes to methods for presenting captured knowledge it's important to understand people's preferences as

---

[38] Bednar (1999).

[39] Ibid. 159.

[40] 10-Step Guide to Knowledge Capture

receivers of knowledge.[41] Some people find reading text a useful way to learn something. Others prefer to learn by listening and observing. So try to develop different channels of knowledge sharing, such as paper, image, video, and so on.

According to 2011 statistics from CPCB, pollution monitoring was conducted by at 2,500 stations nationwide.[42] This translates to one station every 1,315 km^2, which is far below levels in developed nations. Arkansas in the US has monitoring stations per 356 km in comparison.[43] On the Ganga River, there are only 56 monitoring stations or one station every 45 km on average.

In terms of frequency, 32% of the stations have frequency of monitoring on a monthly basis, 28.82% on a half-yearly basis, and 38.64% on a quarterly basis (2009 CPCB figures). Given that pollution levels can peak and spike even on a daily or even hourly basis based on water levels or point-source discharge, the frequency of monitoring is also clearly inadequate.

The CPCB reports that "water samples are being analyzed for 28 parameters consisting of 9 core parameters, 19 other physicochemical and bacteriological parameters apart from the field observations. Besides this, nine trace metals and 15 pesticides are also analyzed in selected samples, and bio monitoring is also carried out on specific locations".[44] However for the Ganga River, only seven values, namely Temperature, pH, DO, Conductivity, BOD, Nitrate, Total Coliform (TC) and Faecal Coliform (FC), are published. Other pollutants such as toxic metals are not represented.

Since 2009, the number of monitoring stations set up has increased (see Figure 5), and there have been efforts to set up real-time monitoring stations. However, only 10 real-time monitoring stations are currently operational and the data is yet to be accessible by the public.[45] The

[41] Ibid. 162.

[42] Central Pollution Control Board (2015).

[43] Rajaram *et al.*

[44] Central Pollution Control Board (2015).

[45] HydroQ — Real-Time Central Monitoring Station (2015).

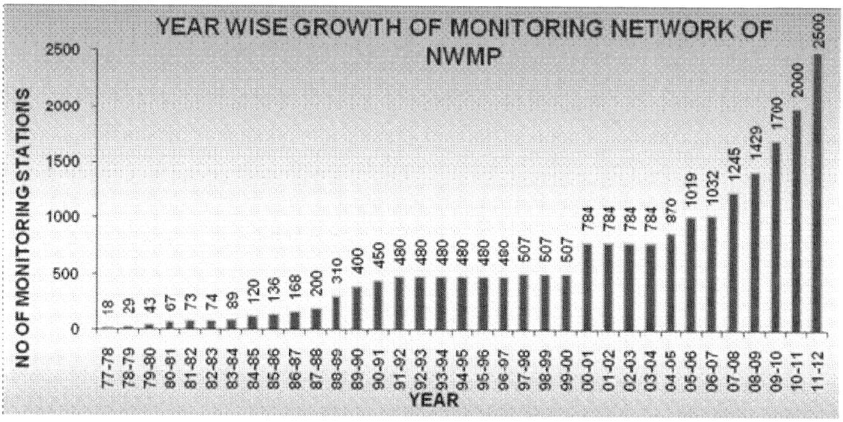

Figure 5: Year Wise Growth of Monitoring Network of NWMP[46]

frequency and number of points of measurement along the Ganga stretch are manifestly inadequate.

From a KM perspective, building more monitoring stations is only one direct way of addressing the knowledge generation gap. It is also possible to find new ways to generate data. For example, new technologies are constantly being invented and tested that open up avenues for pollution data generation. The LIDAR was invented to penetrate dense forests and reveal ground features. Satellite and light-wavelength analysis technologies may yet enable policy makers to generate useful data on a real-time and wide-scale basis. Robotic and miniaturisation technologies may offer the possibility of cheaply manufactured 'micro' monitoring buoys that be released in the river to provide real-time data. New technology also offers the possibility of mass-produced testing kits that can then be used to test water quality at multiple points and increased frequencies, rather expensive monitoring stations. A cheap 'litmus paper' testing kit that could report whether BOD levels are above or below a certain point could revolutionize pollution monitoring and provide health and safety advice for people using the river.

---

[46]Central Pollution Control Board (2015).

On the flip-side, low technology solutions can also be used to collect data at low cost. We observed that there are many communities living near to the river that can be used as resources to collect water samples for testing. It may prove more cost effective and practical to hire and train a team of local personnel to collect water samples than to set up a permanent high-tech monitoring station.

Another possibility is to identify existing data sets that may appear non-related, but yet provide useful information for pollution analysis. For example, water-borne diseases are not effective tracked in rural areas where the existing health and reporting system is inadequate. However, there may be readily available data that can still generate useful information. The sales of antibiotics in rural areas may correlate to general health issues in rural areas, and increased antibiotic sales near to the river may indicate increased water-borne diseases. While the issue of causality between antibiotic sales and water-borne disease is a valid question, non-traditional data approaches may still achieve useful results. What the KM approach can give us is to open the possibility to look at non-traditional data sets to generate useful pollution information and knowledge.

4.4 *Pollution data validation: need for a national water monitoring and testing standard*

Batini and Scannapieco define the criteria for data validation as accuracy, completeness, currency and consistency.[47] Accuracy relates to the relationship between the collected data and the real phenomenon that the collected data aims to represent. Completeness can be explained as having all data sets collected without missing parts. If any data is missing with the reason that it is not existing, existing but unknown or not known if existing, then the collected data set is incomplete. Currency stands for the frequency and length of time since the last data update, whereas the consistency dimension captures the need for a set of data items to have the same units of measurement and to not have variation, especially where data can be represented multiple times in relational tables or records in a file.

[47] Batini and Scannapieco (2006).

Meanwhile, Karadsheh *et al.* explain that knowledge validation should be an evaluation based on value.[48] Knowledge should be relevant to the organizational goal and the knowledge must be accurate and valuable enough to be shared in the next phase. Put in another way, inaccurate knowledge collected from different sources has no value and newly obtained knowledge is useful only when it is assessed as accurate and valuable. Their view, therefore, is that the knowledge generation process must include evaluation and validation as part of the KM process.[49] All data produced by participants in a KM organization should comply with certain defined validation standards, and the KM system should be designed under the consideration of validation.

In the Ganga Rejuvenation Project, all river pollution data generated should, therefore, comply with clearly defined validation standards. Otherwise, the data sets may be inaccurate, incomplete, lacking in currency, or inconsistent. They become valueless and remain raw data rather than useful or practical information that can generate further knowledge or wisdom.

There is some evidence to suggest that pollution data produced by Indian government lacks validity. Shaw Lacy studied the Ganga Action Plan and found that the reporting data from CPCP was not always accurate.[50]

The *Status of Water Quality in India 2008* and the *Ganga Water Quality Trend*, published by CPCB under the Monitoring of Indian National Aquatic Resources Series program, show different pollution data in the same area at the same time. While the former report indicates the BOD in Varanasi D/S (Malvina Bridge) as 14 mg/l,[51] the latter one tells the BOD collected in the same station as 10.5 mg/l.[52]

The pollution data table on Ganga River in the *Status of Water Quality in India 2008* contains pollution data such as temperature, D.O., pH, BOD

[48] Karadsheh (2009).

[49] Karadsheh (2009).

[50] Lacy

[51] Status of Water Quality India (2008).

[52] Ganga Water Quality Trend (2009).

and total Coliform, however, the table has not yet complete in some parameters such as nitrate and nitrite.

As a primary authority for data generation, validation and production, it is not known how CPCB assesses the obtained pollution data from respective sources to test validation criteria. Under NMCG, the GKC has been set up and one of its activities is described as running knowledge repository under the name of collation. A detailed description of the knowledge repository includes computerisation of legacy data, setting up a digital library, and remote sensing and GIS data archiving. GKC does not appear to incorporate data and knowledge validation as part of a KM process and as part of their responsibilities.[53]

In February 2014, CPCB announced that they start to conduct Real-Time Effluent Quality Monitoring System based on the online report from the 17 categories of highly polluting industries. The guideline for this policy explains the parameter validation and the rules of data management. Each pollution parameters such as COD, BOD, pH and TSS has own allowed variability for accuracy between online and laboratory measurement and daily/monthly/periodic check for operation and maintenance of the system is stated. Regulations on human intervention or correction on collecting data has emphasised while raw data should be transmitted directly via online to respective authorities including CPCB.[54] There is also a guide manual published by CPCB outlining a code of practice for the laboratories engaged in carrying out water quality assessment.[55] However, these guidelines and codes of practice do not constitute a unified and codified monitoring standard. It is not clear that testing and monitoring bodies comply with, or are even aware, of these publications.

In South Korea, the Ministry of Environment runs a Real-Time Water Quality Information System and provides online Real-Time Water Quality Index (RTWQI) data to the public.[56] While countries like U.S., Canada, Australia, New Zealand and Malaysia reports the monthly or quarterly water quality, South Korea provides RTWQI in real-time (collected and

---

[53] National Mission for Clean Ganga

[54] Central Pollution Control Board — Final Guidelines (2015).

[55] Central Pollution Control Board — Guideline Manual (2015).

[56] National Institute Of Environmental Research South Korea

analysed in every hour) in five scales (from Excellent to Poor) based on seven parameters such as temperature, pH, DO, EC, TOC, TN, TP. The government operates the 70 stations in the country and the data process system has dual stage of validation process to acquire RTWQI. Data can be provided from July 2012 to current date in numbers and time-series graphs without missing data or error. As for Indian Real-Time Effluent Quality Monitoring System, thc validation issues on pollution data such as accuracy, consistency and currency must be resolved. Only then can the system stand up to objections from industry and justify the expense of system installation.[57]

A codified national water monitoring and testing standard is, therefore, critical in knowledge management of river pollution data, and in achieving success for Ganga rejuvenation. A national water monitoring standard will allow all water monitoring agencies to generate coherent, reliable data, and allow other water monitoring parties to produce reliable data. For example, the Sankat Mochan Foundation collects its own weekly water quality data from three different Ghats along the Ganga River and Varuna confluence and announces it on their website.

A pollution monitoring standard is especially missing in the area of agricultural run-off. There are no standards or legislation relating to the classification, testing, and limits of pesticides or fertilisers, and other household wastes.[58]

Another benefit of a national water monitoring standard is that it can give clarity and validity to data even when there are apparent inaccuracies. The National Water Information System of the United States informs the public why the statistics published by U.S. Geological Survey can be mismatched with other interpretive or data reports.[59] They explain there could be time difference of collecting statistics, difference in treatment of partial record or the mismatch of precision of calculated statistics. It is important for the government to know how to manage the risks related to knowledge validation. Sometimes, pollution data collected from different sources

---

[57]Modi's real-time monitoring of Ganga pollution hits roadblock

[58]Murty *et al.* (2011).

[59]US Water Data

with different method can be different, however, the mismatch can be understood when there is sufficient explanation of the reason.

4.5 *Industrial pollution and law enforcement: KM and a bottom-up approach*

Since enacting the Water Act of 1974 and Emission Regulations of 1984, India has had the necessary environmental legislation for the protection of water resources and the limitation of industrial waste emitting standards. It is clear that industries must be able to meet discharge standards that have been legally set in the country. However, enforcement has been chronically deficient. In UP (Uttar Pradesh), records show that almost all industries inspected by the Central Pollution Control Board in 2013 are in breach of existing standards.[60]

One reason is poor enforcement due to inadequate data. With infrequent checks on individual polluting companies, companies can discharge pollution at night or at times when inspectors are known not to be around. Without more frequent and pinpoint pollution testing, companies will continue to pollute.

Another reason is that it is uneconomical for small-scale industrial units to have effluent treatment plants (ETPs) of their own. Here, technological knowledge such as common-effluent treatment plants (CETPs) and tacit experience implementing these solutions need to be shared. Only with effective knowledge sharing combined with monitoring and law enforcement can industrial pollution be tackled.

Thus far, the emphasis on solutions to river pollution has been based on a centralized governmental, top-down, 'command-and-control' approach. However, M.N Murty and Surender Kumar argue that a bottom-up and market based approach is essential in tackling industrial pollution.[61] The success of centralized government policies in controlling pollution depends upon the capacity of government institutions, the availability of sufficient funds, and quite possibly on the level of endemic corruption. A bottom-up or decentralized regulation system involving civic society

[60]Ganga the river, its pollution and what we can do to clean it, India. 16

[61]Murty *et al.* (2011).

and local communities can avoid these limitations, and bargaining between polluters and the local communities affected by pollution has been shown to be successful.[62]

In India, public interest litigation (PIL) cases filed by organizations in the courts has provided much of the impetus for industrial pollution reduction. Public dissemination of pollution data is, therefore, critical and can be used to hold industrial polluters more accountable to the public and to their financiers.[63] A 'regulatory triangle' can be set up where government, local communities, and industry can cooperate in a self-regulating framework in the area of water pollution. The government will then play a role in R&D, or simply be required to play a catalytic role and provide knowledge such as environmental programme design and clean technologies to companies, and pollution data and some financial incentives to local communities. Researchers have found for example that BOD figures can negatively correlate with the political organization, education, and environmental awareness of local communities.[64] Providing appropriate knowledge to the local communities, as well as to industry, can have a significant impact.

Another important bottom-up approach is the use of Market-Based Instruments (MBI) such as an effluent discharge tax and tradable effluent discharge permits. India is considering to implement MBIs in the control of air pollution, and there is potential to apply these mechanisms in water pollution control.[65] However, there is a need for systematic data collection to estimate marginal abatement costs and regulatory burdens to implement MBIs. Furthermore, implementing these mechanisms require skilled and experienced professionals with knowledge and training in resource and environmental economics, and in analyzing the application of economic instruments to water pollution. KM recognises the need to strengthen institutional capacity though the hiring of staff, as well as to build knowledge sharing between governmental and academic bodies.

---

[62] Ibid. 292.

[63] Ibid. 296.

[64] Ibid. 294.

[65] Ibid. 296.

4.6 *Non-point source pollution and increasing public awareness*

Polluting practices by the public is a smaller but no less significant source of pollution. One area where knowledge is lacking is in the health impact of agricultural runoff, household wastewater, open defecation, and religious and rubbish throwing practices. As stated earlier, there are no clear studies on how pesticides, fertilizers, and household wastewater contribute to pollution and to health or disease. There also has been much criticism, for example, of the religious practices of riverside cremation and throwing of offerings into the river, yet ironically the polluting and health effects of these practices may prove to be insignificant compared to other polluting practices.

The general lack of public knowledge started from the GAP launched by the government of India in the mid-1980s. To the public, the GAP had been acknowledged as a failure resulting from its' narrow focus on 'wastewater interception and sewage treatment' policy and exclusion of viable solution to the non-point sources of pollution. Moreover, the planning of the GAP had been executed by the federal government of India with little, or no, participation from the state and public.[66] Since the GAP only focused on interception and sewage treatment, the state and public weren't provided sufficient funds to promote and share knowledge on the pollution conditions of the Ganga River. Hence, a prominent researcher has quoted that 'the public, even now, does not perceive the current situation [of Ganga River Pollution] as a threat to their existence'.[67] Even now, people living along the Ganga River lack the general knowledge of the current situation and continue to pollute the river increasing the level of non-point sources of pollution. Public awareness is, therefore, one area of knowledge dissemination that is missing. On the flip side, knowledge of local practices is also critical in understanding the most polluting practices and designing measures to counter them.

---

[66] The Ganga and the GAP: An Assessment of Efforts to Clean a Sacred River, India. 1655.

[67] Ganga at Varanasi: Lessons from Environmental Abuse. 66.

4.7 *People as knowledge assets*

For Ganga Rejuvenation Project, one aspect of KM solution is utilizing people and its networks of knowledge. People are a vital knowledge asset and possess valuable local insights and information regarding the Ganga River. In order to utilize people, a knowledge directory needs to be established. A knowledge directory is a list of experts who are knowledgeable in different areas of the Ganga River and its problem of pollution.

Going back to tacit vs. explicit knowledge, we see that it is easier to capture tacit knowledge when people have trust, cooperate with each other, and willing to contribute their knowledge. Trust, cooperation, and sharing of knowledge happens when people who contribute knowledge to the databases are given access with little intervention, recognized, and rewarded for sharing. Hence, people as a knowledge asset increases with cooperation and trust given that they are recognized and rewarded.

At Ernst & Young, incentives to stimulate knowledge sharing are used in performance reviews. Employees who spend a lot of time helping each other and adding to the corporate knowledge base are openly rewarded and recognized.[68] Employees are evaluated on their work contributions and on how well they use the firm's knowledge resources. Bain & Company rewards people for sharing knowledge with others based on how much direct help they give colleagues. The degree of high-quality, person-to-person dialogue a partner has with others can represent as much as one-quarter of his or her annual salary.[69] Reward and compensation are tied and attached to the network of colleague's relationship and directly tied to cooperation and sharing of knowledge.

In the field of KM, the utilization of networks of people is key. Utilization of networks of people can be either centralized or decentralized. In a centralized knowledge assets, the organization has a strong governing core where it has a high degree of interdependence with all its internal functions where knowledge is comes from the core. In a decentralized knowledge assets, all external and internal sources can provide

---

[68] Wah

[69] Hansen *et al.* (1999).

and share knowledge where units are autonomous. Employees are encouraged to transfer knowledge within and outside their areas of expertise. Use of overlapping teams helped ensure continuity of control, joint learning and information sharing. In general, teams that internalize information and knowledge are well equipped to solve problems. Every organization has a slightly different knowledge base and organizational culture. Transferring useful information between organizations is costly in terms of time and effort and strains existing technology. Organizations need to develop fair and equitable reward systems to encourage employees to share tacit and explicit knowledge.

Technology plays a key role in collecting and codifying knowledge for distribution. It is important to have a strong information technology (IT) framework to design and implement the systematic storage and dissemination of information. Information Technology is an enabler, but by itself will not capture tacit knowledge[70]; IT does not provide content. People do.

4.8 *KM and governance and policy-making*

In the area of governance and policy making, critical knowledge is required at several important junctures. Without the required knowledge, policies will not be correctly designed, implemented, and evaluated (see Table 5).

In dealing with the problem of river pollution, one must firstly correctly define the object, scope and magnitude of the problem. This requires proper knowledge about geography and hydrology and how pollution is generated. Inappropriate definition of the problem will lead to failed policies such as GAP which tackled pollution only at the river level rather than the entire Ganga basin. Secondly, one must rely on accurate pollution data to determine the root causes of pollution, and then to develop different sets of possible policies to implement. Misdiagnosing the root causes of river pollution will also lead to misdirected solutions.

Developing good policy alternatives require technical knowledge on possible new technological solutions, as well as knowledge on possible theoretical or practical approaches. For example, sewage treatment can

[70]Wah

rely on many different technologies, and economic instruments or novel solutions by other countries can provide policy alternatives. Local knowledge is also important in generating and choosing between possible solutions. Similarly, the next step of deciding which policy to implement requires an understanding of the policy approaches and their workings. Importantly, policy making experience is also critical in being able to choose the right policy given a multitude of other factors such as institutional capacity, and political, economic, and social considerations.

In the next step of policy implementation, experience and local knowledge is needed to carry out the policy successfully. Placing a good policy in the hands on inexperienced policy implementers will surely lead to the failure of the policy. Finally, good pollution data generation and policy experience is needed for the last step of policy evaluation. One can then tweak parts of a policy that are not working and redesign the policy

Table 5: Overview of Policy Analysis and Knowledge Required

Policy Steps in dealing with river pollution		Knowledge Required
Problem Definition ↓	Defining the object, scope and magnitude of river pollution.	Pollution data, higher level knowledge about river pollution.
Causal Analysis ↓	Understanding the root causes of the pollution	Pollution data
Policy Alternatives ↓	Coming up with different policies choices to implement	Pollution data, technical alternatives, local knowledge
Policy Choice ↓	Deciding which policy to implement	Pollution data, technical and specialized knowledge, local knowledge, policy experience
Implementation ↓	Carrying out the policy implementation on the ground	Policy experience, implementation experience, local knowledge
Evaluation ↓ (Problem definition)	Finding out the impact of the policy implementation, and making changes to the policy as necessary	Pollution data, policy evaluation knowledge and experience

to achieve greater effectiveness in reducing pollution. This section shows that different types of knowledge are critical in the design and implementation of successful solutions to river pollution.

4.9 *Water pollution data dissemination: Ganga knowledge centre as knowledge broker*

The purpose of knowledge dissemination is to deliver the right knowledge assets to the right people at right time.[71] It is an active process to "communicate results to potential users by targeting, tailoring and packaging the message for a particular target audience".[72] The success of knowledge dissemination also "depends on the level of previous knowledge held by the audience and the effectiveness of the channels available to share the knowledge".[73] For Ganga rejuvenation, the "right knowledge" could be "validated knowledge", and the "right people" are the three levels of knowledge demanders. Here the emphasis should be put on the effective system or mechanism and the timing which can ensure the validated knowledge assets to be delivered to the required objectives "at the right time".

In December 2014, through the Governance Study Trip which is organized by Lee Kuan Yew Public Policy School, we did some interviews and investigations with government officials, professors of universities and local communities. We found that there were some problems on Ganga rejuvenation knowledge dissemination. In the first place, there is no nationwide and systematic knowledge asset dissemination system. In the second place, there is a lack of specific regulations on knowledge dissemination. In the third place, staffing is not enough and the responsibilities are not clear. There is only one organization — the GKC — which is responsible for the Ganga rejuvenation knowledge management. And it has been fully operationalized till today. We do not know how many staff will be in GKC. Currently, GKC, which is subordinated to NMCG, seems to be the platform of the knowledge assets dissemination system.

[71] McElroy.

[72] Dobbins *et al.* ().

[73] Debowski (2006).

However, GKC has its own problems: (1). No subordinated organizations or branches to support knowledge assets dissemination, (2). No administrative power, (3). Knowledge assets are not sufficient, systematically, accurate and complete. Currently, users cannot access knowledge assets from its web portal.

In terms of water pollution data dissemination, the main source of water pollution data comes from the CPCB, which publishes data annually in the form of the Status of Water Quality in India yearbook.[74] CPCB publishes the minimum, maximum, and mean figures of 7 water quality values from 56 monitoring stations along the Ganga River. At the time of this writing in July 2015, the latest available yearbook is from 2012[75] while the latest data spreadsheet available on the CPCB website is from 2011.[76]

The main method for the public to obtain water pollution data is via the various websites of the central, state, and municipal water authorities. This patchwork of websites is hard to navigate and searching for relevant data is both tedious and a hit-and-miss affair. For example, one is hard-pressed to find the Status of Water Quality 2012 report from the main CPCB website without resorting to a Being the main source of government pollution monitoring data, it is hard to justify the infrequency and haphazard nature of delivering this data to the public. Why the data from is 2013–14 not available, and why is there no standardised way of publishing the data? "used by Government of India Ministries/Departments (and) their organizations to publish datasets, documents, services, tools and applications collected by them for public use. It intends to increase transparency in the functioning of Government and also open avenues for many more innovative uses of Government Data to give different perspective."

The OGD hosts water pollution data from the Status of Water Quality in India report 2012, and lists the name and contact details of a data controller from the CPCB. This is encouraging, but it is unclear how much mandate the OGD has over agencies and how much cooperation it receives

[74] Central Pollution Control Board (2015).

[75] Status of Water Quality in India.

[76] Central Pollution Control Board (2015).

even though there exists a National Data Sharing and Accessibility Policy (NDSAP) and Section 4(2) of the Right to Information (RTI) Act, 2005 reads:

> *"It shall be a constant endeavour of every public authority to take steps in accordance with the requirements of clause (b) of sub-section (1) to provide as much information suo motu to the public at regular intervals through various means of communications, including internet, so that the public have minimum resort to the use of this Act to obtain information."*

With regards to water pollution data, NMCG must decide how it is to best share pollution data with the public in a consistent and effective way within the frameworks of the NDSAP and RTI. This is especially so given the setting up of the Ganga Knowledge Centre (GKC) which is supposed to "create and manage knowledge resources including analysis and modelling of diverse data sets relevant to Ganga River Basin".[77]

To disseminate the knowledge assets effectively, developing a nationwide dissemination system or mechanism is key. In addition, using the concept of a knowledge broker, and the "push, pull and exchange" knowledge strategies are also required to assist this system.

A nationwide Ganga knowledge assets dissemination system should consist of a series of organizations, including the central level (national level) and its subordinated organizations (state, city and county level) which can effectively connect the knowledge demanders and suppliers. There should be specific regulations which clarify the dissemination mechanism, roles and responsibilities, and incentive and punishment measures. The whole organization should be empowered by law to monitor and inspect the implementation of knowledge dissemination. Also, the system should have enough staff and financial support to fulfil its tasks.

From the stated objectives of the GKC, its current function is mainly as a database and research centre, which is more like a library or archives. However, if GKC or some other authority can be given administrative

---

[77] Ganga Knowledge Centre.

power — which means GKC has monitoring and inspecting power, it will have a more important position on Ganga rejuvenation knowledge management, and could be a national level knowledge dissemination centre for water pollution knowledge. Under this condition, its subordinated organizations: the state and city level water pollution can form a coherent knowledge dissemination system. The vertical dimension of knowledge sharing will form the backbone of this system between the central, state, and municipal water authorities. Currently the Pollution Monitoring website (http://cpcb.rtwqms.com/) and the Environmental Data Bank website (http://cpcbedb.nic.in/) by CPCB are both inaccessible beyond their main pages. The Open Government Data (OGD) Platform under the Ministry of Communications & Information Technology is one possible standardised dissemination platform. Whichever the platform, an integrated administrative system will then be highly effective in disseminating water pollution knowledge.

The horizontal dimension of this dissemination system will be to create sharing mechanisms between knowledge demanders and suppliers. GKC can be an ideal knowledge broker and serve as an intermediary to "facilitate collaborations between Ganga rejuvenation knowledge assets suppliers and demanders, find evidences to shape decisions, assess, interpret and adapt evidences to local context and identify emerging issues that research can help solve".[78] In addition, through coordination between demanders and suppliers, GKC can get feedback of the situation of knowledge dissemination and seek improvement. GKC can serve as the centre of a knowledge sharing system that connects different levels of demanders and suppliers.

Under this system, GKC can function not only as knowledge base and research centre, but also knowledge broker or coordinator, and an administrative inspector. If GKC or similar organisation have both the administrative power and the effective channels, then they are able to utilise legal directives, administrative regulations and efficient tools to disseminate knowledge assets effectively and timely. Firstly, GKC can plan and

---

[78] "Canadian Institutes of Health Research"

implement approaches to push (disseminate) knowledge assets toward demanders (including the superior and subordinates GKC) who GKC believe need or have to receive. For example, GKC can use internet servers to send content to the computers of potential demanders or current demanders at regular intervals, similar to current sport scores or weather reports. Users of push approach may "obtain leads from the information received and then search for more in-depth information".[79]

Using "push" approach, user will take an action due to the disseminated knowledge. In short, "push" approach will disseminate knowledge or action information straight to the targeted users. This action will cause targeted users who are interested in the knowledge to do some action in order to know more about the knowledge. Secondly, GKC can pull knowledge from sources they identify as validated knowledge that is useful to their decision making. They can provide validated knowledge according to different requirements since they have powerful knowledge sources. Thirdly, GKC can be easier to bring "the researcher and knowledge demander communities together through an interactive process (such as seminars, conferences) to 'exchange' knowledge"[80] through its administrative and coordinative role. This face-to-face approach helps the demanders and suppliers know each other's requirements and capacity more directly. In addition, GKC can use multiple methods to ensure the knowledge dissemination timely, such as modern technology (internet, Email, etc.), incentives, punishment and even political orders.

Based on our review of the GKC organisation structure, we noted that the KM's roles and responsibilities are established with the necessary staff to perform the KM functions. The GKC appointed a director to oversee and manage the day-to-day operation of KM. Under the director, the GKC consists of activities such as research, knowledge base, GIS & Remote Sensing, library, and dissemination. The GKC organisation structure is presented in Figure 6. Although GKC has an established organisational structure, we recognise the KM roles and responsibilities with key activities could have been expanded even further.

---

[79] Powis-Dow. "Reconciling Push vs. Pull Technologies"

[80] Shelly *et al.* "Discovering Computers 2009"

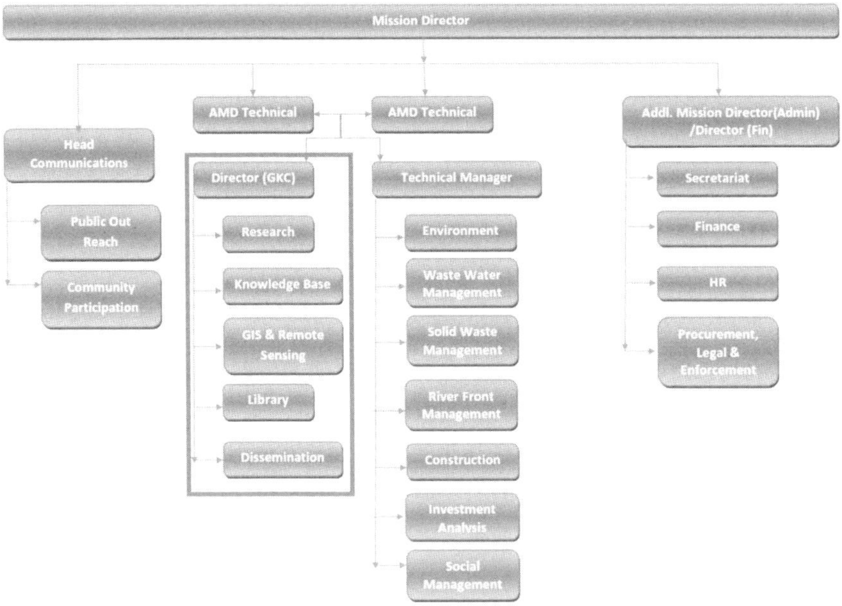

Figure 6: GKC Organizational Structure

For the success of KM, it is key that roles and responsibilities are clearly stated to improve human resource management and performance improvements. In this chapter, it is recommended that KM roles and responsibilities should consist of the following hierarchical structure — establish steering committee, appoint Chief Knowledge Officer (CKO), and establish Knowledge Officers/KM Facilitator.

An effective KM should establish a steering committee where the steering committee functions as the Board of Governance and is responsible for the setting the KM policy including strategy and guidelines. The steering committee should ensure that all senior executives and staffs understands the key KM strategies. It is also responsible for appointing a Chief Knowledge Office, or CKO.

An appointed CKO is required in any organisation wanting to implement a KM policy. CKO needs to have in-depth knowledge of KM and its' day-to-day activities while directly reporting to the Steering Committee.

The CKO also needs to ensure all KM implementation activities are following the set timelines while planning, organizing, leading, and managing the work of Knowledge Office (KO) and/or KM Facilitator.

The KO/KM Facilitators are the workhorse of the KM and reports to CKO. Their roles and responsibilities includes techniques for knowledge capture and/or creation, structuring of knowledge into an appropriate taxonomy, provide advice on systematic storage of knowledge within suitable repositories, access to critical knowledge, sharing of knowledge across all media streams (soft or hard copy), and transfer and application of knowledge.

We propose that GKC is well placed to act as a knowledge broker and to champion the use of KM in the Ganga Rejuvenation Project. However, any government body with the right mandate can perform this role in future.

5. Conclusion

In this chapter, we have shown how poor knowledge management can significantly contribute to policy failures.

By introducing KM concepts and frameworks, we hope to show the importance of identifying and developing knowledge systematically, especially in the area of tacit knowledge. Through knowledge management, we can turn existing data into more useful information, information into more practical knowledge, and finally knowledge into true wisdom in understanding Ganga river pollution and finding a sustainable long term solution.

Through the framework of knowledge generation, validation, and dissemination, we have highlighted some existing knowledge gaps in the Ganga Rejuvenation Project, and hope that these gaps can be effectively addressed in the future. In the areas of pollution data, much more can be done to generate and validate useful knowledge, and to share that knowledge effectively in the efforts against pollution.

We believe that knowledge management is an essential tool in the Ganga Rejuvenation Project, and that NMCG and GKC are uniquely placed to champion the principles of knowledge management to bring about the success of the Ganga Rejuvenation Project.

Glossary/Abbreviations

BOD	Biochemical Oxygen Demand
Cd	Cadmium
CETP	Common-Effluent Treatment Plant
CKO	Chief Knowledge Officer
CPCB	Central Pollution Control Board
Cr+6	Hexavalent chromium
DALY	Disability Adjusted Life Year
DIKW	Data, Information, Knowledge & Wisdom
DO	Dissolved Oxygen
ETP	Effluent Treatment Plant
FC	Faecal Coloform
Fe	Ferrum; Iron
GAP	Ganga Action Plan
GIS	Geographic Information System
GKC	Ganga Knowledge Centre
GRP	Ganga Rejuvenation Project
GSP	Governance Study Project
ICT	Information & Communication Technology
KAD	Knowledge Asset Directory
KM	Knowledge Management
KO	Knowledge Office
MLD	Million Liters per Day
Mn	Manganese
NDSAP	National Data Sharing & Accessibility Policy
NMCG	National Mission for Clean Ganga
MBI	Market-Based Instruments
MEF	Ministry of Environment & Forests
MOF	Ministry of Finance
MoWR, RD & GR; or MWR	Ministry of Water Resources, River Development & Ganga Rejuvenation
NGRBA	National Ganga River Basin Authority

NWMP	National Water Management Project
OGD	Open Government Data
PIL	Public Interest Litigation
R&D	Research and Development
RTI	Right to Information
RTWQI	Real-Time Water Quality Index
SPM	State-level Program Management
STP	Sewage Treatment Plant
TC	Total Coliform
TN	Total Nitrogen
TOC	Total Organic Carbon
TP	Total Phosphorus
TSS	Toxic Shock Syndrome
ULB	Urban Local Bodies
UP	Uttar Pradesh
WHO	World Health Organization

Bibliography

About Open Government Data (OGD) Platform India. Open Government Data (OGD) Platform India. https://data.gov.in/about-us. (Accessed 9 June 2015).

Ackerman, K and Swenson, A. "I-DIAG: From Community Discussion to Knowledge Distillation," Ibid 292–310.

Ahsan, S and A Shah (2006). Data, Information, Knowledge, Wisdom: A Doubly Linked Chain? http://ww1.ucmss.com/books/LFS/CSREA2006/IKE4628.pdf. (Accessed 29 May 2015).

Baporikar, N (2014). Knowledge Management Initiatives in Indian Public Sector. Building a Competitive Public Sector with Knowledge Management Strategy, 53–89. doi:10.4018/978-1-4666-4434-2.ch002.

Batini, C and M Scannapieca (2006). *Data Quality: Concepts, Methodologies and Techniques*. Berlin: Springer.

CENN Knowledge Management Policy. (2011). http%3A%2F%2Fwww.ngoconnect.net%2Fdocuments%2F592341%2F749044%2FKnowledge%2BManagement%2BPolicy%2BManual. (Accessed 2 April 2015).

Central Pollution Control Board Programme/projects Water/Quality Pollution NWMP. Central Pollution Control Board Programme/projects Water/Quality Pollution NWMP. http://cpcb.nic.in/water.php. (Accessed 9 June 2015).

Cleaning of the Ganga (Accessed 16 February 2015). http://webcache.googleusercontent.com/search?q=cache:kXd_zcG9yOAJ:www.geosocindia.org/abstracts/2011/aug/pdfs/r1.pdf+&cd=1&hl=en&ct=clnk&gl=sg.

Das, S (2011). Cleaning of the Ganga. *Journal of the Geological Society of India*, 124–130. doi:10.1007/s12594-011-0073-9.

Davenport, TH and L Prusak (1998). *Working Knowledge: How Organizations Manage What They Know*. Boston, Mass: Harvard Business School Press.

DIKW Pyramid. Wikipedia. http://en.wikipedia.org/wiki/DIKW_Pyramid. (Accessed 10 June 2015).

Dobbins, M, D Ciliska, A DiCenso. Dissemination And Use Of Research Evidence For Policy And Practice: A Framework For Developing, Implementing And Evaluating Strategies. http://gateway.nlm.nih.gov/MeetingAbstracts/ma?f=102237186.html.

Debowski, S (2006). *Knowledge Management*. John Wiley & Sons, Australia Ltd.

Ettore, B (1999). Knowledge management. *Management Review*, p. 8.

Fahey, L and L Prusak (1998). The Eleven Deadliest Sins of Knowledge Management. *California Management Review*, 40(3), 265–76. doi:10.2307/41165954.

Flynn, R and V Marshall (2014). The Four Levers for Change in Knowledge Management Implementation. *Building a Competitive Public Sector with Knowledge Management Strategy*, 227–45. doi:10.4018/978-1-4666-4434-2.ch010.

Ganga Action Plan. Ganga Action Plan. http://www.mppcb.nic.in/gap.htm. (Accessed 10 June 2015).

Ganga Knowledge Centre Website. https://nmcg.nic.in/gkc.aspx. (Accessed 9 June 2015).

Ganges River Pollution. http://www.all-about-india.com/Ganges-River-Pollution.html. (Accessed 16 February 2015).

Ganga Water handbook. http://webcache.googleusercontent.com/search?q=cache:PLoPxswXkv8J:www.downtoearth.org.in/Data-presentations/ganga-water-english.pdf+&cd=1&hl=en&ct=clnk&gl=sg. (Accessed 16 February 2015).

Ganga Water Quality Trend — 2009. Ganga Water Quality Trend — 2009. 2009. http://www.indiawaterportal.org/data/ganga-water-quality-trend-2009. (Accessed 10 June 2015).

Girard, JP and S Mcintyre (2010). Knowledge Management Modeling in Public Sector Organizations: A Case Study. *International Journal of Public Sector Management*, 23(1), 71–77. doi:10.1108/09513551011012330.

Hansen, MT, N Nohria and T Tierney (1999). What's your strategy for managing knowledge?, *Harvard Business Review*, March-April, 106–16.

Hodler, A. Data Centre Advisors. Data Centre Advisors RSS. http://blogs.hds.com/hdsblog/2011/09/is-that-information-and-do-i-care-part-2.html (Accessed 10 June 2015).

HydroQ — Real-Time Central Monitoring Station Website. http://cpcb.rtwqms.com/ (Accessed 9 June 2015).

CPCB (2008) Status of Water Quality in India. Delhi: Central Pollution Control Board Ministry of Environment and Forests .

Intellectual Capital. Wikipedia. http://en.wikipedia.org/wiki/Intellectual_capital. (Accessed 8 June 2015).

Karadsheh, L, E Mansour, S Alhawari, G Azar, and Naser El-Bathy (2009). A Theoretical Framework for Knowledge Management Process: Towards Improving Knowledge Performance. *Communications of the IBIMA* 7 68–79. http://www.ibimapublishing.com/journals/CIBIMA/volume7/v7n7.pdf.

Knowledge Capital Definition | Investopedia. Investopedia. http://www.investopedia.com/terms/k/knowledge-capital.asp. (Accessed 8 June 2015).

Markandya, A and MN Murty (2000). *Cleaning-up the Ganges: A Cost-Benefit Analysis of the Ganga Action Plan. Book, Whole*. New York; New Delhi: Oxford University Press.

Menkhoff, T, HD Evers and YW Chay (2005). Chapter 1 Knowledge of Enterprise: Knowledge Management or Knowledge Technology? *In Governing and Managing Knowledge in Asia*, 23–57. Singapore: World Scientific.

Menkhoff, T, HD Evers and YW Chay (2005). Chapter 3 The Knowledge Gap and the Digital Divide. *In Governing and Managing Knowledge in Asia*, 71–78. Singapore: World Scientific.

Murty, MN and S Kumar. Water Pollution in India An Economic Appraisal. *India Infrastructure Report 2011: Water Policy and Performance for Sustainable Development.*, 285–98.

National Data Sharing and Accessibility Policy (NDSAP) 2012, https://data.gov.in/sites/default/files/NDSAP.pdf.

Norström, P (2014). Knowing How, Knowing That, Knowing Technology. Philos. Technol. Philosophy & Technology. doi:10.1007/s13347-014-0178-3.

Pollution of the Ganges — Wikipedia, the Free Encyclopedia. http://en.wikipedia.org/wiki/Pollution_of_the_Ganges. (Accessed 16 February 2015).

Powis-Dow, DW (2006). Reconciling Push vs. Pull Technologies, http://www.insitepro.com/powis1.htm.

Quinn, JB, T Anderson and S Finkelstein (1996). Managing professional intellect: making the most of the best, *Harvard Business Review*, 71–80.

Tacit Knowledge. Wikipedia. http://en.wikipedia.org/wiki/Tacit_knowledge. (Accessed 7 June 2015).

Tichenor, PJ, GA Donohue and CN Olien (1970). Mass Media Flow and Differential Growth in Knowledge. *Public Opinion Quarterly* 34(2), 159. doi:10.1086/267786.

Sewage Pollution of the River Ganga: An Ongoing Case Study in Varanasi, India — *River Systems* — Schweizerbart Science Publishers. http://www.schweizerbart.de/papers/rs/detail/20/79858/Sewage_pollution_of_the_River_Ganga_an_ongoing_case_study_in_Varanasi_India. (Accessed 16 February 2015).

Schulz, S, K Herrmann, R Kalckloesch and T Schwotzer. Towards Trust-Based Knowledge Management for Mobile Communities, Ibid 150–170.

Shelly, GB and ME Vermaat (2009). Discovering Computers, Course Technology. US.

The Hindu. *Pollution Levels in Ganga Alarming*. http://www.thehindu.com/2003/09/15/stories/2003091502161300.htm. (Accessed 16 February 2015).

Umar, R and M Sami Ahmad (2000). Groundwater Quality in Parts of Central Ganga Basin, India. *Environmental Geology*, 39(6), 673–78. doi:10.1007/s002540050480.

Wah, L (1999a). Behind the buzz. *Management Review*, 17–26.

Wah, L (1999b). Making knowledge stick. *Management Review*, 24–9.

Chapter 10

REGIONAL COLLABORATION IN THE GANGA BASIN: PROSPECTS FOR A CLEANER GANGA AND REGIONAL PEACE

Elijah Lau Maung-Hoe, Chang Lih Kang, Goh Si Mien, Tetsuya Lim Hayashi and Tran Thanh Tung

1. Introduction

Regional collaboration can play an important role in ensuring a cleaner Ganga for India. In the history of India's bilateral relations with Bangladesh and Nepal, there are significant instances of cooperation on the Ganga Basin. Soon after signing the Peace and Friendship Treaty with Nepal in 1950, India cooperated with Nepal on the Kosi and Gandak dam projects to control flooding and aid irrigation in Bihar and Uttar Pradesh.[1] Both countries have also developed several initiatives to develop hydropower in Nepal since the 1960s. But India can do more — specifically in using dams and reservoirs to augment the flow of the Ganga River during the dry season. This can help with pollution control and water supply, for

[1] Nihar, N (2010). India-Nepal Peace and Friendship Treaty (1950): Does It Require Revision? *Strategic Analysis*, 34(4), 579–593.

not only India but also Bangladesh. However, India will need to overcome several policy challenges.

Water management issues have been a source of tension between India and Bangladesh since Bangladesh was formed in 1971, despite the formation of a Joint Rivers Commission in 1972. Bangladesh has long accused India of drawing too much water from the Ganga, which has affected irrigation, navigation, water quality and ecology in southwest Bangladesh. It was only in 1996 that the Indian government, under a coalition led by the Janata Dal party (which took a more generous approach to improving relations with its South Asian neighbours. It adopted a more benign attitude towards the newly-elected Awami League government in Bangladesh and signed the Ganga Water Sharing Treaty.[2] The treaty stipulated the sharing of water flow during the dry season but did not address the more critical issue of augmenting the flow of the Ganga during the dry season. But if India can establish stronger cooperation with Nepal to augment the flow of the Ganga, both India and Bangladesh can derive long-term benefits in reducing pollution in the Ganga and reducing their trans-boundary water problems.

If India can overcome the obstacles to closer cooperation with Nepal, it can improve the Ganga's river flow from its tributaries in the upper riparian state, especially in the dry season. When the river suffers from a reduced flow of water, the Dissolved Oxygen (DO) level in the Ganga, drops significantly, hampering the self-purification ability of the Ganga. Already, the high amounts of untreated sewage, industrial waste and agricultural run-off into the Ganga river is beyond the self-purifying capacity of the river.[3] In 2011, the DO level was 4 milligram per litre (mg/l).[4] As

---

[2] Mohammad, HK (2014). Obstacles to India-Bangladesh Cooperation: An International Relations Theory Perspective. In *International Relations Theory and South Asia: Security, Political Economy, Domestic Politics, Identities, and Images*, *Vol. 1*, Sridharan, E (ed.). Oxford Scholarship Online.

[3] Das, S (2011). Cleaning of the Ganga, *Journal of the Geological Society of India*, 78(2), 124–130.

[4] Kamyotra, IS *et al.* (2013). Pollution Assessment: River Ganga, Central Pollution Control Board, Ministry of Environment and Forests, Government of India.

dissolved oxygen levels in water drop below 5 mg/l, aquatic life is put under stress.

Moreover, improving the river flow, combined with a fresh impetus from India to engage its smaller neighbors through cooperative multilateral frameworks to build better trust and goodwill in the region, can produce harmonious dividends for India's relations with Bangladesh, the lower riparian state in the Ganga Basin. Reduced river flow and pollution has affected Bangladesh, which has not only led to misunderstandings and tensions between the two states, but also generates existing and potential security concerns for India, including border disputes, water disagreements and illegal cross-border migration.

This chapter will explore how the geopolitics of the region and India's historical experience have shaped India's strategic culture and approach to foreign relations with its two smaller neighbors sharing the Ganga River. More importantly, through the analysis of the historical development of India's role in South Asian international relations dynamics, we will identify useful proposals where India can work together with Nepal to achieve a cleaner Ganga and reap the benefits of harmonious relations with both Nepal and Bangladesh. As Bajpai has articulated, India is the largest country in the region and its choices and actions has an influence on the policies of its neighbors.[5] While cleaning up the Ganga River in India requires strong domestic policies, India's cooperation with Bangladesh and Nepal can not only help aid in the cleaning of the heavily polluted Ganga River, it will also help resolve India's security problems with its two neighbors and provide opportunities for closer regional integration in the future. Thus, while the state of the Ganga remains a source of bilateral tensions, it can also be an opportunity for potential cooperation and benefit to India.

Managing the Ganga-Brahmaputra-Meghna (GBM) Basin as a whole is a complex issue — this chapter represents a small part of the puzzle — namely the cooperative management of the GBM Basin among the riparian states which will have benefits for all states in ensuring clean and plentiful water for the inhabitants of the basin.

[5] Bajpai, K (2002). Indian Strategic Culture. In *South Asia in 2020: Future Strategic Balances and Alliances*, Chambers, M (ed.), Carlisle: Strategic Studies Institute, p. 245.

1. Ganga Water Management Efforts by Nepal and India

1.1. *Ganga river in Nepal–India relations*

A major portion of the Ganga's downstream discharge originates from China (through Nepal) or from Nepal itself. According to the Water and Energy Commission Secretariat (WECS) of Nepal, the estimated annual run-off from its more than 6,000 rivers is 220 billion cubic meters.[6] Along with an average annual precipitation of 1530mm and extensive groundwater resources, Nepal contributes 46 per cent of the flow of the Ganga River. During the dry season, Nepal's contribution increases to about 75 per cent.[7] Of the 1,880-km long India-Nepal border, 640 km is delineated by a river.[8] Nepal provides the best, if not the only, option for downstream flood control and dry season augmentation. Hydropower development is another benefit of appropriate water management. Hence, India is keen to cooperate with Nepal to receive the benefits of a lower riparian country.

2. Cooperation in Water Management

Other than abundant water resources, Nepal also has suitable sites to build large dams that are capable of storing up to 77 billion cubic meters of water. This is about 68 per cent of total monsoon flow.[9] Due to its vast water resources, Nepal has proposed cooperating with its lower riparian neighbours, India and Bangladesh, to meet their water demands during the dry season. As early as 1977, the late Nepalese King, Birendra Bir Bikram

[6]WECS (2002). Water Resources Strategy, Water and Energy Commission Secretariat, Kathmandu, Nepal.

[7]Bahadur, PS (2004). *Overview: Conflicts Over the Ganga? In PSA, Disputes Over the Ganga: A Look at Potential Water-Related Conflicts in South Asia*. Panos Institute South Asia.

[8]Dhungel, DN and Bahadur PS (2014). *Nepal-India Relations: Territorial / Border Issue with Specific Reference to Mahakali River*, FPRC Journal.

[9]Poudel, SN (2009). Water Resources Utilization: Irrigation. In *Nepal-India Water Resources Relationship: Challenges*, Dwarika ND and SB Pun (eds.). Springer Science + Business Media B.V. United Kingdom.

Shah Dev, expressed his readiness to share Nepal's water resources.[10] Nepal's offer was turned down by India, although Bangladesh welcomed it with open arms. India responded by implying that water resources management is bilateral in nature and thus should be dealt with bilaterally.[11] In 1987, India's then Prime Minister Rajiv Gandhi's said: "we have not sought to melt our bilateral relationship into a common regional identity, but rather to fit South Asian cooperation in our respective foreign policies as an additional dimension". This is indicative of India's bilateral approach to resolving multilateral issues.[12]

India has been using different strategies and approaches in dealing with Nepal and Bangladesh. It uses its position as a lower riparian state when dealing with Nepal and as an upper riparian state when dealing with Bangladesh. Another possible reason for India's reluctance to engage in regional cooperation is because India wishes to maintain flexibility in its negotiation strategy.[13] If Bangladesh, India and Nepal engaged in multilateral regional cooperation on the Ganga to seek a holistic, Ganga Basin-wide solution to the water sharing problem, India would not be able to use its middle riparian position to play Bangladesh and Nepal off each other.

India is keen to cooperate with Nepal and enjoy the benefits of bilateral cooperation, but not in the form of a regional arrangment. India has set up a Joint Committee of Experts (JCE) with Bangladesh, but has excluded Nepal from this. As of may 2015, it is uncertain if current Indian Prime Minister Narendra Modi's "neighborhood first" policy will serve as a platform for multilateral regional cooperation or if it will remain wedded to the traditional Indian foreign policy model of only bilateral engagements with its neighboring countries. Early signs do not look promising. The Modi government is unlikely to make headway towards stronger multilateral regional cooperation based on the current South Asian Association for Regional Cooperation (SAARC) framework, given Modi's reluctance to engage Pakistan, which was clear in his behavior at

[10] Lohani, M (2013). Harnessing Nepal's Water Resources in National Interest: A Book Review. *Annual Journal by Nepal Council of World Affairs*. Kathmandu.

[11] Upadhyay, S (2013). *International Water Course Law and A Perspective on Nepal-India Cooperation*, EKTA Books.

[12] Ibid.

[13] Ibid.

his first SAARC Summit in November 2014.[14] Modi also did not use the summit as a platform to press for closer water cooperation, and the focus on the summit was on trade and land route connectivity, as well as the India–Pakistan rivalry.

2.1. *Hydropower cooperation*

Despite possessing huge hydropower potential, Nepal suffers from chronic power shortages that could affect its economic development. Theoretically, Nepal's hydropower potential is estimated to be 83,000 mega watts (MW), of which 43,000 MW are financially and technically feasible. As of 2012, Nepal has only exploited 659 MW of hydropower, barely 1 per cent of its hydropower potential.[15] As a result, nearly 85% of the population does not have access to Nepal's national electricity grid. Even in the capital, Kathmandu, it is common to experience power cuts for as long as 16 hours a day during the summer.

International experts have suggested that Nepal work with countries with advanced hydropower capabilities, like India, to capitalise on its hydropower potential and boost its domestic economy. India is the seventh largest hydropower producer in the world. Apart from expertise and proximity, northern India suffers from a power deficit of 50,000 MW which makes it a potential market should Nepal pursue hydropower aggressively.[16]

2.2. *Historical cooperation between India and Nepal*

India and Nepal first began cooperating over water resources in 1920 when their first water management agreement was signed regarding the Sarada Barrage project on the Mahakali River. The Sarada agreement involved an

---

[14] *The Times of India, Narendra Modi Woos SAARC Nations, Pledges Slew of Investments to Counter China*, November 28, 2014.

[15] Bergner, M (2013). Developing Nepal's Hydroelectric Resources: Policy Alternatives. Available at http://www.stimson.org/images/uploads/research-pdfs/Developing_Nepals_Hydroelectric_Resources_-_Policy_Alternatives.pdf. Accessed on 15 May 2015.

[16] Meeking, L (2013). *The Battery of South Asia? The Potential for India-Nepal Hydropower Cooperation*. Available at http://www.futuredirections.org.au/publications/food-and-water-crises/1394-the-battery-of-south-asia-the-potential-for-india-nepal-hydropower-cooperation.html. Accessed on 15 May 2015.

exchange of territory, but not to Nepal's advantage. In addition, water cooperation between the two countries has been plagued by a long-standing disagreement over the interpretation of the Sugauli Treaty signed in 1816. The treaty was signed between Nepal and the British East India Company to delimit the boundary along the Mahakali River in Nepal. The two countries dispute which stream constitutes the source of the river.[17]

In the 1950s, India under Jawaharlal Nehru signed two major water resources agreements with Nepal. In 1954, an agreement regarding the Kosi project was signed, while the Gandak Irrigation and Power Project agreement was signed in 1959. Nehru considered the Himalayas as a main line of defense for India and regarded Nepal as an integral part of India's security system.[18] However, many Nepalese think the two agreements were lopsided, and that India secured disproportionate benefits at the expense of Nepal. For instance, the Indian government built a dam on the Kosi River near the border of Nepal and India after signing the Kosi project agreement. The resulting dam flooded a lot of land in Nepal, but provided irrigation for agriculture in India's Bihar state.[19] Nepal alleged that the project was unfair because it did not get equal benefits from the project. The project only benefited about 29,000 acres in Nepal, even though it had the capacity to irrigate 1.5 million acres. Nepal also felt that India's construction of the dam was an encroachment on Nepal's territory.[20] In April 2008, a devastating flood occurred in the Kosi Basin that displaced about three million people in India and 50,000 people in Nepal. Both India and Nepal blamed each other for causing the flood.[21] Similarly, Nepal also claimed that India had taken advantage on the Gandak

---

[17] Malhotra, P (2010). *Water Issues between Nepal, India & Bangladesh*, IPCS Special Report. Available at http://www.ipcs.org/pdf_file/issue/SR95.pdf. Accessed on 15 May 2015.

[18] Shukla, D (2006). India-Nepal Relations: Problems and Prospects, *The Indian Journal of Political Science*, 67(2), p. 357.

[19] Ranjitkar, SB (2014). Nepal–India Cooperation in Hydropower Development in Nepal. http://www.scoop.co.nz/stories/HL1407/S00160/nepal-india-cooperation-in-hydropower-development-in-nepal.htm. Accessed on 15 May 2015.

[20] Medha Bisht, *Revisiting the Kosi Agreement: Lessons for Indo-Nepal Water Diplomacy*, Institute for Defence Studies and Analyses, September 22, 2008. http://www.idsa.in/idsastrategiccomments/RevisitingtheKosiAgreement_Medha%20Bisht_220908.html.

[21] Meeking, 2013.

Irrigation and Power Agreement to divert the flow of the Gandak River to serve its own interests. Nepal believes that both projects resulted in a substantial portion of scarce agricultural land in Nepal being submerged without the country getting equitable benefits in return; India, on the other hand, believes that Nepal received most of the benefits of the two projects without any investment.[22]

2.3. *Obstacles of cooperation*

The Nepalese also believe that the two treaties signed in the 1990s, the Tanakpur Agreement (1991) and Mahakali Treaty (1996), are also unfair.[23]

The Mahakali treaty, which was signed by then Nepalese Prime Minister Sher Bahadur Deuba and then Indian Prime Minister PV Narasimha Rao, was meant to cover the joint development of the Mahakali River in the areas of irrigation, flood control and power generation. According to the treaty, a project named the Pancheshwar Multi-Purpose Project would produce 6,000 MW of hydropower by 2002. However, Nepal was subsequently unhappy with the treaty and considered it flawed, arguing that the treaty was unclear on Nepal's "water rights".[23]

Nepal is also concerned about the social and environmental impact of future hydropower projects. Civil society groups in Nepal have warned that large-scale hydropower projects would have a disastrous effect on the environment. The construction of dams will cause significant deforestation and soil erosion. This process could also cause severe landslides and flash floods. The Nepalese government is also concerned about the social impact of future hydropower projects. Thousands were forced to leave their homes without ample compensation when the government built dams near their communities in the past. This created many social problems. When displaced families migrate to the cities, it puts pressure on employment opportunities, housing and public health care.

---

[22] Batra, A (2012). *Regional Economic Integration in South Asia: Trapped in Conflict?* Routledge.

[23] Nayak, N (2010). *India-Nepal Peace and Friendship Treaty (1950): Does it Require Revision?* Strategic Analysis, 34(4), 579–593.

Nepal's unhappiness with India goes beyond water. Nepal has long been unhappy with certain clauses in the keystone Peace and Friendship Treaty signed in 1950.[24] More recently, India's trade embargo of Nepal in 1989 as punishment for Nepal's arms deal with China did not help.[25] India also has to contend with domestic Nepalese politics and growing Nepalese nationalism.

2.4. *Recent developments*

Given the internal instability in Nepal in the last 10 to 12 years, bilateral cooperation on water issues has been difficult. However, Nepal's political environment has become more stable and in August 2013, the Nepalese government decided to recommence the Upper Karnali project, which has a capacity of 900 MW. A major Indian company was also given permission to invest in the 44 MW Super Madi Hydropower project in Central Nepal.[26] In November 2014, India and Nepal signed another agreement for the 900 MW Arun III dam, making India the biggest hydropower developer in Nepal.[27]

3. Nepal–India collaboration to clean up the Ganga river

3.1. *Two main approaches*

There are two main approaches to cleaning up the Ganga River.[28] The first is a preemptive approach to stop municipal sewage and industrial effluent from reaching the Ganga River without being appropriately treated. The

[24] Thapliyal, S (2012). India and Nepal Treaty of 1950: The Continuing Discourse. *India Quarterly*, 68(2), 124–126.

[25] Dixit, K (2013). Nepal: Dictated By Geography, *World Policy Journal*, 30(4), pp. 36–40.

[26] Ibid.

[27] Thethirdpole.net, *India and Nepal sign deal to build another mega-dam*, November 27, 2014. Available at http://www.thethirdpole.net/india-and-nepal-sign-deal-to-build-another-mega-dam/.

[28] These approaches are adapted partly from Sunita Narain & Soupamo Banerjee, Ganga: The river, its pollution and what we can do to clean it, Centre for Science and Environment (CSE), India, 2014.

second is a pro-active approach to ensure a minimum level of flow within the Ganga, especially during the dry season, which will allow the river to continuously self-purify.

There is scope for India to collaborate with Nepal to establish sewage and effluent treatment plants along the four main tributaries of Mahakali, Karnali (these two rivers converge to form the Ghaghara in India), Gandak and Kosi. This is similar to the efforts that India has undertaken in recent decades with the Ganga Action Plan (GAP) I and II within its boundary.[29] However, pollution from Nepal is insignificant compared to India, so this approach is not very productive.

However, there is significant potential for India to collaborate with Nepal to establish substantial water storage facilities in the Himalayan state to conserve excess water from the monsoon season (June to November). This can then be used to augment the low flow in the Ganga River during the dry season (December to May) to allow the river to sustain its self-purification capability. This would be a "win-win" situation for both countries. Aside from having a cleaner Ganga River, Nepal would be able to generate hydroelectric energy and India would be able to minimise in Bihar and Uttar Pradesh during the monsoon season.

However, these proposed collaborations can only contribute to a cleaner Ganga River to the east of Patna, where the westernmost Nepalese tributary of Ghaghara flows into Ganga River (See Figure 1). To the west of Patna, the India itself will have to eradicate the pollution problem along the Ganga River and along the Yamuna. In addition, the conserved water released by the Nepalese dams during the dry season might be siphoned off by farmers within the Gangetic Plains for irrigation, which would reduce the flow downstream.

Finally, decades of mistrust have to be overcome. We recommend that India, which will likely benefit more from the collaboration, take a more conciliatory and benevolent stance towards the poorer and less developed state of Nepal. India should also consider enlisting the assistance of Bhutan, given the success of Bhutan–India collaboration in various water

---

[29] See Tare, V *et al.* (2011). SWOT Analysis of Ganga Action Plan, Indian Institute of Technology (IIT), for a discussion on the status of GAP I and II.

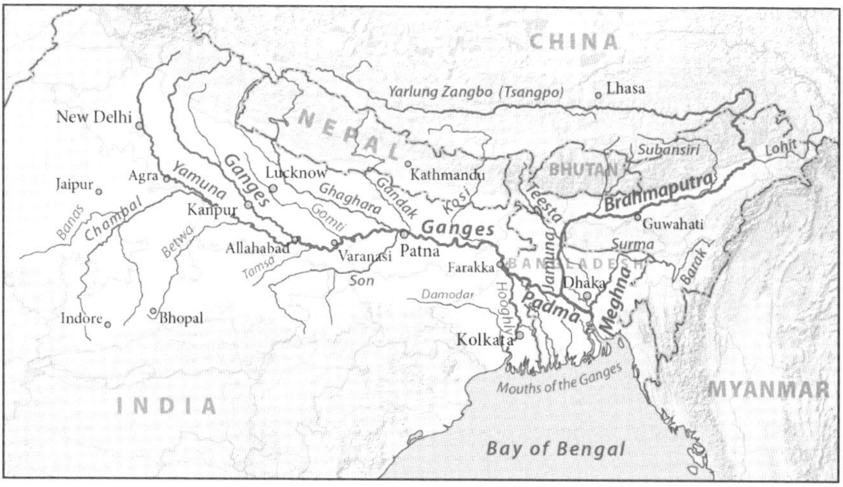

Figure 1: Map of River Ganga.

resource developments that has propelled Bhutan into relative prosperity in recent decades. Given the similarity between Bhutan and Nepal, key success factors that have advanced the Bhutan–India collaboration could potentially be replicated by Nepal.

3.2. *Maintaining acceptable flow in river Ganga during the dry season to sustain self-purification capability*

Although Nepal occupies only 13% of the total drainage of the Ganga Basin, its contribution to the flow of the river is much more significant, amounting to 46% of the average annual flow at the Farakka Barrage and up to 75% of the flow during the dry season. In the World Bank's seminal Ganges Strategic Basin Assessment (GSBA) study[30] in 2012, it was established that the aggregate active water storage along the Nepalese tributaries of River Ganga was about 100 billion cubic meters (BCM), of which about half was already built (56 BCM) with the half being

[30] The World Bank (2012). *Ganges Strategic Basin Assessment (GSBA): A Regional Perspective on Risks and Opportunities*. Washington DC: The World Bank.

proposed (44 BCM).[31] This was about 20% of the 500 BCM average annual flow of River Ganga.

In the GSBA study, a simulation was conducted to investigate the effect of completing the other half of the proposed Nepalese dams (another 23 dams) on the main stem of the Ganga River at the Hardinge Bridge in Bangladesh during the dry seasons. The simulation assumed that "*Indian abstraction would increase to the full off-take channel capacity until no more water could be drawn from the system without additional investment in dredging or expanding existing infrastructure*".[32] Even under such a scenario, low flow at Hardinge Bridge in December would increase from 8,000 million cubic meters per month (MCM/mth) to 12,000 MCM/mth; in January from 3,000 MCM/mth to 9,000 MCM/mth; in February from 2,000 MCM/mth to 5,000 MCM/mth; and in March from 2,000 MCM/mth to 4,000 MCM/mth.[33] However, the simulation revealed that April and May would see negligible low flow augmentation (2,000 to 3,000 MCM/mth), and that riparian states should explore other solutions, such as pumping underground water from aquifers in eastern Uttar Pradesh and Bihar.

3.3. *Resolving bilateral mistrust between Nepal and India*

To benefit from low flow augmentation from December to March in the River Ganga east of Patna, India will need the cooperation of Nepal to build the proposed dams. However, getting Nepal on board will be difficult. The previous treaties — the Kosi Treaty and the Gandak Treaty — are generally considered to unfair by the Nepalese. Although the latest treaty — the Mahakali Treaty signed in 1996 — was regarded as consistent with international good practice, it was still met with widespread resistance in Nepal.

Nepal stands to profit substantially if the proposed dams are built. Based on the GSBA study, Nepal has an estimated 40,000 MW of

---

[31] GSBA, p. 41.

[32] GSBA, p. 52.

[33] These flows are estimated from Figure 46 in GSBA, p. 52.

potential hydropower in the Himalayan headwaters of the Ganga River, of which only 2% has been developed.[34] With the construction of the proposed 23 dams, it is estimated that Nepal will have an installed capacity of 25,000 MW, which can produce approximately 53 TWh of electric power annually. This is worth around US$5 billion annually, 26% of Nepal's 2014 GDP of US$19.6 billion.[35]

On its own, Nepal is not able to exploit its water resources, which are plentiful, annually renewable and relatively environmentally safe. Large-scale water development projects require huge capital investment and sophisticated technologies, which the Nepalese lack. Usually, the most attractive of these water development sites in the Himalayas are not easily accessible and additional costs of building access roads need to be incurred to connect them to the existing transportation network. As a land-locked country, Nepal will require substantial assistance from its neighbors to develop its water resources. However, bound by the Himalayas to the north, Nepal is effectively an "India-locked" state with access to the world only through its long southern boundary with India. Nepal thus has little alternative to collaborating with India to fully harness its water resources.

India's motivation for collaborating with Nepal is equally strong. Besides the benefit of having a cleaner Ganga River east of Patna, India stands to profit on three other major fronts. First, flood control and management in India can potentially be enhanced.[36] Second, as the power that will be generated by the proposed Nepalese dams will far outstrip domestic demand in Nepal, there is huge scope for excess electricity to be exported to India to urgently meet its energy needs. Third, irrigation

---

[34] GSBA, p. 59.

[35] Ibid.

[36] The GSBA's observation is that many floods in the Ganga Plains are caused by rising waters on rivers that do not have significant upstream dam sites or by heavy rains that find no drainage, and as such, building dams will contribute little to flood prevention. However, our team opines that though relatively insignificant, upstream storage might help in some cases. This GSBA observation was disputed by academics from Nepal, which argued that the benefit can be significant. See Narendra Man Shakya, *Ganges Strategic Basin Assessment: Consultative Workshop Report*, Jalsrot Vikas Sanstha, Nepal, 2012 for an example.

during the dry season east of Patna can be better met with low flow augmentation, allowing arable lands to be productive in the dry months.[37]

To bridge the gap of mistrust between Nepal and India, we propose a two-pronged approach for India's consideration. First, India should critically examine the impact of the existing water treaties with Nepal to appreciate from the Nepalese perspective. While India is entitled to dispute the validity of the Nepalese concerns, the Indian government should attempt to take a more conciliatory and benevolent stance towards addressing these concerns in order to move future collaboration forward. The Indian government should not only consider offering Nepal more favorable terms, it should also signal visibly to Nepal the importance that India places on water cooperation with Nepal. Prime Minister Narendra Modi's visit to Nepal in August 2014, the first by an Indian Prime Minister in 17 years, is one such important signal but India needs to follow up on such visits to deepen cooperation, not just at the political level, but also at the working level.

Second, India should consider using Bhutan as a showcase . Similarly landlocked and blessed with an abundance of water resources like Nepal, Bhutan's successful collaboration with India can offer Nepal a viable blueprint to chart its own water resource development. In 1961, hydropower cooperation between Bhutan and India started with the signing of the Jaldhaka agreement. The development of the first major hydroelectric project started in 1974 when a bilateral agreement was signed for the construction of the Chukhahydel project across the River Wangchu in western Bhutan. The project proved to be mutually beneficial to both countries and catalyzed the development of other collaborative hydropower projects in other parts of Bhutan. The impact of these developments on Bhutan was significant. Its per capita electricity consumption in 1980 was 17 kWh, the same as Nepal. By 2003, it has increased 13 times to 218 kWh, whereas the comparative figure for Nepal only increased five times

[37]The GSBA's observation is that this improved irrigation during dry seasons might be economically insignificant given the current low marginal benefit associated with surface water irrigation in the Ganga Plains. Similarly, this observation was disputed by academics from Nepal, which argued that India will benefit substantially from such low flow augmentation.

to 91 kWh. Bhutan's GDP per capita in 1980 was the lowest among all South Asian countries. By 2008, it had the second highest GDP per capita in South Asia of US$1,933, a figure that was nearly five times that of Nepal. While Nepalese are justified in bristling at unfair comparisons between their country and Bhutan in terms of economic and social development given that both countries have different political histories and demographic distribution, having better diplomatic relations with their Indian neighbor is a worthwhile example to emulate from Bhutan, with the potential for Nepal to reap similar benefits as Bhutan in hydropower cooperation with India.

3.4. *Initiating proposal to advance Nepal-India collaboration*

To rejuvenate Nepal–India collaboration using this two-pronged approach, we propose the establishment of a formal and suitably high-level Indo-Nepalese Joint Rivers Commission (INJRC), co-opting Bhutan as a third-party observer. The INJRC can build on and re-energize the past efforts of the Joint Ministerial Commission on Water Resource (JMCWR) at the level of Ministers of Water Resources of Nepal and India. The initial task of the INJRC would be to build long-lasting trust between water bureaucrats and politicians of both countries to provide a platform for meaningful bilateral discussions. This will involve, among other things, the transparent sharing of information from both sides, especially with regard to benefit projections from any water development projects. Having established a certain level of trust, the INJRC can then embark on drafting a mutually-beneficial framework based on international best practices that can be applicable to further Nepal–India collaboration.

In retrospect, the evolution of the series of past water treaties and amendments between Nepal and India from Kosi to Gandak to Mahakali has revealed an encouraging trend towards international best practices in water agreement and benefit sharing. With additional experience from the successful Bhutan–India collaboration, the INJRC should strive towards an amicable water development framework based on transparent cost-benefit analysis and extra goodwill gestures from India to facilitate the process. With this framework in place, the INJRC can then prioritise resources to develop projects that will yield the highest return for both countries. This will eventually lead to, among other benefits, low flow augmentation during the dry season in the Ganga River east of Patna for the river to sustain its self-purification capability.

3.5. *Current window of opportunity and existing challenges*

With the election of Prime Minister Modi in India and his emphasis on cleaning up the River Ganga, a real window of opportunity has opened up for meaningful Nepal–India collaboration on water development projects. Prime Minister Modi's visit to Nepal in August 2014, the first by an Indian Prime Minister in 17 years and his conciliatory address to the Nepalese legislature, the first by a foreign leader, can provide the much needed momentum for more meaningful collaboration. The extension of a US$1 billion line of credit to Nepal at a concessional rate of interest of 1% by India will provide a boost to future hydropower projects. The establishment of the proposed INJRC will be a catalyst to formalise and operationalise these efforts sustainably for the benefit of both countries.

Despite the warming between Nepal and India since the appointment of Prime Minister Modi, there are existing challenges that might derail progress. First, while there is strong political will to advance bilateral collaboration in India, the same cannot be said for the Nepalese, where legislative acrimony is a constant. A 10-year insurgency ended in 2006, but a resulting Constituent Assembly failed after four years of effort to write a Nepalese Constitution. As such, political unity in Nepal is needed for any Nepal–India collaboration.

Second, in recent years, Nepal has been more proactive in its engagement with China as a counterweight to its prior dependence on India. While this move might have contributed to Prime Minister Modi's eagerness to re-engage Nepal to preempt China's substantial involvement in South Asia, it might potentially be destabilizing for Indo-Nepalese relations if not handled properly.

Third, the devastating earthquakes in April 2015 are a cause for concern. The prospect of dams in the earthquake-prone Himalayas being damaged by seismic activities and inflicting uncontrollable destruction downstream is a nightmarish scenario.

3.6. *Overcoming the challenges*

What the GSBA study revealed is that the low flow situation in the River Ganga can be partially ameliorated by the construction of 23 additional dams in Nepal, bringing low flow to a more acceptable minimum rate of 4,000 to 5,000 MCM/mth for 10 out of the 12 months in the River Ganga east of Patna. To augment the low flow in the other two months, and to increase low flow throughout the dry seasons in River Ganga west of Patna, an alternative will be required. A potentially feasible option would be to systematically tap the vast aquifers of eastern Uttar Pradesh and Bihar, which have approximately 30 to 50 BCM of underground water according to the GSBA study.[38] This is comparable to the 44 BCM from the construction of additional Nepalese dams.

A complementary project to pump water from these aquifers into the River Ganga east of Patna in April and May would increase the low flow in those months to a more acceptable 4,000 to 5,000 MCM/mths. A more substantial project to pump underground water into the River Ganga west of Patna from December to May would ameliorate the severe pollution between Delhi and Varanasi during the dry season.[39] Although this option lies entirely within Indian control, the electricity needed to pump the underground water could potentially be supplied on a sustainable basis by the hydropower generated by the proposed Nepalese dams. Thus, a Nepal–India collaboration could help clean up the River Ganga both east and west of Patna.

4. Ganga Water Management Between India and Bangladesh

4.1. *The Ganga river in Indo-Bangladesh relations*

India and Bangladesh have a shared history and culture; India played an important role in the creation of Bangladesh and was one of the first countries to recognise its independence. However, Indo-Bangladesh relations

[38] GSBA, p. 55.

[39] See Kamyotra, *et al.* (2013) for a discussion on the most polluted stretches of River Ganga.

have become embroiled in a number of issues, such as illegal immigrants, border disputes and the sharing of trans-boundary rivers.[40] Rivers are vital natural resources to the two countries. Water is an integral part of their social, economic and spiritual lives. Given the growing population, the effects of climate change, and poor management of water resources in both countries, the need for a comprehensive approach to water management is more imperative than ever. Nevertheless, water management and sharing of trans-boundary rivers have been a source of concern in bilateral relations since the formation of Bangladesh in 1971. The two states have concluded only one agreement on water sharing of one of their 54 common rivers. Although the Ganga Water-Sharing Treaty reached by the prime ministers from both sides in December 1996 is a landmark achievement, it still cannot resolve the root cause of the bilateral water dispute. The agreement manages the water sharing at the Farakka Barrage but fails to augment the Ganga flow for utilisation of both countries in the dry season.

Under the Indo-Bangladesh Treaty of Friendship, Cooperation and Peace of 1972, the two countries recognised each other's rights to use the Ganga River and established the Joint Rivers Commission (JRC) to maintain communication and facilitate efforts to maximise the benefits from shared rivers. However, the Commission excluded the issue of sharing of the Ganga water and left it for discussion between the two prime ministers.[41] Bilateral relations subsequently chilled following India's construction of the Farakka Barrage just ten miles from the border of Bangladesh to transfer water from the Ganga to the Hooghly River to improve the navigability of the Port of Calcutta. Bangladesh maintained that the construction of such a barrage would have serious consequences on downstream Bangladesh. In 1974, the prime ministers of India and Bangladesh realised that the flow during the lean season at the Ganga River was not sufficient for both Calcutta's port and use by Bangladesh. As such, the two sides agreed on the need to augment the Ganga flow but differed on how

[40]Thota Narasimha Rao, *India-Bangladesh Relations: Issues, Problems and Recent Developments*. Undated. http://www.academia.edu/4365028/India-Bangladesh_Relations_Issues_Problems_and_Recent_Developments.

[41]Khan,TA (1996). Management and Sharing of the Ganges. *Natural Resources Journal*, 36, p. 462.

to do so. While Bangladesh suggested constructing storage facilities in the upper Ganga areas in India and Nepal to reserve monsoon flow for the lean months, India recommended building canals to divert surplus water to the Ganga from the Brahmaputra, another shared river between the two countries. Neither proposals were acceptable by the other party. India opposed the idea of involving Nepal in the discussion and argued that building such reservoirs in Nepal was outside the purview of the JRC. Bangladesh, on its part, feared that India's canal-linking proposal would have disastrous consequences for Bangladesh.[42]

The Ganga water management became an international issue in 1976, when Bangladesh brought the case of India's continuous and unilateral withdrawal of water at the Farakka Barrage to the Islamic Foreign Ministers Conference, the Colombo Summit of the Non-Aligned Movement, and the United Nations General Assembly.[43] Bangladesh complained that India had been unilaterally withdrawing as much as 60% of the flow at the Farakka Barrage after the agreed trial run of the barrage ended on 31 May 1975.[44] The smaller nation believed that the unilateral action of its bigger neighbour resulted in a significant reduction of the Ganga river's flow to the country and adverse effects on vast areas of Bangladesh. Following recommendations by the UN Special Political Committee, India and Bangladesh reached a five-year Agreement in 1977 in which Bangladesh was allocated 60% of water volume at the Farakka Barrage and a guarantee clause of 80% of the water in case of low flows. Upon the expiration of the 1977 agreement, the two sides agreed to another two MOUs in 1982 and 1985, respectively, without a guarantee clause, but share on a 50:50 basis when the flow falls below 75% of the standard flow.[45] However, from 1988

---

[42] Tiwary, R (2014). *The Ganga Brahmaputra Basin: A Geopolitical Analysis of Transboundary Water Resource Management in South Asia*, Jawahalal Nehru University. Available at http://shodhganga.inflibnet.ac.in/bitstream/10603/16725/12/12_chapter%204.pdf. Accessed 10 June 2015.

[43] Tiwary (2014).

[44] Burns, JF (1997). Sharing Ganges Waters, India and Bangladesh Test the Depth of Cooperation, *New York Times*, May 25. Available at http://www.nytimes.com/1997/05/25/world/sharing-ganges-waters-india-and-bangladesh-test-the-depth-of-cooperation.html. Accessed 10 June 2015.

[45] Tiwary (2014).

to 1996, the both countries could not reach any settlement on the Ganga water and Bangladesh alleged that India conducted unilateral withdrawals of Ganga water during that period.[46]

In May 1992, the prime ministers of India and Bangladesh agreed in to reach a comprehensive and long-term solution to the Ganga water-sharing dispute. Nonetheless, a number of ministerial and secretary-level talks could not agree on how to augment flow. During this period, Bangladeshi experts highlighted the negative impact of the Farakka Barrage on Bangladesh. It was only when Mr HD Gowda and Mr Inder Kumar Gujral became Prime Minister and Foreign Minister of India, respectively, in May 1996, and when the India-friendly Awami League party came back to power in Bangladesh with Madam Sheikh Hasina as Prime Minister, that the Ganga issue have a chance for resolution. The two prime ministers signed a landmark 30-year Treaty on Ganga water management in New Delhi on 12 December 1996.[47] The Treaty recognised Bangladesh's rights as a lower riparian nation and specified the amount of water released by India at the Farakka barrage in the dry season. More specifically, the sharing of water is based on 10-day periods and on a 50:50 basis if the total flow reaches 70,000 cubic meters per second (cumecs). The amount will vary slightly, if the flow is more.

The Treaty of 1996 is an important landmark and a great achievement in the history of India–Bangladesh relations. The Treaty has been praised as a peaceful resolution of water-sharing friction between the riparian states, with the two prime ministers having upheld the principles of "equity", "fairness", and "no harm rule to either side."[48] Furthermore, the Treaty has paved the way for further cooperation between the two nations in other areas of mutual concern, such as sharing of Teesta and other shared rivers, illegal immigration and border demarcation.[49]

However, the Treaty did not resolve a number of contentious points. First, according to the Treaty, the amount of water allocated to India and

---

[46] Ibid.

[47] Ibid.

[48] Pandey, P (2014). Bangladesh, India, and Fifteen Years of Peace: Future Directions of the Ganges Treaty. *Asian Survey*, 54, p. 651.

[49] Tiwary (2014).

Bangladesh at the Farakka Barrage is based on data between 1949 and 1988. However, the current flow through the barrage is significantly lower compared to that period. Bangladesh has alleged that India's over-withdrawal of water in the upstream tributaries is the main reason for the significant reduction in flow at Farakka.[50] Second, the Treaty failed to solve a long-standing problem of augmenting the Ganga River flow during the dry season for the use by both countries. The two sides still disagree on how to enhance the augmentation of the Ganga's flow for mutual benefit. Third, the Ganga Water-Sharing Treaty covers just one of the 54 shared rivers. Hence, a comprehensive approach on the river and water management between India and Bangladesh.

4.2. *Negative impacts of the Farakka barrage on Bangladesh and its implications for India's security*

India's unilateral extraction of Ganga water at the Farakka Barrage has been a major bugbear in relations between the two countries. It is alleged that India unilaterally withdrew water from the Ganga River at the Farakka Barrage for the period from 1975–1977 and from 1988–1996.[51] There was no water-sharing agreement between the two countries during those periods. Given that the Ganga system and tributaries provide water to 37% of Bangladesh, the reduction of the Ganga's flow to Bangladesh has resulted in numerous adverse environmental, ecological and economic consequences.[52] The harmful impact of the Farakka Barrage operations can be found in Bangladesh's hydrology, groundwater, salinity, irrigation, industry, fisheries, navigation, ecology, and health. Bangladesh reported that the Ganga's flow at Hardinge Bridge, 170 km downstream, reached a record low at 23,000 cusecs in 1976 in comparison with its historic level of 65,000 cusecs. The resulting increase in salinity level is believed to cause a deterioration in groundwater quality, agricultural production, and fisheries resources in Bangladesh.[53] Bangladesh also complained that the

---

[50] Ibid.

[51] Ibid.

[52] Ibid.

[53] Ibid.

diversion of the Ganga's flow has dried out its rivers and jeopardised the livelihoods of millions of Bangladeshis. At least seven of the 15 tributaries in Bangladesh are suffering due to flow reduction.[54] It is estimated that Bangladesh has been deprived of about US$600 million annually as a result of its decreasing share of the Ganga.[55] Ashok Swain, at Uppsala University in Sweden, summarised the harmful impact of the Farakka Barrage and the reduced flow of Ganga as follows: "Farakka changed the river's hydrology, disrupting fishing and navigation, brought unwanted salt deposits into rich farming soil … and caused an annual loss estimated at 2–2.5% of GDP."[56] In response, India has said that these adverse effects are overestimated and not scientifically proven.[57]

The Farakka barrage and its diversion of water also has security implications for India. As a result of degradation of the Ganga's dependent areas in Bangladesh, its people are forced to leave their homes.[58] They then see India as a convenient and suitable neighbouring destination, given the proximity between the two countries and India's loose migration policies. It is estimated that were about 20 to 30 million illegal Bangladeshi immigrants in India in the 1990s.[59]

India's unilateral withdrawal of Ganga water has been a source of anti-India sentiments among the Bangladeshis and exploited by successive Bangladeshi governments and parties to gain political advantage. The Ganga water issue is a popular theme in Bangladesh's public debate and domestic politics. The general feeling is that India is utilising Ganga water

---

[54] Ramachandran, S (2006). India, Bangladesh Fight Against the Current. *Asia Times Online*, June 8. http://www.atimes.com/atimes/South_Asia/HF08Df04.html. Accessed 10 May 2015.

[55] Khan, TA (1996). Management and Sharing of the Ganges. *Natural Resources Journal*, 36, p. 473.

[56] Islam, S (undated). Bangladesh-India Water Sharing Disputes: Possible Policy Responses. Available at http://www.academia.edu/3204779/Bangladesh-India_Water_Sharing_Disputes_Possible_Policy_Responses. Accessed 10 May 2015.

[57] Tiwary (2014).

[58] Khan (1996), p. 473.

[59] Joyeeta Bhattacharjee, India: Resolving the Bangladesh Immigration Issue, *The Diplomat*, May 27, 2014. http://thediplomat.com/2014/05/india-resolving-the-bangladesh-immigration-issue/. Accessed 10 May 2015.

to the detriment of Bangladesh.[60] As a result, India is frequently blamed for all water-related problems in Bangladesh, whether it is flood or drought.[61] The ruling government in Bangladesh is supposed to protect water benefits and is frequently criticised by opposition parties for how it has dealt with the water issue with India.[62] It is, therefore, not unusual to find that India is caught between a confrontation between the Bangladesh Nationalist Party (BNP) and the Awami League, the two main political parties in Bangladesh. The BNP has condemned the Awami League regime for signing the 1996 Treaty that allocated an unequal share of water to Bangladesh.[63] These political maneuvers by Bangladeshi parties have, to some extent, fueled nationalistic sentiment against India and portrayed India as a big bullying neighbor. For India, this is certainly not a favourable situation for its security and prestige because the anti-India feelings of a neighboring country may develop beyond the control of the two governments and adversely affect bilateral relations.

4.3. *Evaluation of Indian policy on Ganga water management*

The 1996 Treaty ushered bilateral relations into a new period of cooperation, development and more effective management of trans-boundary water resources. Although the Treaty failed to find a long-term solution to facilitate the Ganga's flow, it removed a major irritant in India–Bangladesh relations and paved the way for further cooperation in other areas. However, as a bigger, upstream and more powerful country, India seems to have ignored the harmful environmental and ecological effects as well as economic losses on the Bangladesh side by allegedly carrying out unilateral withdrawals of Ganga water from 1975–1977 and from 1988–1996. In addition, to gain maximum advantage in dealing with the smaller and weaker country, India has been consistent in its policy of

[60] Tiwary (2004).

[61] Rao (Undated).

[62] Tiwary (2004).

[63] Ramachandran (2006).

bilateralism and opposed the idea of involving other regional countries.[64] As a result, it has been challenging for both sides to reach a consensus on augmenting the Ganga. India's policy of bilateralism has hindered the concerned parties from finding a long-term solution for the Ganga issue. India also seems to have violated international law by unilaterally withdrawing water at the Farakka Barrage from an international and trans-boundary river. It is the legitimate right of Bangladesh to have an equitable share of the Ganga. No nation has the right to unilaterally divert the natural flow of an international river.[65] India's policy of neglect and disregard for interests of the lower downstream country has provoked distrust among Bangladeshis.

Given the growing population in both India and Bangladesh, the adverse effects of climate change, and the lack of a regional cooperation framework, trans-boundary rivers management will continue to be an important and complex issue in the relations between India and Bangladesh. If the two states fail to come up with a comprehensive and durable approach to the problem, it may run the risk of triggering bilateral tensions and threatening regional peace and security.[66]

5. Indo-Bangladesh Cooperation on the Ganga Pollution and Clean-Up

5.1. *Three main approaches*

Climate change and population growth in both countries will lead to more resource scarcity in future as demand for water increases. In addition, in 2026, the 1996 Water-Sharing Treaty between India and Bangladesh will expire. As such, the Ganga River water disputes need to be settled over the next decade to avoid increasing bilateral tensions.

A key factor in the dispute between the two countries is that both suffer from water shortages during the dry season (December to May). The following three policy proposals will help mitigate the water dispute

---

[64] Tiwary (2004).

[65] Islam (1987).

[66] Ibid.

between India and Bangladesh and reduce the water pollution through cleaning up the Ganga. They involve cooperating with Nepal, organising a multilateral mechanism to manage water use, and the participation of neutral third parties in the new framework.

5.2. *Creating cooperation with Nepal*

Water supply from Nepal is inextricably linked to relations between India and Bangladesh. As such, Nepal must be a key stakeholder in the Ganga water problem. As elaborated earlier, one of the main proposals is the construction of dams in Nepal with the cooperation of India in order to improve the water quality and reduce the water shortage in India and Bangladesh.

In this proposal, dams will store water during the monsoon season (June to November) and release the accumulated water into the river during the dry season to increase the low flow.[67] Nepal's contribution to the flow of the river accounts for 46% of the average annual flow and up to 75% of the flow during the dry season. The increase in the water flow in the Ganga delta by the dams can be expected to augment river flow at the Farakka Barrage as well. This is a spillover effect of improving river flow through Nepal–India collaboration by constructing the dams in Nepal. As a result of the flow augmentation, the countries downstream, India and Bangladesh, should be able to overcome their water shortage problem, especially if a robust and impartial monitoring framework can be established to assure Bangladesh that India is not extracting water from the Ganga beyond the 1996 Treaty. The increase in water during the dry season would allow the river to sustain its self-purification capability, as well as enable the lower riparian states to reduce the water conflict caused by water shortages. For Bangladesh, reducing pollution in the Ganga River will improve the health and environmental conditions of the Bangladeshi people given that 30% of the population rely on the river for their domestic use. Nepal, in turn, can benefit from its untapped hydropower potential, which could meet its domestic demands with the surplus sold to the region. In this scenario, market-based water transfers from Nepal should be also discussed. One of

[67] GSBA (2014), p. 73.

the possibilities is that Bangladesh buys excess water from Nepal, and that India also pays for an additional proportion of water from Nepal. This transfer system would also help to resolve the water sharing conflict.[68] Needless to say, proper site selection and assessment are necessary before the dams are constructed to ensure that any negative impact on Nepal is minimised. With careful planning and implementation, this policy option can make a win-win situation not only for the India–Nepal relationship but also for India–Bangladesh relations.

5.3. *Organizing a multilateral mechanism*

Given the difficulty of conducting bilateral diplomacy in the face of domestic politics and historical baggage, both India and Bangladesh should consider establishing a new multilateral framework to help resolve their water disputes.

The 1996 treaty has a problem because the flow of the river at Farakka Barrage was much less than expected.[69] One year after the treaty was signed, the water released to Bangladesh was less than the quantity fixed by the treaty. A similar problem occurred in 1998.

In addition, if the flow at Farakka reduces substantially due to upstream abstraction, it is said that India is not under any obligation to protect the flow.[70] Hence, without any clear obligation to control the upstream withdrawal of the Ganga flow, and no minimum guaranteed flow for Bangladesh in such situations, water sharing becomes very much dependent on the goodwill of the users upstream from the Farakka Barrage. The Joint River Commission (JRC), a bilateral working group established by India and Bangladesh to work for the common interests and sharing of water resources, has been held several times since the signing in 1972, measures to resolve this issue have failed.

---

[68] Kamruzzaman, M, Beechama and GM Zuppi (2012). A Model for Water Sharing in the Ganges River Basin. *Water and Environment Journal*, 26(3), 308–318. p. 315

[69] Kolås, Å, L Barkved, L., Bhattacharjee, J., Edelen, K., Hoelscher, K., Holen, S., Jahan F, Jha HB, & Miklian, J. (2013). Water Scarcity in Bangladesh: Transboundary rivers, conflict and cooperation. *Oslo: Peace Research Institute*. p. 41.

[70] Islam, M. (1987). The Ganges Water Dispute: An Appraisal of a Third Party Settlement. *Asian Survey*, 27(8), 918–934. doi:10.2307/2644864

Disagreements between India and Bangladesh over the impact of the Farakka Barrage have contributed to the overall difficulty in forging not only a water-sharing agreement but also a flow augmentation agreement. Nevertheless, despite the 1996 agreement, the lack of coordination and acrimony fanned by domestic politics would still be a main obstacle for Ganga River issues. As long as both sides insist on only their own interest

To solve the water problem and to strengthen governance of the Ganga River, we suggest a new framework organised at a multilateral level as a second step. One example of this is the Mekong River Committee (MRC) in Southeast Asia. The organisation was established in 1995 to deal with Mekong River issues by four countries: Cambodia, Laos, Thailand and Vietnam. The central purpose of the committee is to control the developments along the river and flow of water. Another example of a regional framework on trans-boundary river management is the Nile Basin Initiative (NBI) involving ten countries around the Nile basin including Egypt, Kenya and Ethiopia. The NBI, started in 1999, focusses on dam construction and water resource sharing and plays an important role in sustainable social and economic development in the region.

A multilateral framework prevents an unbalanced condition where an upper riparian or a large country holds bargaining power in negotiation, and encourages member nations to deepen mutual understanding without conflict. If India and Bangladesh accept a multilateral scheme, the Ganga River issues would take a big step towards finding workable solutions.

If the multilateral framework is established for the Ganga River, calling for the participation of Nepal and China, which occupy 16% of the Ganga basin in total, is vital. Therefore, participation by all stakeholders in the multilateral scheme is a pivotal step towards solving the Ganga's water problems.

The creation of a multilateral mechanism in which all the co-riparian countries have equal power is important. Ideally the body should have transboundary legislative powers pertaining to the basin and its water.

The World Bank has expressed its willingness and readiness to negotiate and cooperate with the parties in the region for water cooperation.[71]

---

[71]The creation of a multilateral mechanism in which all the co-riparian countries have equal power is important. Ideally the body should have transboundary legislative powers pertaining to the basin and its water.

In this respect, the most hopeful reference for the Ganga River is the Indus water dispute between India and Pakistan that was negotiated and concluded under the auspices of the World Bank. Based on the experience of the Indus water dispute, India and Bangladesh should collectively approach The World Bank for assistance. It was also the World Bank which issued a study in 2012 that called for significantly enhanced regional cooperation in water, weather and climatic information which is essential for the sustainable management of the basin and the safety and prosperity of the people.[72]

The United Nations Development Program (UNDP) is another potential partner. It once attempted to initiate cooperation between India and Pakistan for the development of all major rivers common to India and Pakistan.[73] The UNDP also organised a meeting in 1992 in order to discuss the Mekong River. Moreover, the UNDP has also called for regional cooperation by enhancing governance among stakeholders in its report of 2003 titled "Water Strategy of UNDP — Approach towards Effective Water Governance."

Another reason for the need for a neutral third-party organisation is that China and India are key stakeholders of the Ganga River. As a matter of fact, India and China are less interested in adopting international or regional frameworks that restrain them from unilateral actions in developing commonly shared water resources. For example, China voted against the adoption of the UN Convention on the Law of the Non-Navigational Uses of International Watercourses (UNCIW) as the primary basis for river sharing and use in 1997.[74] Likewise, India objected to the mandatory settlement of the dispute in the law. Moreover, China,

---

[72] Dhungel, D. (September 2013). Regional Cooperation on the Ganga Basin: Yet a mirage? Intra-Regional Ganga Initiative. Issue Brief. The Asia Foundation. Accessed on 1 May 2015. Available at http://asiafoundation.org/resources/pdfs/ORFIssuebrief61DwarikaN.Dhungelformail.pdf

[73] Islam, M. (1987). The Ganges Water Dispute: An Appraisal of a Third Party Settlement. *Asian Survey*, 27(8), 918–934. doi:10.2307/2644864

[74] Chintan, Gopal Siwakoti (2011). Trans-boundary River Basins in South Asia: Options for Conflict Resolution. Berkeley: International Rivers. Accessed on 15 May 2017. Available at https://www.internationalrivers.org/sites/default/files/attached-files/transboundaryriverbasins.pdf

which occupies one-fifth of the basin of the Mekong River, does not take part in the MRC. Therefore, when discussing Ganga River issues, a third-party neutral international organisation which could convince China and India to compromise and help regional cooperation is necessary. As can be seen from the past examples and experiences, the participation of the United Nations and the World Bank as a neutral third party to the multinational scheme for Ganga River would be valid and they can play an active role in creating a multilateral framework and promoting cooperation in managing the Ganga River to resolve water conflicts and mitigate river pollution.

7. Conclusion

Better cooperation between India, Nepal and Bangladesh, on managing the use of the Ganga River will have not only environmental, but also political, economic and security benefits for all three riparian states. Specifically on reducing pollution in the Ganga River, international cooperation can play a meaningful role in achieving the goal of sustaining an acceptable level of river flow in order to maintain the self-purifying capacity of the Ganga River.

As the upper riparian state, Nepal is well-positioned to help India increase the Ganga River flow from Nepalese tributaries that feed the Ganga, especially as these tributaries contribute more than two-thirds of river flow to the Ganga in the eastern part of India during the dry season.

Dams serve the interests of both India and Nepal. Dams for water storage in Nepal can benefit India by increasing river flow, while Nepal stands to benefit economically from the sale of electricity generated from hydropower. India could even buy such electricity to pump underground water to augment river flow for other parts of the Ganga to further reduce pollution in the river. Given Nepal's economic challenges in funding such large infrastructure projects, the Indian government should deepen its engagement with its Nepalese counterpart through a strong bilateral mechanism such as a Joint Rivers Commission.

India should seize the current opportunity as Nepal emerges from its political chaos and build on the positive bilateral relationship generated by

Modi's ground-breaking visit to Nepal. India's strong relationship with Bhutan can serve as a useful template in furthering a positive Nepal–India relationship.

If India takes the effort to improve the Ganga's flow, it will improve its relations with Bangladesh and help reduce India's security problems, which have been caused by water stress in Bangladesh. Given India's role in helping Bangladesh achieve independence in 1971, relations between both countries have been a string of missed opportunities. If the 1996 Indo-Bangladesh Water-Sharing Agreement was the most significant development in water resource management between both countries after 20 years of negotiation, it was only possible by the fortuitous circumstance of the two governments then in power having the inclination and will to successfully ink an agreement. And even then, the agreement could be signed only because the tricky issue of augmenting the Ganga River flow was excluded from the agreement.[75]

Given that augmenting the Ganga River flow has remained unresolved between India and Bangladesh since the inception of Bangladesh in 1971, progress in this area will be a long overdue development in regional harmony. However, it is unlikely that augmenting the flow alone will produce the necessary peace dividends between India and Bangladesh, given the history of political mistrust and the political football of internal politics. Having a multilateral framework to discuss cooperation or the involvement of an international organisation might also help insulate India and Bangladesh from domestic fallout in making decisions on international cooperation.

As the largest and more developed state in the South Asian region, India needs to take the lead in forging closer cooperation with Nepal and Bangladesh, not in the sense of exercising hegemony over the region but more in the Nehruvian sense that states and peoples can come to understand each other better and sustain peace through negotiations, compromise and restraint.[76] This will create not only domestic benefits for India in having a cleaner Ganga River, but also develop more secure relations

[75] Pandey (2014), p. 673.

[76] Bajpai (2002), pp. 251–252.

with Nepal and Bangladesh and reduce India's security problems with these countries.

Bibliography

Bajpai, Kanti (2002). Indian Strategic Culture. In *South Asia in 2020: Future Strategic Balances and Alliances*. Carlisle: Strategic Studies Institute.

Bhattacharjee, J (2014). India: Resolving the Bangladesh Immigration Issue, *The Diplomat*, May 27. Available at http://thediplomat.com/2014/05/india-resolving-the-bangladesh-immigration-issue. Accessed 1 May 2015.

Biswas, AK (2011). Cooperation or Conflict in Transboundary Water Management: Case Study of South Asia. *Hydrological Sciences Journal*, 56(4), pp. 662–670.

Burns, JF (1997). Sharing Ganges Waters, India and Bangladesh Test the Depth of Cooperation. *New York Times*, May 25. Available at http://www.nytimes.com/1997/05/25/world/sharing-ganges-waters-india-and-bangladesh-test-the-depth-of-cooperation.html. Accessed 1 May 2015.

Chintan, GS (2011). Transboundary River Basins in South Asia: Options for Conflict Resolution, *International Rivers*, 2011. Available at http://www.internationalrivers.org/resources/transboundary-river-basins-in-south-asia-options-for-conflict-resolution-2445. Accessed 1 May 2015.

Das, Subhajyoti (2011). Cleaning of the Ganga. *Journal of the Geological Society of India*, 78(2), 124–130.

Dhungel, DN (2013). *Regional Cooperation on the Ganga Basin: Yet a Mirage?* Observer Research Foundation (ORF).

Ghosh, P (1995). *Cooperation and Conflict in South Asia*. Manohar Publishers and Distributors.

Islam, M (1987). The Ganges Water Dispute: An Appraisal of a Third Party Settlement. *Asian Survey*, 27(8), 918–934.

Islam, S (undated). *Bangladesh-India Water Sharing Disputes: Possible Policy Responses*. Available at http://www.academia.edu/3204779/Bangladesh-India_Water_Sharing_Disputes_Possible_Policy_Responses. Accessed 1 May 2015.

Kabir, MH (2014). Obstacles to India–Bangladesh Cooperation: An International Relations Theory Perspective. In *International Relations Theory and South Asia: Security, Political Economy, Domestic Politics, Identities, and Images, Vol. 1*, Sridharan, E (ed.). Oxford Scholarship Online. Retrieved from http://libportal.nus.edu.sg/. Accessed 1 May 2015.

Kamruzzaman, M, S Beecham, GM Zuppi. (2012) A model for water sharing in the Ganges River Basin. *Water and Environment Journal*, 26(3), 308–318.

Kamyotra, JS (2013). *Pollution Assessment: River Ganga*, Central Pollution Control Board, Ministry of Environment and Forests, Government of India.

Khan, TA (1996). Management and Sharing of the Ganges. *Natural Resources Journal*, 36.

Kolas *et al.* (2013). *Water Scarcity in Bangladesh*. Peace Research Institute Oslo Report, January 2013. Available at https://www.prio.org/Projects/Project/?x=969. Accessed 1 May 2015.

Malhotra, P (2010). *Water issues between Nepal, India and Bangladesh: A Survey of Literature*. Institute of Peace and Conflict Studies (IPCS).

Narain, S and S Banerjee (2014). *Ganga: The River, its Pollution and What We Can Do To Clean It*. Centre for Science and Environment (CSE), India.

Nayak, N (2010). India-Nepal Peace and Friendship Treaty (1950): Does It Require Revision? *Strategic Analysis*, 34(4), 579–593.

Nepal, KD (2013). Dictated By Geography. *World Policy Journal*, 30(4), 36–40.

Pandey, P (2014). Bangladesh, India, and Fifteen Years of Peace. *Asian Survey*, 54(4), 651–673.

Rahaman, M (2006). The Ganges Water Conflict. A Comparative Analysis of 1977 Agreement and 1996 Treaty. *Journal of International & Peace Studies*, 1(2), 195–208.

Ramachandran, S (2006). India, Bangladesh Fight Against the Current. *Asia Times Online*, June 8. Available at http://www.atimes.com/atimes/South_Asia/HF08Df04.html.

Rao, TN (). *India-Bangladesh Relations: Issues, Problems and Recent Developments*. Undated. Available at http://www.academia.edu/4365028/India-Bangladesh_Relations_Issues_Problems_and_Recent_Developments.

Seth, BL (2013). *Status Paper for River Ganga: Past Failures and Current Challenges*. Centre for Science and Environment (CSE), India.

Shakya, NM (2012). *Ganges Strategic Basin Assessment: Consultative Workshop Report*. Jalsrot Vikas Sanstha, Nepal.

Shukla, D (2006). India–Nepal Relations: Problems and Prospects. *The Indian Journal of Political Science*, 67(2), 355–374.

Tare, V *et al.* (2011). *SWOT Analysis of Ganga Action Plan*. Indian Institute of Technology (IIT), 2011.

Thapliyal, S (2012). India and Nepal Treaty of 1950: The Continuing Discourse. *India Quarterly*, 68(2), 119–133.

The Times of India (2014). *Narendra Modi Woos SAARC Nations, Pledges Slew of Investments to Counter China*. November 28.

The World Bank (2012). *Ganges Strategic Basin Assessment (GSBA): A Regional Perspective on Risks and Opportunities.*

Tiwary, R (2004). *The Ganga Brahmaputra Basin: A Geopolitical Analysis of Transboundary Water Resource Management in South Asia.* Doctor of Philosophy Thesis. Centre for International Politics, Organization and Disarmaments Studies. Jawahalal Nehru University. New Delhi.

Chapter 11

MANAGING COMPETING DEMAND FOR WATER IN THE GANGA: LESSONS FROM THE YELLOW RIVER

Han Mingyu, Cao Dongdong, Li Na, Panji Ruky

1. Introduction

The River Ganges suffers from severe pollution from urban sewage, industrial effluents and agricultural run-off as well as nonpoint source pollution such as religious waste and cremated bodies. Over the decades, the Indian Government has adopted many policies and programs, including the Ganga Action Plan (GAP) I and II, to mitigate and remove these sources of pollution. But the quality of water has not improved, and in fact, has possibly deteriorated.

Previous clean-up efforts that concentrated on controlling sources of pollution have been largely thought to be failures; hence, restoring river flow has become an increasingly attractive option. There are now urgent calls for the government to widen its focus from pollution abatement to addressing the issue of environmental flow (hereby abbreviated as *e-flow*) into the Ganga. In fact, pessimists argue that the plan to rejuvenate the Ganga led by Prime Minister Modi, will meet the same fate as past attempts because "the plan fails to list concrete steps to restore the flow of the river — essential for washing away pollutants" (Chaudhary, 2014). Proponents of the water flow remediation theory also contend that rivers have a self-cleansing ability which allows for the assimilation and treatment of biological waste.

This chapter offers a discussion on the inadequate water flow in the Ganga, specifically in the upper and middle reaches of the river, specifically from Uttarakhand to Uttar Pradesh. We argue that unmanaged and competing demands for water use in agriculture, industrial development, hydropower generation and human settlements along the Ganga and its tributaries has led to significant withdrawals from the river. Moreover, the combination of such withdrawals and extreme seasonality of rainfall has resulted in insufficient flow to help wash away pollutants during the dry season, which then contributes to the high level of pollution found in cities in the aforementioned river section such as Kanpur and Varanasi.

The primary motivation for this research is to urge the National Ganga River Basin Authority (NGRBA) to take concrete steps to realise one of its mission statements, i.e. to maintain the minimum e-flow of the Ganga. In this chapter, we use the comparative case study as the main research methodology, drawing lessons from the Yellow River in managing competing demands for water so as to achieve e-flow. Since there is lack of comparison with similar initiatives between the Ganga and other major river basins in Asia so far, we believe the information here will help water policy-makers and practitioners in Asia to understand the importance of developmental context to proposed solutions. It is worth pointing out that both e-flow and water-sharing are crucial components in the Integrated Water Resource Management (IWRM) framework, ubiquitous in many river basin rejuvenation programs. Hence, we make the comparison by analysing three important governance issues, i.e. the institutional framework, the enabling environment as well as policy instruments, through IWRM-related dimensions.

1.2. *Comparing the Yellow River to the Ganga*

Both the Yellow River and the Ganga share some key similarities, including geographic features, diverse water use and importance to national development (see Figure 1).

	The Ganga	The Yellow River
Location	Northern India	Northern China
Length	2, 525 km	5, 464 km
Drainage Area	1, 080, 000 km^2	752, 400 km^2
Annual Precipitation	300 mm (Monsoon season from June to August)	452 mm (Monsoon season from April to October)
Annual Mean Runoff	550 billion m^3	58 billion m^3
Passing Localities	5 states (plus Nepal and Bangla-desh)	7 provinces and 2 autonomous regions
Serving Population	400 million	140 million
Serving Farmland	561, 600 km^2	160, 000 km^2
Cultural Significance	Holy River for the Hindus	Mother river of the Chinese nation and cradle of the Chinese civilisation
Diverse Water Usage	Agricultural irrigation, domestic use, industrial production, hydropower, religious practice and tourism	Sediment flushing, agricultural irrigation, industrial production, domestic use and afforestation
Focal Problems	Water over-extraction and pollution	Water scarcity, soil erosion and sedimentation, flooding and water pollution

Figure 1: Comparison of The Ganga and The Yellow River

Both rivers both face severe water stress in certain stretches. Consisting of only 2 percent of national water resources, the Yellow River has to accommodate 15 percent of China's farmlands, 12 percent of the national population, more than 50 large and medium-sized cities, and dozens of large-scale energy bases. The increasing imbalance between demand and supply exacerbated the problem of water scarcity, resulting in no-flows downstream from 1972 to 1998. Reduced water volume also adversely affected the river's self-dilution function and caused 71 percent of the river to be categorised as highly polluted (YRCC, 1998). On the other hand, the Ganga is considered to have sufficient water compared with the Yellow River, especially during the monsoon season. Nevertheless, the shortage of water in the Ganga is still acute in the first two states — Uttarakhand and Uttar Pradesh, during the dry season. From this perspective, this section of the Ganga corresponds with the middle and lower reaches in the Yellow River, i.e. Inner Mongolia, Shaanxi,

Shanxi, Henan and Shandong. And because the size of overall population dependent on water from the Ganga is nearly three times than that of the Yellow River, the average water usage per capita within the two river basins is similarly limited as a result.

Another comparable point is the developmental context that the two rivers are have to contend with. Both India and China are developing countries with huge populations. The total population of the two countries make up more than one-third of the world population in 2014, with 1.27 billion in India and 1.39 billion in China.[1] The two countries have also successfully pursued economic growth. Between 2010 and 2013, the annual GDP growth in India was 10.4 percent, 9.3 percent, 7.7 percent and 7.7 percent. During the corresponding period, China's growth rate was 10.3 percent, 6.6 percent, 5.1 percent, 6.9 and percent.[2] Both governments have attached great emphasis on economic development and have more or less sacrificed environmental concerns on many occasions. The intense pursuit of economic gain has made all common resources contested, including rivers that can provide water. Consequently, nature takes its revenge, as manifested by problems in rivers such as heavy pollution and no-flows, which then teaches people to realise the importance of establishing a balance between environmental and socioeconomic needs. In China recently, GDP growth is no longer the key criterion to assess the performance of localities; ecological indicators have been taken into account instead. In India, there has been a strong outcry from both government and civil society to restore the ecological integrity of the Ganga.

Given the similarities in terms of severe water stress, stage of economic development, and the national, social and cultural features of the two rivers, we believe an examination of how the Yellow River Conservancy Commission (YRCC), a river basin management authority, has evolved to incorporate administrative, legal and policy measures to address water scarcity will provide useful insights for NGRBA and the Ganga Rejuvenation program. We will provide a historical perspective so that

[1]World Meters (2014). Countries in the World (ranked by 2014 population). Retrieved from http://www.worldometers.info/world-population/population-by-country/.

[2]The World Bank (2014). GDP growth. Retrieved from http://data.worldbank.org/indicator/NY.GDP.MKTP.KD.ZG.

readers may better understand the underlying dynamics that YRCC faced in its water governance journey. The study shows that YRCC's successful practice in managing competing demand for water actually lies in three critical sequential steps: first, set a volume appropriation for e-flow and human socioeconomic use based on scientific research; second, define water rights and separate use rights from ownership rights; and third, establish a water allocation and quota mechanism to ensure sound socio-economic development while maintaining the intrinsic value of the river. It also reveals a complementary and interdependent relationship between e-flow and water sharing, as simply emphasising one while disregarding the other will not be conducive to creating a sustainable governance eco-system. On the other hand, NGRBA is still in the early stage of agenda setting in terms of managing diverse demand for water. Although several research reports commissioned by NGRBA and the National Water Policy 2012 have identified a clear need for a better definition of water rights and sustenance of minimum e-flow, debates are still ongoing and consensus on the way forward has not been reached.

In the sections below, we will first conduct a review of the theoretical framework to be employed in our comparative analysis. Then we elaborate on the current situation of water flow in the Ganga, the policies taken by NGRBA, and offer an analysis related to key IWRM pillars found in the Ganga Rejuvenation program. After that, we will explore YRCC's policies and programs and discuss the same IWRM pillars that made YRCC's policy interventions successful. Analytical comparisons against major IWRM concepts and pillars are presented afterwards so as to identify differences and similarities of governance issues between the two rivers. In the end, we propose policy lessons, following which potential challenges and opportunities are touched upon in the conclusion.

2. Theoretical Framework

Our research is heavily influenced by the Integrated Water Resource Management (IWRM) approach to water governance. The concept has become the leading paradigm in managing river basins since its formal adoption by countries in global water councils (Unver, 2008). There are a number of versions on what constitutes IWRM. The broadest definition states that IWRM is "a facilitated stakeholder process to promote coordinated activities in pursuit of common goals for multiple objective development and management of water aligned with the sustainable water resource system criteria" (Davis, 2007). Although there are various versions, the common themes are emphasis on *coordination, meeting multiple objectives,* and *defining water as economic good.*

IWRM specifically focuses on the coordination of efforts among agencies, national and state governments, legal jurisdictions and participants representing the general public who have a stake on the river. It also recognises the multitude of interests and needs that river basins have to fulfill from basic human demand, economic livelihood, social and cultural practices to ecosystem requirements for flora, fauna and a healthy environment. As such, water governance influenced by IWRM would generally attempt to meet these multiple objectives, which is similar to managing competing demands for water. In order to achieve this, it brings up the key dimension of treating water as economic good that introduces market-based principles to resolve trade-offs between limited supply and excessive demand for water. There have been critiques of IWRM that the approach is too encompassing and ambitious and very difficult to implement in practice (Biswas, 2004). Therefore, we argue that IWRM should be viewed as a process rather than an outcome.

We employ IWRM as the theoretical framework in making a comparative analysis across river basins. In a departure from fragmented approaches to water governance, IWRM endeavors to address not only the hydrological aspect, but also the socioeconomic, cultural, administrative and political issues in river management. According to the 2002 Global Water Partnerships, the three pillars of IWRM are to have an *Enabling Environment*, an *Institutional Framework* and a set of *Policy or Management Instruments*. By using this framework, we avoid outcome comparisons (which may not be visible in the short term) and focus instead on the capacity and quality of processes achieved within the three pillars in our research.

Enabling Environment. To achieve success, IWRM requires an environment conducive to the implementation of strategies and policies for managing the river basin. Such an environment is enabled through the combined will from the nation's leadership and public, supportive legislation, laws and regulations defining water rights, and cooperative water governance between national and sub-national governments with agreements settled among riparian countries, states or localities.

Institutional Framework. The decisive role of the river basin organisation in water governance must be clearly identified. Institutional arrangements address the river basin organisation's administrative discretion, whether a centralised or decentralised model is used, and the extent of the interaction between it and other government agencies and local authorities. Furthermore, the organisation needs to be authorised and have adequate resources in manpower, funding, knowledge and the appropriate skills to plan, design, implement and monitor relevant policies.

Policy Instruments. If an enabling environment and a supportive institutional framework are in place, the river basin organisation can then deliver a set of policy instruments to actively manage the river ecosystem. Many tools and best practices in the IWRM inventory are available to be implemented in the river basin, depending on the trimmed application for the specific river or country. Tools include using hydrological statistics to assess, define and monitor e-flow and using economic measures to develop a mechanism for water allocation and sharing.

Environmental Flow (E-flow). A concept frequently visited in IWRM is the e-flow of a river. E-flow has a number of different definitions used by scholars and governments. In this chapter, we take the definition from the Brisbane Declaration in 2007 which defines e-flow as "the quantity, timing and quality of water flows required to sustain freshwater and estuarine ecosystems, human livelihoods and well-being that depend on these ecosystems" (Richter *et al.*, 2012). Hence, e-flow offers a scientific basis to determine the quantity of water that should be left in the river after taking multiple sources and uses of water into consideration.

Over-extraction of water from a river body significantly affects the ecosystem and creates environmental problems. In addition, such problems are often magnified by the seasonality of rainfall or other variations of water sources. When water supply contracts, the level of oxygen drops and water quality deteriorates (Jones *et al.*, 2007). By ensuring e-flow during dry periods, the river would be able to sustain the ecosystem (including transporting sediment to downstream) and still meet the needs of human society. In reality, to achieve e-flow during the dry season would mean storing surplus water in reservoirs and dams during the rainy season and releasing the stored water into the river when the river flow falls below levels set as e-flow. Other practices include imposing limits on water withdrawal for irrigation or industrial purposes during the dry season (Jones *et al.*, 2007).

Water Sharing.[3] A primary goal of IWRM is to manage competing demand for water in the river basin. IWRM attempts to help address this by treating water as an economic good (Davis, 2007). In order to turn water into an economic good, access to water (or rights to water) is usually defined in water laws or even constitutions. Many environmental issues in rivers suffer from the classic problem of the commons. Lack of water rights or definition of water as a common resource with oversight contributes to ineffective policies in managing competing demand for water. Explicitly defined water rights are also a premise for water sharing. By attaching economic value to water, the river basin authorities may implement market-based mechanisms to allocate water to various users upon agreement, and mitigate the negative externalities of water as a common resource.

Water sharing is broadly defined as "control and allocation of water from headwaters to sea or sink" (Davis, 2007). It may include determination of and restrictions on the supply of water, demand for water (extraction) and allocation of water for ecosystem and human needs. For instance, once e-flow is set for the river, the amount of water available for extraction is also determined. In river basins with a water sharing mechanism, that portion of water flow is the supply to the "market" for water. Water allocation may differ between dry and wet seasons, particularly in the dry season, and may operate according to riparian rights (given to land owners adjacent to river) or appropriative rights (based on historical occupation or shared equity) (Davis, 2007). In a sense, e-flow also represents the demand from the ecosystem as a water user. Any loss incurred by human beings that leads to lessened water for use is also shared in accordance with a water rights system. Another key component of water sharing is a mechanism to mediate and resolve conflicts, either

[3] Matthew Davis (2007) wrote a comprehensive synthesis on IWRM and water sharing. This section heavily draws from Davis' article.

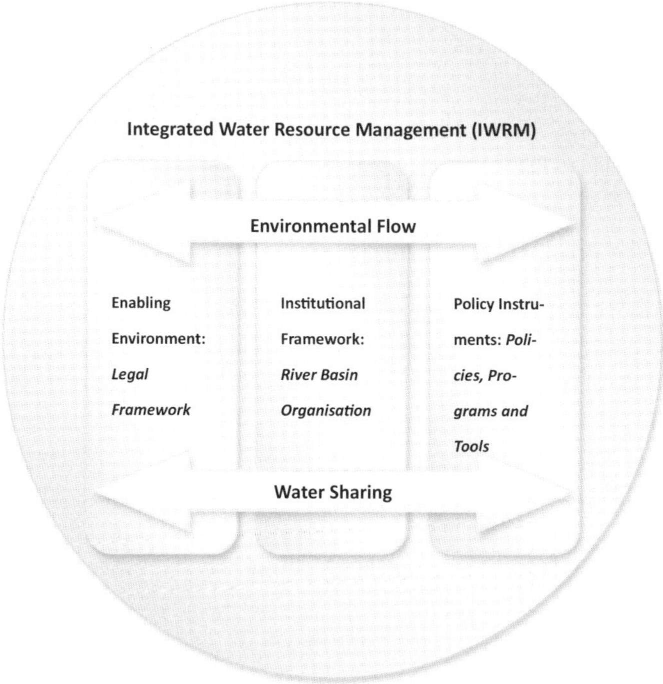

Figure 2: Umbrella Framework of IWRM and Relevant Concepts and Pillars

through specialised courts or specific water authorities set up by the government that oversee the management of water as a common resource in the river basin (Davis, 2007).

The umbrella framework of IWRM is supported by three pillars and two key concepts and is illustrated by Figure 2. This will be systematically incorporated in the following case narratives, comparative analysis as well as policy lessons.

3. Evolving Water Governance in The Ganga River Basin

The Ganga faces the challenge of inadequate water flow in the upper and middle reaches for many reasons, among which severe water extraction and competing use of water are critical. Based on the above theoretical framework, an examination of the current state of flow in the Ganga will be carried out and then followed by a discussion of three governance issues pertaining to IWRM pillars, i.e. *river basin organisation, the legal framework* and *the policy context.*

3.1. *Water flow in the Ganga*

Water flow is another important issue for the Ganga clean-up initiative in parallel with pollution control. But there has been a dearth of reliable information regarding the flow and health of the river (Das, 2011). According to statistics released by the Central Water Commission in 2004, the annual flow of the Ganga amounts to 525,023 MCM (replace with million m^3?)whereas the live storage capacity of barrages completed and under construction totals 63,736.84 MCM. In addition, there have been discussions to add 30,617.15 MCM of storage by the previous administration. Hence it gives the impression that the river flow is in abundance. In reality, however, 90 percent of runoff is diverted for irrigation and hydropower generation before reaching Kanpur, leaving the Ganga with hardly any water during the dry season in the middle stretch.[4]

Many reasons contribute to the reduction of water flow in the upper and middle reaches of the Ganga. Firstly, the Himalayan glaciers that are estimated to supply 30 to 40 percent of the water volume to the Ganga are shrinking because of global warming and climate change. Coupled with the over-exploitation of aquifers, the Ganga (during December to May) prior to the monsoon season has threatened to run dry (Das, 2011; Depraz/WWF). Secondly, hydropower projects with dams and barrages have had a serious impact on the state of water flow. It is reported that "the large hydropower projects in Uttarakhand including Tehri Dam are critically depleting discharge of the river and its tributaries" (Das, 2011). In the upstream as a whole, there are nine hydropower projects in operation, 12 under construction and another 33 proposed (IIT, 2010), all mainly

[4]Chaudhary, J (2014). India's Plan to Clean up the Ganga River is Flawed, Say Experts. *China Dialogue*, Oct. 16. Retrieved from https://www.chinadialogue.net/article/show/single/en/7395-India-s-plan-to-clean-up-the-Ganga-River-is-flawed-say-experts.

located in the states of Uttarakhand and Uttar Pradesh. They hoard huge amounts of river flow in dams and barrages, which makes the Ganga dry up even more than usual (Mcdermott, 2011). When the river flow is unable to fill hydrologic projects due to effects of climate change and further degradation of catchment areas, dams like the Tehri dam recycle utilised water, storing it for power generation instead of discharging it into the mainstream. Thirdly other, competing uses for water significantly dampen water flow. Apart from hydropower projects, irrigation through canal systems use a substantial amount of water. The Upper Ganga Canal network, for example, diverts huge amounts of water that reduces water flow to 15 billion m^3 annually at Balawali (Gupta, 1984). The Lower Ganga Canal system transfers water between Hardar and Aligarh, leading to a meaningful drop of water flow downstream in the dry season followed by deteriorating water quality (CPCB, 2013). Controlling all the tributaries in the river, barrages shift roughly 60 percent of water to large-scale irrigation programs.[5] As a result, the mean water flow in the dry season between Kannauj and Kanpur, coupled with heavy discharge of pollutants, has increased the pollution level in this section of the river. Upstream of Varanasi, the river flow disappears in many regions and the water depth at Varanasi decreases to 30 m in some places (Das, 2011).

The outcome of reduced water flow can be disastrous. Water scarcity caused by excessive agricultural extraction "has increased dependence on ground water, the loss of water-based livelihoods, and the destruction of habitat for 109 fish species, and other aquatic and amphibian fauna".[6] According to the WWF, agricultural productivity has declined due to the deficiency in the soil's organic content incurred by lowering water levels. Furthermore, over-exploitation of groundwater has seriously affected water quality and "impairs the natural cleansing of arsenic which becomes water soluble when exposed to air, and threatens the health of 75 million people who are likely to use water contaminated with up to 2 Mg/L of arsenic".[7] Uncontrolled water use poses a critical threat to natural flows and severely damages people's livelihoods in the Ganga river basin.

---

[5] Dépraz, M, "Threat of Water Extraction on the Ganges." wwf.panda.org.

[6] Ibid.

[7] Ibid.

3.2. *River basin organisation*

Over the years, various authorities have been put in charge of cleaning up the Ganga. Constituted in 1974, the Central Pollution Control Board (CPCB) is a statutory organisation designated for improving water quality by the prevention, control and abatement of pollution. After CPCB issued two comprehensive reports on pollution, the Ganges Action Plan was set up in 1985, with the goal of controlling pollution along the Ganga (IIT, 2011). GAP's overall design and implementation was under the jurisdiction of the Ministry of Environment and Forests (MoEF). One year later, the Central Ganga Authority (CGA) headed by the then Prime Minister was set up to finalise the relevant policy framework and supervise the implementation of GAP I. An additional agency entitled the Ganga Project Directorate (GPD) was also established to put projects under GAP I into practice with empowered financial and administrative discretion. CGA and GPD were transformed into the National River Conservation Authority (NRCA) and the National River Conservation Directorate (NRCD) respectively after the installation of GAP II in 1995.

The lack of tangible results from earlier GAP efforts prompted the setting up of the NGRBA by the Singh administration in 2009,[8] a time during which e-flow started to draw attention from the authority. As a planning, financing, monitoring and coordinating water governance body between the central and state governments, the NGRBA endeavors to ensure the effective abatement of pollution and conservation of the river Ganga through adopting a river basin approach for comprehensive planning and management with both regulatory and developmental functions (MoEF, 2015). Its mission includes developing a river basin management master plan, regulating water affairs, maintaining minimum e-flow, developing schemes to control water pollution, collecting, analysing and disseminating information, conducting investigation and research on pollution and conservation issues, promoting water conservation practices, and monitoring and reviewing implementation of all relevant programs. In addition to these functions, the NGRBA also oversees the power vested with

---

[8]Yadav, S (2015). Namami Gange: In The Name of Ganga. *The Indian Express*, May 18. Retrieved from http://indianexpress.com/article/explained/in-the-name-of-ganga/.

riparian state governments and their water institutions, whilst the latter's role is relatively vague in terms of authorised execution. According to the MoEF, riparian states may choose to turn to the Ganga River Conservation Authority for coordinating and implementing water governance at the state level, but is not yet an imperative for them to comply. Besides, state governments are requested to act for "comprehensive management" within their jurisdiction, again a somewhat ambiguous definition.

The National Mission for Clean Ganga (NMCG) was established in 2011 as the implementation arm of NGRBA and enforce its mandates as well. Both NGRBA and NMCG come under the Ministry of Water Resources, River Development and Ganga Rejuvenation (MoWR, RD and GR). The Secretary to MoWR, RD and GR chairs the governing council of NMCG at present. Moreover, its Mission Director is held by a Joint Secretary (JS) in the Government of India. The NMCG exerts administrative authority in the entire Ganga river basin, including all regions that the river flows through and the National Capital Territory of Delhi. The mission of NMCG is in line with that of NGRBA: to ensure the effective abatement of pollution and the rejuvenation of the river Ganga by adopting a river basin approach to promote inter-sector coordination for comprehensive planning and management; and to maintain minimum e-flow in the Ganga with the aim of ensuring water quality and environmentally sustainable development. The rejuvenation of the Ganga is defined as ensuring "*Aviral Dhara*" ("continuous flow"), "*Nirmal Dhara*" ("unpolluted flow"), and geological and ecological integrity. While *Namami Gange*, the latest integrated Ganga development project, was announced and laid onto the track (unsure what this means) after Prime Minister Modi took power last year, NMCG functions as the coordinating organ at the national level and much of its work still relies on support from the State Level Project Management Groups (SPMGs), which consists of senior officials from states along the river.

3.3. *Legal framework*

Water rights have been absent in Indian laws and hence many uncertainties lie with water usage (Siddiqui, 2008). Traditionally speaking, water rights in India are closely associated with the property right of land inherited from colonial law. But there is no clear definition or statement assigned to it in the 1987 and 2012 national water policies (Cullet, 2007). In practice, water is widely regarded as a common resource, also "the basic need for the survival of the human beings and part of right of life and human rights" (Cullet, 2007). The government provides universal and free access to water for its predominantly poor and largely rural population.

In addition, no national umbrella legislative framework is in effect to regulate water affairs in all dimensions. This has led to the situation that legal interventions and principles set at the central and state levels may not necessarily align with each other (Cullet, 2007). Above all, water is defined as a state subject by the 1935 Government of India Act, empowering the local governments to legislate water-related decrees. It is due to this that states have been authorised with exclusive power in managing water supply, irrigation and canals, water storage and hydropower, etc. (Cullet, 2007). The central government's discretion in legislating water laws is constrained by the constitutional institution. There is virtually no restriction on water extraction from rivers that flow across states. Although the 1956 Inter-State Water Disputes Act introduced a channel for addressing water conflicts among states, it has never been applied to the Ganga. Also, the 1956 River Boards Act empowered the central government to set up river boards for advising states on the regulation and coordination of inter-state river issues such as control and use of water, water supply, conservation, irrigation and so on, but again the act has not been put into practice yet (Cullet and Gupta, 2009).

The 1987 and 2002 national water policies came into place mainly because of disputes among states owning absolute power and the lack of a unified national water governance framework. The two documents both anchored the principles that water, as a valuable natural resource, can help enhance economic productivity, and that domestic use of water has the overriding priority in water allocation in terms of allocation; also "incentives" shall be cultivated to promote efficient and productive use of water (Cullet and Gupta, 2009). Most importantly, they proposed to introduce water rights, which can be the premise for participation from varied entities in water resources management and potential trading of entitlements (Cullet and Gupta, 2009).

Nevertheless, some fundamental questions still remain to be addressed under the current legal framework. For instance, what water uses should be prioritised in which parts of the river system and under what circumstances? When can socioeconomic uses of water from human society transcend environmental interests? And how can rivers flowing across states be managed fairly within a legal framework that authorizes water rights onto the state government? It is obvious that a simple distribution of water resources through engineering technology that takes the river as a pipeline will not bring sound and sustainable solutions.

3.4. *Policy context*

The Ganga clean-up initiative at present features two points: first, pollution abatement is still the top priority and as a result, great attention is placed on constructing Sewage Treatment Plants (STPs); second, e-flow is identified in the national agenda, yet appears to be a secondary priority. The government directs the majority of resources on controlling sources of pollution. When Prime Minister Modi took office in 2014, restoring the sound and clean flow of stream in the Ganga was mentioned by him on many occasions. But when he launched *Namami Gange*, its focus was "on urban sewage and industrial effluents in order to check pollution at its source".[9] The program is envisaged to be a grand integrated development project with a total budget of around USD 334 million. It incorporates elements such as removing pollution, maintaining e-flow, building ghats at selected points, and making a navigable river route from Allahabad to Haldia. However, many critics have commented that the program is rich in rhetoric but fails to outline detailed measures under each objective and internal correlations among them.[10]

The program continues to focus on adding more STPs, and even includes the erection of 16 new dams over a 1,600 km long stretch between Varanasi and Hooghly, a project that has been criticised as

[9] Down to Earth (2015). 'Namami Gange', May 22. Retrieved from http://www.downtoearth.org.in/content/namami-gange.

[10] Ramachandran, S (2014). Clean up the Ganges. *The Diplomat*, Sep. 15. Retrieved from http://thediplomat.com/2014/09/cleaning-up-the-ganges/.

converting the Ganga into 16 large ponds (Ramachandran, 2014). The World Bank has appropriated 1 billion USD in financing to support a portfolio of investments in sewage networks, waste treatment and river-front development. Meanwhile, in many people's views, adding STPs is simply a repetition of past failures in GAP I and II, since many finished STPs are not operating at full capacity due to power shortages. According to a CPCB report in 2013, less than 60 percent of the installed STPs ran at their full capacity and 30 percent were not in operation at all (CSE, 2014). Another illustration of the narrow minded approach to solving the complex issue is the India–Australia Water Science and Technology Partnership set up in 2012, where the modeling framework used for the Murray-Darling basin is applied to the Ganga river basin. It "allows for evidence-based decision making in the regulation of a complex water market and the allocation of environmental flows according to climatic conditions" (Hanasz, 2014). Nevertheless, this time the emphasis has been put on technology, e.g. data collection and information processing, while the more fundamental issues of defining water rights and competing demand for water are still low on the agenda. Hence the data gathered is not able to help bring governance to a higher level.

The government has realised the importance of maintaining e-flow, but the supporting theoretical arguments and the empirical evidence are still being debated. Ambiguity over hydrology, use of water and questions surrounding storage infrastructure may also explain the barriers to formulating concrete policies on e-flow (Wu and Whittington, 2013). At the current stage, the argument centers on how much water should be left in the river for environmental needs, and there are primarily three viewpoints regarding this. For hydropower engineers, 10 percent of average runoff is believed to be enough for maintenance in the water body and will not hinder the huge benefit of energy generation (CSE, 2014). According to the Wildlife Institute of India (WII) however, 20 to 30 percent of river flow in different seasons should be kept for the ecosystem and biodiversity requirements. Last but not least, the 2013 Inter-Ministerial Group (IMG) report recommended an e-flow volume of 25 percent for eight months and 30 percent for four months (IMG Report, 2013), whilst the Center for Science and Environment (CSE) believes this is far from adequate and proposes 50 percent for mean season flow and 30 percent for rest time periods (CSE, 2014). The latter contends that the figure is mandatory particularly when taking both critical ecological functions and societal needs into consideration. In summary, no consensus has been obtained on the required e-flow quantity set for the Ganga, which further contributes to the lack of realistic regulatory action. It is thus understandable that against such a background, e-flow has been regarded as a secondary priority on NGRBA's agenda (IIT, 2011).

4. Controlling Diverse Water Use in the Yellow River Basin

4.1. *Basics of the Yellow River*

Originating from the Bayankala Mountain and pouring into the Bohai Gulf, the Yellow River is the second largest river in China with a total length of 5,464 km and a drainage area of 795,000 square km. It can be divided into three reaches featuring distinctive landforms. The upper reach, from its source to Hekouzhen in Inner Mongolia, wanders through the Qinghai–Tibetan Plateau, the Inner-Mongolia Plateau, the Hetao plain and drops 3,800 m in elevation (Wu *et al.*, 2004). The middle reach lies

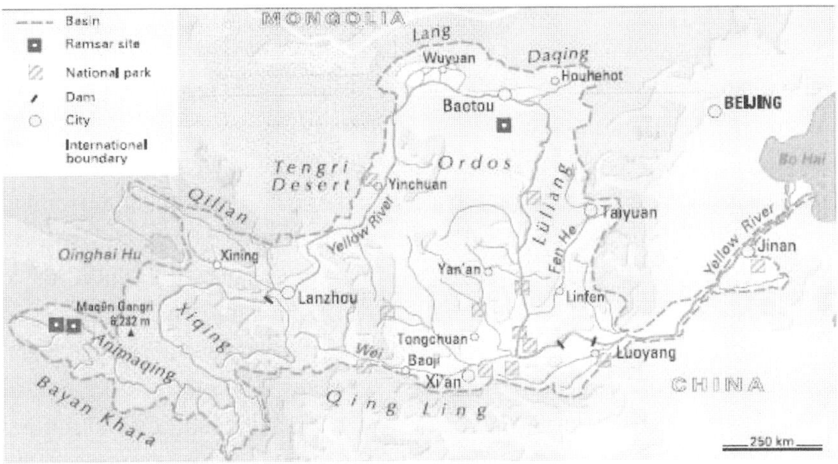

Map of The Yellow River Basin (*Source*: *UN World Water Development Report 3: Water in a Changing World, 2009*)

between Hekouzhen and Taohuayu in Henan Province. It passes the Loess Plateau and descends from 1,000 m to 100 m above sea level. The lower reach flows across the Huabei alluvial plain with geomorphic channel patterns transiting from being braided to meandering.

The Yellow River basin covers nine localities (seven provinces and two ethnic autonomous regions) and nurtures about 140 million people in China. It has long been regarded as the cradle of the Chinese civilisation. Different cultural relics dated back to the Stone Age and inscription-laden tortoise shells recording human activities have been widely found along the river. The Chinese nation emerged in the mid- and downstream area more than 4,000 years ago after the legendary ruler, the Yellow Emperor, united scattered tribes there. Many ancient dynasties established their capitals in the basin, making it the political, economic, technological and cultural center in China for a period of over 3,000 years. Although the economic power gradually shifted towards the Southern China after the Northern Song Dynasty ended, the Yellow River basin still enjoys an irreplaceable status in the nation's development course in terms of historical and contemporary contexts.

4.2. *Water availability vs usage: Contested issues and the negative impact*

Precipitation, snowmelt, river flow and groundwater storage make up the major sources of water availability in the Yellow River basin, measured on the basis of hydrologic records from 1956 to 1979 is 73.5 billion m^3 annually after taking feasible groundwater exploitation into consideration (Kusuda, 2010). However, since the majority of the river basin is situated in arid and semi-arid regions, it has only 2 percent of the nation's surface water resources. These water resources vary significantly in terms of space and time, with more seasonal rainfall and average annual runoff upstream and downstream. The annual runoff has been decreasing since the 1950s in particular, which was 25 percent lower than long-term mean estimation of 58 billion m^3 per year in the 1990s (Kusuda, 2010). In addition to natural variations, water withdrawal and flow control conducted by human society has seriously affected natural flow in the middle and lower reaches.

The overall water shortage caused by uneven geographical and inter-annual distribution leads to various problems, including soil erosion and sedimentation, flooding, competing extraction and water pollution. While sedimentation and flooding remain as a constant threat downstream, water scarcity exacerbated by an increasing imbalance between water demand and supply in recent decades has had a negative impact on the ecological environment and socioeconomic production in the entire river basin, endangering its sustainable development in the long run.

4.2.1. *Natural challenges*

The Yellow River flows through the vast Loess Plateau in the whole mid-stream and part of the upper reach, an area of 640,000 km^2 within which 439,000 km^2 of land suffers from severe soil erosion. It is estimated to carry an average sediment load of 1.6 billion tons annually, ranking first among all rivers worldwide (Wu *et al.*, 2004). Only about 25 percent of the sediment load is conveyed to the sea, the remaining 75 percent being deposited in the lower reach and the estuary. As a result, the riverbed in the downstream area has been rising at a rate of 5 to 10 cm per year, and flood control embankments have been raised accordingly. The perched river is in some places up to 10 metres higher than its surroundings so the region is thus under threat from breached levees and flooding.

Rainstorms from the midstream with high intensity and short duration used to bring floods to the lower reach. Floods have breached the down-stream embankment 1,590 times and caused a change in the course of the river 26 times from 602 BC to 1938 (Wu *et al.*, 2004). Although the flood has been tamed through the integration of structural and non-structural measures such as building detention and diversion projects and establishing a forecasting system, some tough issues need to be addressed in a critical manner. They include the lack of a solution to substantially reduce the sediment load, unstable braided river reach, a shrinking main channel, and a secondary perched river between the low inner embankment and the levees.

4.2.2. *Diverse socioeconomic use*

Water usage has become more diversified and intensified due to rapid urbanisation and socioeconomic development. First and foremost, a considerable amount of water is diverted to support agricultural irrigation.

The river basin has been a strategic granary in ensuring food security in China for years. For example, nearly 60 percent of the nation's wheat and 40 percent of its corn is produced from the downstream North China plain (Webber, *et al.*, 2008). To replenish the water moisture deficit in dry land, irrigation has been used for over a millennium and since 1949, it has been considerably expanded. Statistics show that the current irrigated area in the basin is 7.51 million ha, an increase of more than nine times between 1950 and 2000, and the total amount of diverted water for irrigation has also increased from 7.9 billion m^3 in 1949 to 30.8 billion m^3 in 1997 (Kusuda, 2010). In general, agricultural use accounts for 80 percent of water withdrawal in the basin.

Industrial production accounts for another big share. Along with swift industrialization, industrial water use in the upper and middle reaches increased 1.8 times (from 7.34 billion m^3 to 13.17 billion m^3) from the 1950s to 1990s, while in the more populated and economically developed lower reach, the figure rose to 5.7 times during the same period (Kusuda, 2010). In 1998–2000, 12 percent of the extracted water was utilised by the industrial sector. And according to China's strategic plan, four specialised economic centers have been envisaged to be built in the river basin, including hydropower and non-ferrous metal base near Lanzhou, a comprehensive economic and high-tech center in Xi'an, an energy industry cluster along the midstream, and a downstream base for petroleum production and marine exports. Industrial water demand is still increasing exponentially because of economic growth.

Domestic use also plays an important role in water consumption. Home to more than 50 large and medium-sized cities nowadays, the river basin has witnessed remarkable urbanisation during the past decades, with an urbanisation rate of about 26 percent in 2000. Not only do affluent urban citizens consume more water than before, but the income growth and improved water supply system have also enhanced the rural households' capacity to use more water. It is estimated that domestic water use increased from 2.49 billion m^3 in 1990 to 3.87 billion m^3 in 2001, about 8 percent of total water withdrawal (Kusuda, 2010). The rate is increasing at about 0.2 percent annually, almost identical to that of industrial water use.

4.2.3. *Fragmented governance and problem magnitude*

Before 2000, a fragmented administration system was a problem. It was vividly described by the Chinese as "multiple dragons managing the water". Nine ministries including the Water Resources, Environmental Protection, National Development and Reform Commission, Land and Resources, Agriculture, State Forestry Bureau, Public Health, Construction and Commerce, shared management authority and made different policies at the central level. Meanwhile, nine riparian localities were separately responsible for the river sections passing through their jurisdiction at the local level. Due to water scarcity especially during the dry season, localities in the upper and middle reaches usually over-extracted large amounts of water to sustain their socioeconomic needs.

The lack of coordination between upstream and downstream, coupled with less annual runoff and more competing water usage among varied sectors, resulted in a no-flow situation in the lower reach for the first time in 1972. Ever since then, no-flows have increased in frequency and became almost a seasonal event in the 1990s. The most serious incidence occurred in 1997, with a record-high drying-up length of 704 km and 295 days altogether. Hundreds of bird and fish species became extinct consequently. Needless to say, no-flows not only affected the eco-system but also caused a disastrous economic loss to agricultural and industrial production in the downstream region, which was measured at 21.64 billion RMB during 1990–1996 (Leong, 2011). In addition, the reduced water volume adversely affected the river's ability to dilute effluents and made pollution more acute. Five billion tons of wastewater was discharged into the river in 2000, which was 2.3 times of that in the 1980s (Kusuda, 2010). And 71 percent of the river's length (plus some main tributaries) was categorised as grade IV to V^+ in terms of water quality level in 1998, meaning highly polluted.[11]

[11]Yellow River Conservancy Commission (1998). Yellow River Water Resources Bulletin 1998. Note: The Ministry of Environmental Protection promulgated *The National Environmental Standards for Groundwater* in 2002, which divided groundwater resources in China into five quality categories, grades I, II, III, IV and V. I refers to good quality, II slightly polluted, III suitable for swimming, IV not applicable for direct body contact, and

4.3. *Water allocation and quota mechanism*

Critical water stress with consequent environmental deterioration and constraints on socioeconomic growth forced the Chinese government to shift its perception of the river from unlimited resources to an ecosystem that has intrinsic value and sustainable necessity in the end of the 1990s. After several years' exploratory practice, an integrated water resources management strategy was officially introduced in 2004 to fulfill the ultimate goal of "maintaining the health of the Yellow River" and meet four strict standards, i.e. no levee breaching, no cut-offs in the river flow, no pollution beyond criteria and no riverbed ascending. At its core is the water allocation and quota mechanism, which not only puts the maintenance of e-flow as top of the agenda, but manages diverse socioeconomic water usage through holistic approaches as well. To be more specific, the mechanism has been built upon three pillars, namely administrative reform, legal empowerment and policy innovation. Infrastructure and technology support have also been given great emphasis, to function as the backbone for the operation of the whole system.

4.3.1. *Administrative reform as guarantee*

YRCC was founded in 1949 under the leadership of the Ministry of Water Resources. Though nominally promulgated as a "basin-wide management institution", in its early years the commission did not have much authority to exert administrative control in the upper and middle reaches, nor was it able to resolve disputes among ministries and riparian provinces. During the same period, the ultimate goal of river management was identified by *the 1955 Report on The Comprehensive Planning of Disaster Elimination and Water Resources Development of The Yellow River* as eliminating floods and harnessing water resources through "storing water and retaining sediment in various sections, i.e. from plateau to valley and from mainstream to tributaries, taking unremitted efforts in advancing water utilisation for industrial, agricultural and transportation purposes, and

---

V meeting basic agricultural and normal landscape requirement. Water measured beyond V is considered to have no function for utilisation anymore.

conserving loess and rainwater in farmland" (State Council, 1955). In March 1999, however, the State Council granted YRCC the power to be in charge of water resource management and allocation in the whole basin. A centralised water allocation framework has been established with YRCC taking the leading role as in collaboration with other institutions: YRCC is authorized to make the annual water diversion master plan among riparian localities after consultation with their water authorities and reservoir management bureaus; then it facilitates the provincial water authorities on water allocation and supply to their subordinate cities, counties, towns and villages. Thus management at both basin-wide and local levels is organically integrated with all agents acting at its full play and complementing others as well. Apart from gaining autonomy and greater jurisdiction, YRCC also set up branch offices in all basin provinces to better coordinate the work on the ground (see Figure 3). For example, local teams are usually dispatched to do site inspections and spot checks on policy implementation during the water allocation period. There are more than 30,000 staff, engineers and technicians working for the commission and its affiliated organs.

4.3.2. *Legal empowerment as basis*

A water sharing system cannot be fulfilled without defining various water rights at the initial stage. In general, water rights in China include the right to own, manage and use water resources (Li, 2009). The Chinese Constitution and 1988 Water Law explicitly stipulates that the State owns all water resources in the national territory, and the Water Law further mandates that water administration authorities at different levels exercise management rights. Opposed to the two rights, water use right is not attributed to any specific entity in either the Civil Law or Water Law, and as a result, it can be divided according to regions, conditions, time and needs (Li, 2009). Therefore, an ecosystem of interplaying water rights built on their definitions is formed.

A series of decrees have been enacted to consolidate YRCC's authoritative role and provide guidelines to allocate water in an evolving manner. Among them, the milestone ones are: first, the 1987 Yellow River Water Allocation Scheme, which legalised the right of water withdrawal

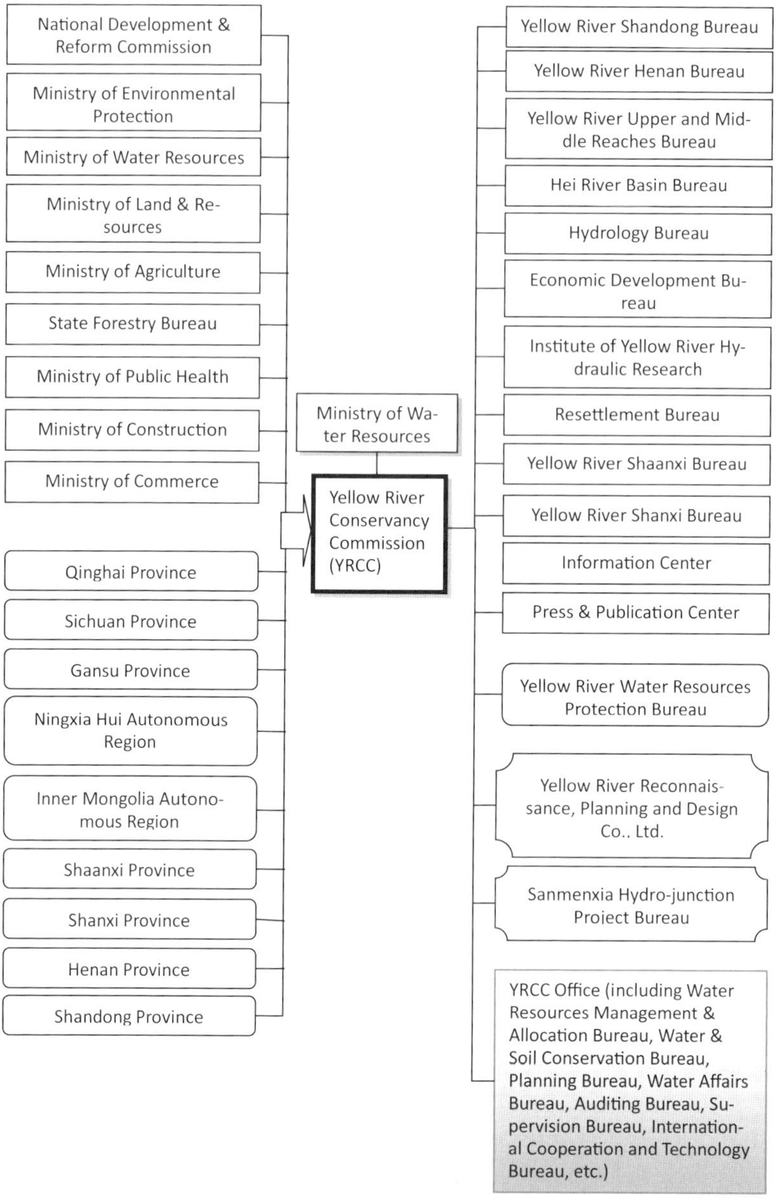

Figure 3: Administrative Reform of the Yellow River Basin Management and YRCC's Organization Chart

at the national level and divided the long-term average annual runoff of 58 billion m^3 into two portions — 37 billion m^3 for socioeconomic uses and 21 billion m^3 for sediment flushing. It was the first time that the river's imperative need for sustainability was taken into account. The figure of 21 billion m^3 has become widely recognised as the necessary environmental flow for the Yellow River (Liu *et al.*, 2008). The second milestone was the 1988 Water Law which stipulated a Water Withdrawal Permit (WWP) system. YRCC was authorised to administer the WWP and given total amount control in riparian provinces. The law also offered a basis for local legislatures to make related water regulations and rules accordingly. But it only specified the water use quota each year while ignoring the uncertainty around inter- and intra-annual natural variations. The third milestone is the 1998 Administrative Measures for Yellow River Water Diversion which addressed these concerns. It not only added set the monthly quota within the annual quantum, but also established the important principle of increasing and reducing a fixed proportion of allocation to all water users in high and low flow years respectively. The principle is also applied to the volume of environmental flow. The fourth milestone is the 2006 Yellow River Water Diversion Bill which marked the first official regulation in enforcing water allocation at the basin scale and strict compliance by riparian provinces, whose governors bear responsibility for possible violations.

4.3.2. *Policy innovation as instrument*

On the basis of total amount control and the WWP system, policies applying market instruments were piloted to promote water use efficiency and realize trade-offs across sectors and regions. The most noteworthy one in this regard is the water rights trading policy, which aims to tackle the challenges of limited water supply, inefficient water utilisation in irrigation, and increasing demand from industries together. The background is as follows: making up 80 percent of total water withdrawal, agricultural irrigation often suffers from poor efficiency due to out-of-date infrastructure and methods. In the meantime, provinces like Ningxia and Inner Mongolia boast rich energy resources including coal, petroleum and natural gas and thus have committed to constructing energy bases to be necessarily supported by water from the Yellow River. However, the water quota limit and non-transferrable license restricted further development of those large-scale energy programs. To cope with this, YRCC devised a scheme for allowing eligible industries to invest in irrigation infrastructure in exchange for the use of saved water from farmers. Many detailed regulations and rules were issued to define the implementation process: initially, applications for water rights exchange should be approved by YRCC; then provincial authorities would arrange for the two parties to sign a contract and put the real transfer into place; meanwhile, YRCC appropriates a related fund, organises an open tender for the construction of those transfer projects, sets evaluation criteria and supervises the implementation; performance monitoring and assessment reports have to be submitted by project participants to YRCC. Pilot projects were first launched in Ningxia and Inner Mongolia in 2003, where five enterprises including Ningxia Daba Power Plant and Inner Mongolia Dalad Power Plant were paired with groups of water-saving farmers in Ningxia Qingtongxia Irrigation Region and the Inner Mongolia South River Bank Irrigation Region respectively. By the end of 2008, the number participating enterprises had increased to 26 entities in the two localities, with a total investment of 1.226 billion RMB. Saved water from the agricultural sector amounted to 257 million m^3 and 228 million m^3 of this water was transferred over. (YRCC, 2008). As for project participants, it was a win-win outcome since farmers spent less on irrigation and enjoyed increased yields, while industries increased their water quota and sustained profitable production.

Economic incentives have also been used in reforming the water pricing system and initiating basin ecological compensation mechanisms. The former, water pricing reform, is conducted to balance water supply cost and usage charge. In urban areas, for example, progressive water pricing methods have been tested in Nanyang, Henan Province, since 2012 and have achieved positive results. Justified rational water rate is adopted in part of the lower reach to mitigate water waste in irrigation. A high price and/or progressive rate for excessive water use is employed to effectively combat waste water discharge. The basin ecological compensation mechanism, on the other hand, economises environmental value by rewarding those who protect and having those who benefit pay in return (Li, 2009). Responsible for providing sufficient clean water to the downstream, provinces like Qinghai located in the upstream usually sacrifice certain development opportunity to make room for the lower reach. Although Qinghai has access to more than half of the annual runoff, it was only allocated with 1.41 billion m^3 water according to the 1987 quota scheme and the water demand deficit by 2020 will amount to 1 billion m^3.[12] Under such circumstances, the lower reach localities need to pay an eco-compensation tax to a central compensation fund and the central government will redistribute the fund to upstream regions, offsetting their contribution cost and encouraging further protection efforts.

4.3.4. *Infrastructure and technology support*

Water use control and allocation is built on infrastructure and technology support. In terms of infrastructure, the development of reservoirs plays a crucial role in adjusting the seasonal flow and retaining water resources for usage in dry periods. In 2012, there were 196 large and medium-sized reservoirs along the river with a water storage volume of 40.2 billion m^3. The overall storage capacity of all hydro facilities in the basin actually hit 70 billion m^3 in 2005, far exceeding the mean annual runoff of 58 billion m^3. Other facilities such as sluices, canals and gauging stations have also been built to make up a comprehensive water allocation and management network.

More importantly, Digital Yellow River, a concept advocated by the leadership of YRCC since the beginning of this century, has been developed to provide solid support for scientific decision-making and practical implementation. It includes the construction of six application systems, i.e. flood control, water allocation, water resources monitoring and protection, water and soil conservation, hydraulic engineering administration and e-governance. New techniques such as remote sensing (RS), distributed control system (DCS), global positioning system (GPS)

[12] Economic Observer (2013). Entangled Interests of Water Allocation in the Yellow River Basin, Jan. 5. Retrieved from http://www.eeo.com.cn/2013/0105/238352.shtml.

and geographic information system (GIS) have been widely deployed in collecting basin-wide information, processing data, simulating models and monitoring operation in real time. Through this technical platform, the Yellow River Water Allocation Center, for example, is capable of controlling remote sluices located 1,000 km away. Also, when emergencies like low water flow are forecasted by the alert system, it may react accordingly by shutting down some upstream water-diverting gates so as to regain normal flow and prevent cut-offs in advance.

4.4. *Policy outcome and practical challenges*

Through ration control and unified allocation of water resources in the whole basin, the basic demand on socioeconomic water usage has been met and the no-flow phenomenon in the lower reach has been eliminated since 1999. Water quality in the main stream has been steadily improved since 2005, with the proportion of the river length measured at grades I to III rising from 45.8 percent in 2005 to 88.2 percent in 2012 whilst water of grades IV to V decreased from 51.5 percent to 11.8 percent in the same period (YRCC, 2012). No V^+ water has been detected since 2009. The ecosystem in the Yellow River delta has witnessed obvious progress, as both plant and bird species have increased in quantity.

The water allocation and quota mechanism is in essence an issue of interest redistribution. Its complexity inevitably ushers in many challenges in implementation. The most difficult task is to make a widely acceptable annual water allocation master plan and have all riparian provinces comply with it. Tense conflicts often arise during the consultation stage that requires central government intervention. In 2002, for example, downstream Shandong Province acquired an additional 0.8 billion m^3 water from the State Council at the expense of reduced water allocation from two upstream provinces after lobbying (Xia and Pahl-Wostl, 2012). Besides problems such as conflicts between regulations issued by YRCC and local governments, heavy dependence on government and industry administration, the lack of a channel for information dissemination and public participation, and weak capabilities in enforcement and supervision also combine to hinder the sound operation of the mechanism.

5. Comparison of The Ganga and The Yellow River

After examining the evolving water governance discourses in the two river basins mainly through perspectives of e-flow and water sharing, this section will focus on the institutional framework, enabling environment and policy instruments further, comparing the similarities and differences against the three IWRM pillars so as to better facilitate the formulation of insightful lessons later.

5.1. *Institutional framework*

YRCC is the sole river basin organisation for the Yellow River, while its counterpart for the Ganga is in stratified form, with NGRBA being the governance body and NMCG as the implementation agency. Both parties have adopted an integrated river management strategy and recognized e-flow and water sharing issues in managing the river basin, but the administrative focus and capacity in terms of the two aspects (what are the two aspects?) still vary to a large extent due to complex reasons.

On one hand, path dependency has led to divergence on current work objectives. According to NGRBA's Objectives, Approach and Functions, announced in 2009, the holistic approach it adopts to manage water in the river basin will "take into account the competing demands on water and seek to ensure minimum ecological flows" (CPCB, 2009). It reflects that the authority has realised the importance of maintaining e-flow and admitted competing water extraction has great influence on water flow. Nevertheless, the majority of the work done by agencies under the oversight of NGRBA such as NMCG, CPCB, CGA (later as NRCA) and GPD (later as NRCD) until now has been focused on pollution abatement. The NGRBA's function is defined as "...regulation of activities aimed at prevention, control and abatement of pollution, to maintain water quality and take measures relevant to the river ecology in the Ganga basin states" (NMCG, 2015), which implies that e-flow is more or less the secondary concern when compared with control of pollution. In the meantime, unlike the already matured understanding on mitigating sources of pollution, there are still ongoing debates on how to augment e-flow in the Ganga. For instance, which proposed e-flow scenario is more feasible?

Is it 25 percent of river flow for eight months and 30 percent for four months as proposed by IMG or 50 percent for six months and 30 percent for another six months as championed by CSE (CSE, 2014). Hence the issue of e-flow is still at the stage of agenda setting for NGRBA and requires more research to help turn it from ideas and rhetoric into real governance actions. YRCC has also evolved as an administrative body. In the early years, its mission dwelled largely on combating floods and harnessing water resources for national development. But the consequent environmental deterioration as typically manifested by no-flows in the lower reach from 1972 to 1998 shaped the consensus that the intrinsic value of the river should be respected and safeguarded over human needs. In line with this, the 1987 Yellow River Water Allocation Scheme set explicit amounts for e-flow and socioeconomic uses in the long-term average annual runoff. And after several years of operation, YRCC was authorised to manage basin-wide water sharing in 1999. Nowadays, its functions has become quite extensive, except for pollution abatement and water conservation, which include ensuring the rational exploitation and utilisation of water, managing and supervising water resources, coordinating water for livelihood, production and ecological purposes, protecting water resources in the river basin, preventing flood and drought, and instructing hydrology projects. (YRCC, 2015). Although its role has been enhanced, its core is still focused on maintaining the sound flow of the Yellow River.

Varied institutional design significantly affects the capacity of policy coordination and implementation. Huge differences in power distribution can be observed between the central and local governance bodies, though both NGRBA/NMCG and YRCC have constructed a network to exercise their functions. In general, we define the model in the Indian context as "partial devolution", with NGRBA having authority to draw up the master plan, appropriate a budget, and approve and monitor projects whereas the states have a wider discretion in executing real programs. Much of NMCG's work has to seek support from the five riparian State Ganga River Conservation Authorities (SGRCAs) chaired by their Chief Ministers respectively. However, due to their often competing demand for water, when states demonstrate "moral hazard" (e.g. develop overlapping water schemes, over-extract water at the expense of others' interests)

NGRBA/NMCG does not appear to have the power to control this behaviour. As a matter of fact, since its inception, NGRBA has only had three meetings on inter-state coordination with the latest one held in April 2014, and almost all these meetings have ended in heated debate. The Chinese governance structure, is in essence a centralised framework with YRCC taking the leading role and interacting with provincial water authorities proactively. As best illustrated in the water allocation and quota mechanism, YRCC not only takes the responsibility of making master plan based on consultation with all involved actors but assists provinces to allocate water to their subordinates as well. Localities are encouraged to fully exercise their governance competence, but whenever their interests are in conflict, it is imperative for them to comply with adjustment and arbitration mandated by YRCC. In this sense, management at river basin and regional levels is aligned in an organic manner. Furthermore, YRCC's pervasive authority on coordination and implementation is enhanced through the branch offices set up in each riparian province. Those offices play a decisive role in routinely liaising with local water bureaus, and also conducting much practical work by themselves to put programs envisaged by YRCC into place.

5.2. *Enabling environment*

Although both India and China have developed a series of decrees to support water governance, there is still a significant difference in a number of aspects including legislative systems, definition of water rights and legal empowerment of river basin organisations. Firstly, varied legal basis has led to diverging legislative mechanisms and outcomes. In India, since the entitlement of water body is determined by adjacent land ownership, riparian states naturally own river sections within their boundaries and hence the river becomes a state subject. Based on this fact, states are entitled to enact their respective water laws and regulations. As the central government does not have ownership of rivers, that deprives it of the right to legislate water law at the national level. Consequently, conflicts arise among states without an umbrella regulation framework to prevent and mitigate them. The 1956 Inter-State River Water Disputes Act was passed with the goal of helping solve inter-state river disputes, but it has been only applied to other river basins for a couple of times, not yet to the Ganga. Chinese legislation on water governance is a centralized model featuring a top-down approach. The 1988 Water Law marked the official establishment of a holistic legal system for water resources management. Within it, the Water Law firmly holds the guiding role, and other rules such as the Water Pollution Abatement Law, Flood Prevention Law, Water and Soil Conservation law and so on have been gradually enacted to complement it in more specific fields. Provincial legislatures are mandated to draft and enforce water regulations based on guidelines set by the Water Law. In this way, water rules and regulations at the national and local levels are framed into a unified structure, which helps to avoid potential inconsistency and contradiction.

Secondly, a gap in the definition of water rights can be observed in the two countries. All in all, there is no clear separation between ownership right and use right for water resources in India. While states legally own sections of the river flowing through their areas, use right of water has not been well defined in any official document and thus it becomes the default for people to extract water without limit. It also results in the prevailing free use of water nationwide, particularly in the vast rural areas, which has virtually imposed insurmountable obstacles to the installation of any water pricing system. Compared with India, China has made an explicit distinction among rights to own, manage and use water resources. It is stipulated in the Water Law that "water resources are owned by the State… The State encourages units and individuals to exploit and use water resources in accordance with law, and protects their legitimate rights and interests… For water resources, the State applies the systems of Water Withdrawal Permit and Pay-for-use…" (Water Law, 1988, 2002) Also in it, water governance organs at various levels are authorised to exercise management rights. Hence an ecosystem is built on the definitions: while the state maintains the supreme right of all water resources, water administration authorities, e.g. YRCC and local water bureaus, shall manage the possession, use and disposal of water resources by various users within the legal framework. More importantly, it laid a solid foundation for the establishment of the water sharing mechanism, where all involved actors agree that valuable water resources are not free of charge and there ought to be control on competing demands in place.

Thirdly, legal empowerment on river basin organisations especially in terms of e-flow and water sharing varies significantly. YRCC's powers have evolved over time. The 1987 Yellow River Water Allocation Scheme first set the baseline of quantifying separate water volume for e-flow and human needs, then the 1988 Water Law authorized YRCC to manage total amount control and WWP on socioeconomic uses, and subsequent regulations further addressed issues of natural variations and collaboration with localities. Hence, viewed retrospectively, the underlying logic in YRCC's administration is to safeguard the natural livelihood of the river first and then manage the share for human development needs through measures concentrating on allocation, implementation, adjustment and enforcement. In comparison, NGRBA in India is relatively young, with only a history of six years. The concept of e-flow was put on the agenda from the time of the establishment of the NGRBA in 2009 but compared to pollution abatement, e-flow lacked explicit identification and methods in legal documents. Statements on water sharing have also been rare. Because of this, it can be concluded that there is no legal support for NGRBA/NMCG to realize policies to achieve e-flow and water sharing in the Ganga river basin so far.

5.3. *Policy instruments*

Due to the varying administrative focus and legal framework, control of pollution still dominates NGRBA's policy agenda whereas YRCC targets its attention on e-flow and water sharing. On the Ganga side, both the government and public in general have taken pollution as the most serious focal problem in the Ganga clean-up initiative for decades. Almost all fiscal resources and manpower were spent in designing and implementing projects that handled pollution, e.g. building STPs and sewage networks, upgrading hygiene facilities in urban and rural regions, etc.. With the establishment of NGRBA, maintaining minimum e-flow has drawn increasing attention. The idea to manage competing demand for water, a premise that will lead to water sharing, has emerged in some official documents. But they are both secondary priorities in realistic policy formulation. As showcased in the latest *Namami Gange* Program, 21 comprehensive visions covering seven thrusts, i.e. wholesome river, clean river, river front development, capacity building, research and monitoring, protection of aquatic flora and fauna, and awareness creation, are proposed by NGRBA (NMCG, 2015). In this, the focus is still largely on pollution abatement. Determining and maintaining e-flow is mentioned, and no policy can be found on water sharing.

On the Yellow River side, both e-flow and water-sharing are not only anchored in policy agenda but also implemented on the ground. The operation of the water allocation and quota mechanism complements and safeguards the maintenance of e-flow as the top priority among all governance work. Since water sharing involves interest redistribution, YRCC has applied regulatory and economic approaches to deal with disputes among localities and sectors. For the regulatory method, YRCC is responsible for making the annual water allocation master plan based on inclusive consultation with riparian provinces and reservoir management bureaus. Water users are granted WWP from YRCC after the master plan is set, in turn, supervise the extraction of water and punishes possible violations. As for economic tools, water rights trading, ecological compensation and progressive water pricing systems have been piloted to promote efficiency and ensure equity in water utilisation. For example, water rights trading helps transfer surplus water from irrigation to industry where water use is in deficit; ecological compensation policy assists to realize the trade-off between economic development and environmental protection.

The Yellow River case implicitly reveals an interplay between e-flow and water sharing. Both are indispensable components of IWRM, water sharing is employed to control competing demand for water so as to ensure sound water flow in the river body, whilst the continuous and abundant water volume conversely consolidates the operational basis for water sharing. Mere emphasis on e-flow will not provide a concrete and sustainable solution to complex problems that river basins normally face, as the unconstrained extraction of water resources from different regions and sectors will devastate the quantity and quality of water body in the long run. Since the vast middle and lower reaches of the Yellow River have been witnessing severe water scarcity for decades, e-flow and water sharing have been assigned top priority on the policy agenda for some years. But in the Ganga situation, over-extraction of water appears to be more acute in the first two riparian states especially during the dry season and the river on the whole is still considered by many to have enough water. This might be part of the reason to explain the insufficient attention from the government and general public on e-flow and water sharing issues so far.

6. Policy Lessons from The Yellow River

The Yellow River case provides useful inputs for the Ganga and its river basin authority on its rejuvenation program. Based on our comparative analysis, the following governance lessons in fields of administration, legislation and policy can be applied to the river Ganga at its current developmental stage.

Firstly, build consensus on e-flow in parallel with pollution abatement on NGRBA's policy agenda and align the collaboration between NGRBA/NMCG and states in a more integrated way. A group of Indian scholars were commissioned to produce a report on e-flow, which nonetheless has not been adopted by the government. To generate a widely accepted research outcome, an inclusive committee comprising of experts from various backgrounds, e.g. specialists from home and abroad, international organisations and local practitioners, should be formed. We assume it may take the committee at least one year to complete a convincing report with sufficient evidence from measuring stations along the Ganga. Once the ideal quantities of e-flow at varied sections and seasons are defined, NGRBA/NMCG should pass it on to all relevant agencies and make it become the clear administrative focus in their governance work. Meanwhile, water commissions under the direct leadership of NGRBA/NMCG should be established in each riparian state, which should work closely with local water authorities and implement programs initiated by NMCG on the ground. In this way, a vertical governance network inside NMCG apart from its semi-horizontal interactions with states will be created.

Secondly, set up an umbrella water governance legal framework in the short run and explore the possibility of legislating a national Water Law in the long run. At the very beginning, a debate on water rights needs to be launched to promote people's awareness of ownership right and use right of water. It is hoped that a clear definition of different water rights can be generated from this process. Based on that, an umbrella water governance legal framework at the national level should be established, through which NGRBA/NMCG should be empowered with more autonomy and jurisdiction in policy making, implementation, enforcement and supervision. The key is to have the scientifically proven quantity of e-flow inked into real articles, as this will provide the crucial regulatory basis for NGRBA/ NMCG to ensure e-flow and exercise water sharing afterwards. Ultimately, it would be imperative to enact a national Water Law that will not only integrate issues of water rights, e-flow and water sharing in an organic way, but align the often conflicting state legislation into the umbrella framework as well. States still maintain independent legislative powers, but they also have responsibility to comply with the national Water Law.

Thirdly, set an explicit amount control on e-flow and human socio-economic use respectively and launch a pilot experiment of water allocation system. Should e-flow statistics produced by expert committee be acceptable to all, NMCG should start to formulate a practical plan, e.g. a water release process from existing dams and barrages, to ensure the maintenance of e-flow. In the meantime, a basin-wide water sharing mechanism should also be taken into consideration. To accumulate more empirical data, it is suggested that pilot experiments be first launched in the two riparian states — Uttarakhand and Uttar Pradesh — where there is more competing demand for water and greater pollution in the water body. Initially, water users from various sectors should be gathered together to be consulted and to determine how much water they can extract from the Ganga. The State water authority should take the leading role in this process, and make a cross-sector water allocation scheme to be operated within their jurisdiction. NGRBA/NMCG should collaborate closely with the two states particularly in aspects of policy formulation, enforcement and supervision. NMCG should then apply its experience to draft an inter-state water sharing program for operation in the entire basin.

6. Conclusion

This chapter is a departure from the dominant attempt to reduce pollution in the Ganges — instead, it makes an effort to highlight another underlying root cause of pollution against the aforementioned background, i.e. lack of natural flow in the Ganga particularly in the middle sections, which are also, not coincidentally, one of the most polluted parts. A healthy water flow helps sustain the river ecosystems and wash away pollutants — it is, however, profoundly influenced by mankind's withdrawal of water. Water extraction for irrigation, hydropower generation and settlement needs represent the competing demand for water in the Ganga. Managing competing demand for water is a more substantial challenge than control of pollution since there is a need to resolve socioeconomic trade-offs among various water users and hence requires a more consensus from the river's stakeholders. The often repeated belief that the Holy Ganga has an abundance of water is also a source of cultural and political inertia to address inadequate flow. The significance of this chapter, therefore, is to identify and advocate necessary steps on the governance side towards building a consensus that may ultimately contribute to resolution of the trade-offs through operating a water sharing program.

From this perspective, the experience from the rejuvenation of the Yellow River offers meaningful insights for the Indian government as it embarks on the *Namami Gange* program. The Chinese government instituted the key concepts and three pillars of the IWRM framework in an attempt to revitalise the Yellow River in the late 1980s but achieved more success only at the beginning of this century, following a series of legal and administrative reforms. The reforms enabled YRCC to put basin-wide policies in place so as to achieve e-flow and water sharing, which are equally important in terms of managing the competing demand for water in the river. Using a comparative approach, this chapter has identified several gaps between the strategies of NGRBA and YRCCC, the most salient being the absence of an enabling environment that allows effective policies to be implemented by NGRBA in India. This includes the necessary legislation that (1) differentiates ownership right and use right of water in the river body and (2) strengthen NGRBA's authority and capacity to secure cooperation from all relevant agencies and enforce policies across states.

While it may take years to establish the scientific foundation for e-flow in the Ganga, build political and societal consensus on water rights, and have riparian states agree with the allocation of water during lean times, it is worth pointing out that a stronger political will has been demonstrated since the Modi administration took office last year. There is a supportive civil society that proactively advocates and participates in the Ganga clean-up initiative. Both these factors provide a greater impetus for the whole rejuvenation program. It is believed that as a potential viable option, sharing of water may help policy-makers in India devise a road map to a more sustainable and healthy flow in the Ganga.

References

Bansal, S (2014). National River Linking Project: Dream or Disaster? *India Water Portal*, September 14. Retrieved from http://www.indiawaterportal.org/articles/national-river-linking-project-dream-or-disaster.

Barnett, JM Webber, M Wang, B Finlayson and D Dickinson (2006). Ten Key Questions about the Management of Water in the Yellow River Basin. *Environmental Management*, 38(2), 179–188.

Bhowmik, NG (2009). River Basin Management: An Integrated Approach. *Water International*, Rutledge, 23(2), 84–90.

Biswas, A (2004). Integrated Water Resources Management: A Reassessment. *Water International*, 29(2), 248–256.

Biswas, A and M Mukherjee (2014). Ganga Clean Up a Herculean Task. *The Business Times*, September 11. Retrieved from http://lkyspp.nus.edu.sg/iwp/wp-content/uploads/sites/3/2014/09/Ganga-cleanup-BT.pdf.

Cai, X and M Rosegrant (2004). Optional Water Development Strategies for the Yellow River Basin: Balancing Agricultural and Ecological Water Demands. *Water Resources Research*, 40(8). doi:10.1029/200,WR002488.

Center for Science and Environment (2014). Ganga: The River, its Pollution and What We can do to Clean it. Available at mission-ganga.thewaternetworks.com

Central Pollution Control Board (2009). NGRBA: Objectives, Approach and Functions. Retrieved from http://cpcb.nic.in/ngrba/about.html.

Central Pollution Control Board (2013). Pollution Assessment: River Ganga. cpcb.nic.in/upload/NewItems/NewItem_203_Ganga.report.pdf.

Chaudhary, J (2014). India's Plan to Clean up the Ganga River is Flawed, say Experts. *China Dialogue*, October. 16. Retrieved from https://www.chinadialogue.net/article/show/single/en/7395-India-s-plan-to-clean-up-the-Ganga-River-is-flawed-say-experts.

Chinese State Council (1955). Report on The Comprehensive Planning of Disaster Elimination and Water Resources Development of The Yellow River, July 18. Retrieved from http://www.gov.cn/test/2008-03/06/content_911471.htm.

Cullet, P (2007). Water Law in India: Overview of Existing Framework and Proposed Reforms. International Environmental Law Research Center (IELRC), Working Paper, Geneva.

Cullet, P and J Gupta (2009). India: Evolution of Water law and Policy. In *The Evolution of the Law and Politics of Water*, pp. 159–175. Dordrecht: Springer Academic Publishers.

Das, G (1984). The Ganga Basin, Part II The Ganga Basin. Center for study of Man and Environment, Central Board for The Prevention and Control of Water Pollution, New Delhi.

Das, S (2011). Cleaning of the Ganga. *Journal of the Geological Society of India*, 78(2), p. 124.

Davis, MD (2007). Integrated Water Resource Management and Water Sharing. *Journal of Water Resources Planning and Management*, 133(5), p. 427.

Dépraz, M, Threat of Water Extraction on the Ganges. *World Wide Fund for Nature*. Retrieved from http://wwf.panda.org/about_our_earth/about_freshwater/freshwater_problems/river_decline/10_rivers_risk/ganges/ganges_threats/.

Down to Earth (2015). '*Namami Gange*', May 22. Retrieved from http://www.downtoearth.org.in/content/namami-gange.

Economic Observer (2013). Entangled Interests of Water Allocation in the Yellow River Basin, January. 5. Retrieved from http://www.eeo.com.cn/2013/0105/238352.shtml.

Gosain, KA (2014). Presentation: Cleaning up the Ganga and the Role of An Independent Scientific Voice. India Institute of Technology. Presented at SOURCE 2014, 18–19 September, Canberra.

Gupta, D (2008). Implication of Environmental Flows in River Basin Management. *Physics and Chemistry of the Earth*, 33(5), 298–303.

Gupta, J (2012). Unquiet Flows the Ganga. *Thethirdpole.Com*, June 15. Retrieved from http://www.thethirdpole.net/unquiet-flows-the-ganga/.

Hanasz, P (2014). Can Ganga Governance Problems be Engineered Away?. *Australia India Institute*, September 30. Retrieved from http://www.aii.unimelb.edu.au/blog/can-ganga-governance-problems-be-engineered-away.

Hooper, BP (2005). *Integrated River Basin Governance: Learning from International Experience*. London: IWA Publishing.

Huang, R (2013). Study on Administrative and Legal Management of Water Resources in the Yellow River Basin. China Northwest Agriculture and Forestry University.

Huntjens, P, C Pahl-Wostl, B Rihoux, M Schlüter, Z Flachner, S Neto, R Koskova, C Dickens and IN Kiti (2011). Adaptive Water Management and Policy Learning in a Changing Climate: A Formal Comparative Analysis of Eight Water Management Regimes in Europe, Africa and Asia. *Environmental Policy and Governance*, 21(3), 145–163.

IANS (2014). Need states' cooperation for a clean Ganga: Center to SC. *Mathrubhumi.Com*, September 14. Retrieved from http://www.mathrubhumi.com/english/news/latest-news/need-states-cooperation-for-a-clean-ganga-centre-to-sc-152246.html.

Indian Institute of Technology (2011). SWOT Analysis of Ganga Action Plan. GRB EMP: Ganga River Basin Environment Management Plan. Available at http://nmcg.nic.in/writereaddata/fileupload/50_006GEN.pdf . Accessed on 1 May 2015.

Indian Institute of Technology (2011). Environmental Flows: State-of-the-Art with special reference to Rivers in the Ganga River Basin. GRB EMP: Ganga River Basin Environment Management Plan. Available at http://gangapedia.in/?q=content/environmental-flows-state-art-special-reference-rivers-ganga-river-basin . Accessed on 8 May 2017.

Indian Institute of Technology (2010). Report on River Ganga at a Glance: Identification of Issues and Priority Actions for Restoration. Available at Ministry of Environment, Forestry, and Climate Change website http://www.moef.nic.in/sites/default/files/ngrba/001_GEN_DAT_01.pdf. Accessed on 8 May 2017.

Indian Institute of Technology (2011). SWOT Analysis of Ganga Action Plan. GRB EMP: Ganga River Basin Environment Management Plan. Available at http://nmcg.nic.in/writereaddata/fileupload/50_006GEN.pdf . Accessed on 1 May 2015.

Inter-Ministerial Group (2013). IMG Report on Issues Relating to River Ganga. Government of India.

Jeuland, M, H Nagaraja, E Jorge, D Blackmore and C Sadoff (2013). Implications of Climate Change For Water Resources Development in The Ganges Basin. *Water Policy*, 15(1), 26.

Jones, R, J Crean, P Aluwihare, and R Letcher (). Economic Cost of Environmental Flows in an Unregulated River System. *The Australian Journal of Agricultural and Resource Economics*, 51, 305–321.

Kaushal, N (2014). Environmental Flows for River Ganga. In *Our National River Ganga: Lifeline of Millions*, 223, Cham: Springer.

Kusuda, T (2010). *The Yellow River: Water and Life*. Singapore: World Scientific Publishing.

Leong, C (2011). Dragons Managing Water: Integrated Water Resources Management in the Yellow River. National University of Singapore.

Leong, C and M Mukherjee (2015). Managing the Socio-Ecology of Very Large Rivers: Collective Choice Rules in IWRM Narratives. *Global Environmental Change* (upcoming publication).

Li, Guoying (2005). Maintaining the Healthy Life of the Yellow River. Zhengzhou: Yellow River Conservancy Press.

Li, Guoying (2009). Questions and Answers about the Yellow River. Zhengzhou: Yellow River Conservancy Press.

Li, X (2011). Study on Measures to Enhance Integrated Water Resources Management of Yellow River. China Water Resources, (7): 021. Available at http://en.cnki.com.cn/Article_en/CJFDTOTAL-SLZG201107021.htm. Accessed on 1 May 2016.

Liu, X, Y Lian, J Huang, S Ke, J Zhang and X Wang (2008). Environmental Flows of the Yellow River. *Science and Technology Review*, 26(17).

McDermott, M (2011). River Ganga Running Dry — Climate Change, Hydropower, Unsustainable Water Use To Blame. *TreeHugger.Com*, June 9. Retrieved from http://www.treehugger.com/natural-sciences/river-ganga-running-dry-climate-change-hydropower-unsustainable-water-use-to-blame.html.

Mijic, A (Water Resources in India — the Story of Ganges Basin). Imperial College London. Retrieved from https://www.imperial.ac.uk/grantham/our-work/impacts-and-adaptation/ipcc-working-group-ii/water-security-and-flood-risk/the-ganges-river-basin-in-india/.

Ministry of Environment and Forests (2015). Function and Power of NGRBA. Retrieved from http://www.moef.nic.in/sites/default/files/ngrba/ngrbaread.html.

National Mission for Clean Ganga (2015). Function and Power of NRGBA. Retrieved from https://nmcg.nic.in/ngrbaread.aspx.

National Mission for Clean Ganga (2015). PPT — World Bank Assisted National Ganga River Basin Project and 'Namami Gange Program. Ministry of Water Resources, River Development and Ganga Rejuvenation. Retrieved from http://wrmin.nic.in/forms/list.aspx?lid=1262andId=4.

Ramachandran, S (2014). Clean up the Ganges. *The Diplomat*, September 15. Retrieved from http://thediplomat.com/2014/09/cleaning-up-the-ganges/.

Richter, BD, MM Davis, C Apse and C Konrad (2012). A Presumptive Standard for Environmental Flow Protection. *River Research and Applications*, 28(8).

Siddiqui, MS (2008). Water Policies and Legal Framework in India, 569–585. International Water Management Institute. Available at publications.iwmi.org/pdf/H042921.pdf. Accessed on 1 April 2016.

Singh, B (2013). Sufficient Waterflow can Help Solve Ganga's Pollution Woes. *The Times of India*, October 26. Retrieved from http://timesofindia.indiatimes.com/home/environment/pollution/Sufficient-waterflow-can-help-solve-Gangas-pollution-woes/articleshow/24718097.cms.

Sun, G and Y Pei (2004). Practices and Outlook on Integrated Management and Allocation of Water Resources in the Yellow River. *Yellow River*, 26(5).

The World Bank (2014), 'GDP growth'. Retrieved from http://data.worldbank.org/indicator/NY.GDP.MKTP.KD.ZG.

Trivedi, RC (2014). Water Quality Challenges in Ganga Basin, India. In *Our National River Ganga: Lifeline of Millions*, pp. 189. Cham: Springer.

Tripathi, BD and S Tripathi (2014). Issues and Challenges in River Ganga. In *Our National River Ganga: Lifeline of Millions*, pp. 211. Cham: Springer.

Uncer, O (2008). Global Governance of Water: A Practitioner's Perspective. *Global Governance*, 14(4) 409–417.

Wang, X (2011). Comparative Study on Watershed Management between Foreign Countries and China. *Environmental Science Survey*: 003.

Water Law of the People's Republic of China (1988, 2002). Articles 3, 6 and 7.

Webber, M, J Barnett, M Wang, B Finlayson and D Dickinson (2008), The Yellow River in Transition. *Environmental Science and Policy*, 11(5), 422–429.

World Meters (2014). Countries in the World (ranked by 2014 population). Retrieved from http://www.worldometers.info/world-population/population-by-country/.

Wu, B, Z Wang and C Li (2004). Yellow River Basin Management and Current Issues. *Journal of Geographical Sciences*, 14(1), pp. 29–37.

Wu, X and D Whittington (2013). Water Resource Development on the Ganges: Moving Beyond Ambiquity. *Water Policy,* 15(1), 1–8.

Xia, C and C Pahl-Wostl (2012). The Development of Water Allocation Management in the Yellow River Basin. *Water Resources Management*, 26: 3395.

Xinhua News (2003). China will construct the 'Three Yellow Rivers' in An All-round Way. January 6. Retrieved from http://news.xinhuanet.com/fortune/2003-01/06/content_680207.htm. Accessed 6 January 2015.

Yadav, S (2015). Namami Gange: In The Name of Ganga. *The Indian Express*, May 18. Retrieved from http://indianexpress.com/article/explained/in-the-name-of-ganga/. Accessed 1 May 2015.

Yao, Z, J Liu and R Su (2006). Effects of Centralized Management and Strategies for Integrated Management for the Yellow River. *Resources Science*, 28(2).

Yellow River Conservancy Commission (2015). Institutional Function. Retrieved from http://www.yrcc.gov.cn/zwzc/zjhw/znjg/201108/t20110810_26215.html.

Yellow River Conservancy Commission, 'Yellow River Water Resources Bulletin 1998, 2008 & 2012. Retrieved from http://www.yellowriver.gov.cn/.